TRENDS IN ELECTROCHEMISTRY

TRENDS IN ELECTROCHEMISTRY

Edited by

J. O'M. Bockris
The Flinders University of South Australia
Bedford Park, South Australia, Australia

D. A. J. Rand
CSIRO Division of Mineral Chemistry
Port Melbourne, Victoria, Australia

and

B. J. Welch
The University of New South Wales
Kensington, New South Wales, Australia

PLENUM PRESS•NEW YORK AND LONDON

The Library of Congress cataloged the first conference of this series as follows:

Royal Australian Chemical Institute.

Australian Conference on Electrochemistry.
  Plenary lectures & abstracts of original papers.
  Sydney, Butterworths.
  v. illus. 26 cm.
  At head of title   : Royal Australian Chemical Institute.
  1. Electrochemistry—Congresses I. Royal Australian Chemical Institute.
QD551.A87   541'.37   79-18318

ISBN-13 978-1-4613-4138-3   e-ISBN 13 978 1 4613 4136 9
DOI 10 1007/978-1-4613-4136-9

Plenary and Invited contributions presented at the
Fourth Australian Electrochemistry Conference
held at The Flinders University of South Australia,
February 16-20, 1976

© 1977 Plenum Press, New York
Softcover reprint of the hardcover 1st edition 1977

A Division of Plenum Publishing Corporation
227 West 17th Street, New York, N.Y. 10011

THE ELECTROCHEMISTRY DIVISION
THE ROYAL AUSTRALIAN CHEMICAL INSTITUTE

FOURTH AUSTRALIAN ELECTROCHEMISTRY CONFERENCE

ORGANIZING COMMITTEE

J.O'M. BOCKRIS, *Chairman*

T. BIEGLER, *Honorary Secretary*

A.J. PARKER	D.A.J. RAND	L.J. WARREN
E.C. POTTER	J.H. SHARP	B.J. WELCH

BENEFACTORS

The Division wishes to acknowledge the following companies and institutions for financial assistance towards organizing the Conference:

The Broken Hill Associated Smelters Pty. Ltd.
The Broken Hill Pty. Co. Ltd.
Broken Hill South Ltd.
Chloride Batteries Australia Ltd.
C.S.R. Ltd.
Education Fund, Royal Australian Chemical Institute.
Electrolytic Zinc Company of Australasia Ltd.
Imperial Chemical Industries of Australasia Ltd.
John Lysaght (Australia) Ltd.
The Government of South Australia.
Utah Electronics.

Foreword

This volume presents plenary lectures and invited papers that were delivered during the Fourth Australian Conference on Electrochemistry held at The Flinders University of South Australia, 16-20th February 1976.

Electrochemistry for a Future Society was selected as the Conference theme since the organising committee were mindful of the rapid change in technological perspective which the world now faces. We no longer have a prospect of uncontrolled spontaneous expansion and change as the result of technological enterprise. Rather, we face the task of attempting to reach a state of very restricted growth. In the next few decades special accent must be placed on minimizing pollution and maximizing the efficient utilization of all available energy sources.

With this in mind, the Conference organisers considered that a conventional electrochemistry symposium, with its divisions into the various academic aspects, would be less relevant than a meeting devoted to aspects of electrochemistry which may underlie parts of the new and necessary technology for the future state of affairs. What has actually been achieved by the Conference organisers is a balance between the ideals expressed and the resulting response from electrochemists. This response has a bias which reflects the dominance of certain resources, e.g. metallic minerals, within Australia. Consequently, the papers included in *Trends in Electrochemistry* cover subjects which are of both global and local concern.

EDITORS   J.O'M. BOCKRIS

Melbourne   D.A.J. RAND

June 1976   B.J. WELCH

Contents

IV.   SURFACE AND COLLOID ELECTROCHEMISTRY

V.   ADVANCES IN ELECTROCHEMICAL TECHNIQUES

VI.   MINERAL SULPHIDE ELECTROCHEMISTRY

VII.   ELECTROCHEMISTRY AND METALLURGY

ELECTROCHEMISTRY FOR A FUTURE SOCIETY

John O'Mara Bockris

School of Physical Sciences
The Flinders University of South Australia

PART 1   DIFFICULTIES OF THE PRESENT SOCIETY

Indigestion: The Pollutional Danger

    Pollutional difficulties arise from the automobile and from
factories. Effluents from the latter can be dealt with, and have
been in Japan and England. Effluents from cars are dealt with well
in laboratories but the introduction of catalytic converters into
cars lessens their performance and the catalytic purification is
expensive and short-lived. Removal of lead from petrol increases
pollution from unsaturates. If growth of the economies continues,
the pollution rate will increase again in the 1990's (Fig. 1).*

Starvation: The Exhaustion of the Fossil Fuels before the
Abundant Clean Energy Sources have been Engineered

    Estimation of the time of exhaustion of the present fuels is
not difficult to make if a given growth rate of the economy is
assumed. A cessation of growth in our economy brings depression;
growth in energy demand is, in any case, difficult to stop,
particularly as the population of most areas of the world is still
growing. If we assume a continuance of a 5% growth rate in energy
need (the U.S. one has been 5.8% for some years), there will be
exhaustion** of world oil by 1995. However, this assumes continued

---

* This figure was given to the author by Dr. C. Heath, then of
  Esso Research, Engineering, Linden, N.J.

** "Exhaustion" is taken arbitrarily as the situation where scarcity
   has forced a (real) price rise 4 times that of 1976 (i.e. ten
   times that of 1973).

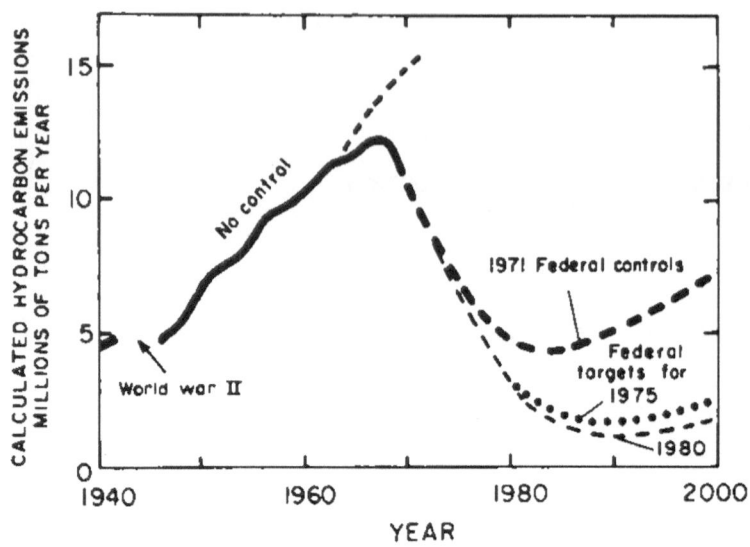

Fig. 1 Estimated effect of federal controls in the U.S.A. on hydrocarbon emissions from passenger vehicles. A rising trend is
expected after 1985 because of the increasing numbers of cars.

export of oil from the Middle East. Saudi Arabia, the main
supplier, plans to cease exports in 1990 and divert the rest of its
resources to a planned indigenous industry (1).

After the exhaustion of gasoline and natural gas, synthetic
fuels from coal are planned to bridge the gap until atomic and
solar sources have been researched, developed and built.

Gross over-estimates of the supply of coal have been published
(2); they omit the effect of growth and the fact that the coal
production extrapolation line (from which is calculated the
exhaustion date) is based upon the use of coal largely for
electricity production (15% of need). When coal products take over
from oil the use-rate would increase many-fold. Turner and Elliot
(3) have allowed for this in Fig. 2, (see also Table 1 taken from
Linden (4)).

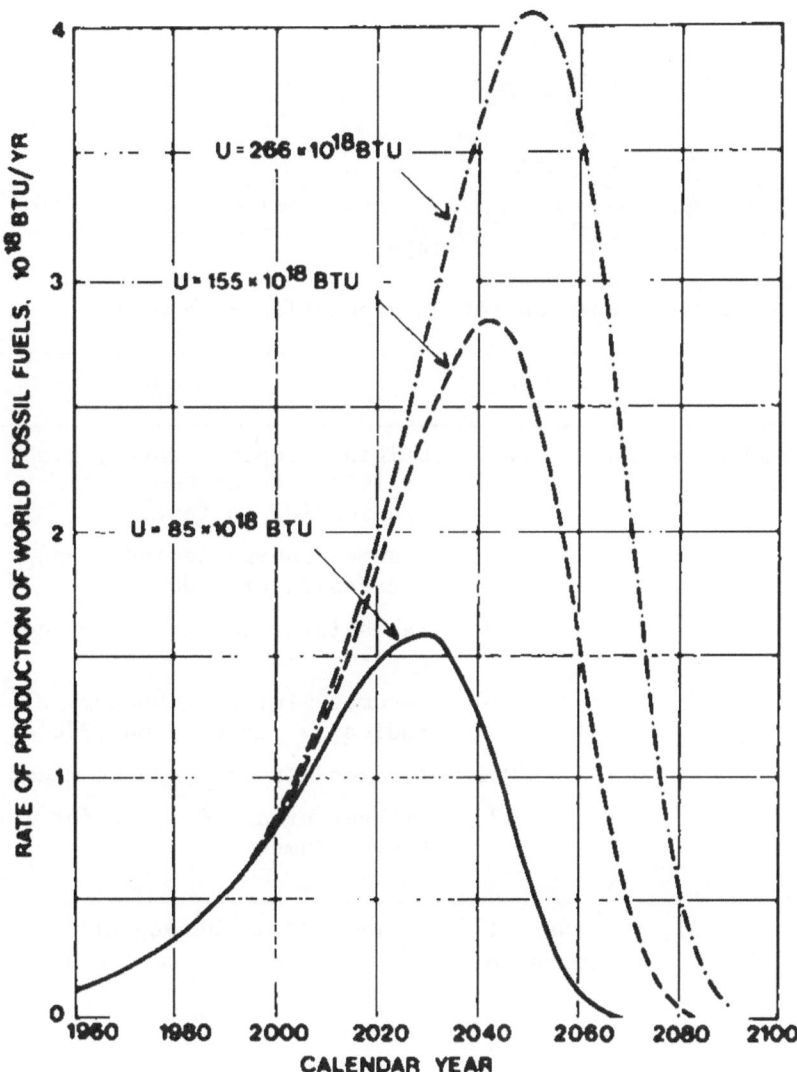

Fig. 2 Effect of the resource base on the projected rate of world fossil fuel yields. (Turner and Elliot (3)).

TABLE   1

Estimates of Year when Maximinum Production Rate of Coal could Occur*

Source	Year	Comment
Chase Manhattan Bank, elucidated.	2033	Original report states, 1,500 years at 1972 rate of consumption. Elucidation in text.
Brennan	2000	Assumes energy demand stops increasing at 2000.
Elliott and Turner	2040	Takes into account use of coal to replace oil.
Linden	2010	Assumes mining technology not radically improved on 1970's.
Linden	2052	Assumes all coal can be mined.
Linden	2073	Highest credible limit for resource base assumed.

*   In general, the assumption is made that the logistical
     difficulty of mining coal *at a sufficient rate* has been solved (4).

A Latter-Day Coal Age?

The use of coal to replace oil on a large-scale seems improbable for the following reasons:-

1. There are difficulties of extracting coal at the rate needed. These have been stressed by Arthur (5). *Five* new mines of average size per day in the U.S. would have to be completed between now and 2000.

2. The pollutional problems which would be provided by the use of dirty American coal (much of it 5% sulphur) are great. Although it may be feasible to reduce the sulphur in coal in laboratory experiments, the economics of doing this on a large scale (massive milling to small particle size) and the degree of removal create difficulties. Were the rate of burning of coal increased not only seven times - needed to replace oil - but three times more to allow for energy growth to 2000, a disagreeable increase in air pollution seems probable.

3. Amortisation: a massive development of the coal industry is unlikely because it would need a guarantee of continuity of use through 30-40 years. This will not be possible except by Government Fiat.

A population of 10^{10} people would need 2000 reactors, each 40,000 MW. The population of the United States would need 40 large reactors, built in about 40 years to have an all breeder atomic energy economy by 2016. This is not an impossible amount of investment capital per year, but the breeder technology - after 20 years research - is not yet ready and the pollutional problems of the wastes are regarded by some as insoluble.

Conservation

Although some contraction in energy spending per person can be achieved by caution, a decrease of energy per person of > 10% would lead to deprivation (6) and hence political unrest. The population of most countries continues to grow inexorably. A cutback in energy demand, or even stopping of growth, is an impractical expection with democratic (i.e. low control-power) governments. That political party will be elected which makes most credible its promise of the highest standard of life in the *immediate* future.

Democracy must therefore drive an economy at full speed to exhaustion of its resources. Of course, whilst the capital lasts, the standard of life in the "spend what we have" democratic society will be greater than that in the Economy which holds off consumer spending so as to obtain an industrial base to build military forces and energy production, i.e. Long Term Strength.

Media Of Energy

Fossil fuels are sources and media of energy. However, the *new* sources will give electricity or heat at source.

Heat is not a suitable medium of energy because it cannot be transported. One possible medium is electricity, but it is not suitable for long distance transport of energy nor for carrying out tasks now carried out by natural gas.

Thus, were we to use electricity, even though there were no difficulties in respect to the cost of sending it over large distances (as needed with solar sources), one would have to research and develop electrochemical technology. Such technology is only at the point of projection from the fundamental stage.

There is need for a fuel easily produced by heat or electricity and easily transportable. Hydrogen is likely to be useful. It could interface with electrochemical devices, viz. fuel cells. The alternative is methanol. Internal combustion devices driven off methanol at 30% efficiency would have to compete with the fuel cell driven cars which function off hydrogen at 60% efficiency.

Electrochemical Consequences

One consequence of the new energy sources is a rebuilding of much technology toward utilising electricity, hydrogen and methanol. At source, the cost of these fuels will be electricity < hydrogen < methanol; hydrogen will always be cheaper than the product of the reaction with CO_2, methanol. A greatly increased component of electrochemical technology therefore arises on economic (and of course, ecological) grounds.

PART II   ELECTROCHEMISTRY

Definitions of Electrochemistry

There are misunderstandings of what electrochemistry is.   It was originally discussed in terms of electrode process-chemistry as exemplified by Faraday and Nernst.   The European concept of electrochemistry ended with Tafel (1903) and the birth of electrode kinetics.

In this early era, there was less concentration upon ionics. This was awakened by Arrhenius.   The parallel development of the electrode and ionic theme within electrochemistry was thus begun.

The Nernstian hiatus occurred between 1900 and 1950.   Ionics prospered because of the application of statistical mechanics to the properties of solutions and a fruitful approximation which made ionic systems the centre of physico-chemical discussions from about 1925 to 1945.   Ionics became asymptotic by the 1960's and by 1970 most of the remaining fundamental interest went into the study of the ionic environment.

The 1950's saw the rebirth of the electrode process interests, sparked by vigorous British and German schools.   An International Society for Electrochemistry was formed.   The phenomenological theory of electrode kinetics became established (7).

During the 1960's the terms Ionics and Electrodics (8) were coined.   It was a time of excitement in electrochemistry for the field played a part in space technology and was seen by some* to offer an interesting alternative technology.   This brilliant phase saw the birth of the concept of Electrochemical Science (9): electrochemistry is no longer a part of Chemistry but stands as an interdisciplinary subject, see Table II.

---

* For a few years in the mid 60's, electrochemistry was a subject of interest in Washington Government circles.   Fuel cells and electrochemical developments became something of a political football because of the considerable effects on the automotive industry.   Thus, the field and its leading exponents, worldwide, was subject to study by the American Security Services.

TABLE II

Electrochemistry, a Perspective from Afar

Period of Early Discovery	Nernstian Hiatus	Rise of Ionics	Electrochemistry Becomes Predominantly Concerned with Interfaces	Electrochemistry Becomes an Interdisciplinary Area
Muscles move when currents pass.	Over-thermodynamical approach prevents development of charge transfer kinetics at interfaces.	Study of ions in solutions, adjunct to electrochemical studies, becomes identified with electrochemistry itself in England and America.	The kinetics of electrical reactions at interfaces is formulated, and some basic progress in investigating its mechanisms is made.	Electrochemistry plays a role in space power, stability of materials, functioning of biological cells, nylon synthesis, vehicular transportation.
1791	1910–1940	1920–1940	1950's	1960's

Relevant Aspects of Electrochemistry

1. Many problems of syntheses, materials science, engineering and biology are associated with electrochemistry. Interfacial properties are electrochemical in the sense that all interfaces are charged, and the situation at them is electrochemical.

2. Electrochemical reactions can go up and down the free energy gradient, depending on the interfacial monolayer charge.

3. The range of fields in science in which the important step is electrochemical is comparatively great. Examples of the interdisciplinary character of electrochemistry have been given in a well-known book (10). Concepts of electrode processes underlie much Analytical Chemistry. In Metallurgy, stability is based upon electrochemical kinetics. In Engineering, much energy production and storage is electrochemical in prospect. In Biology, the applications of electrochemistry are legion, the most well-known one being the transmission of electric currents in nerves, and the most recent, the influence of potential on the growth of bones.

4. Electrochemistry is "another chemistry"? There seem to be usually two ways of doing things - the normal thermochemical way or by an electrochemical alternative. The Van't Hoff Law relates cell potential and free energy. Many reactions, for example Kroll's synthesis of Ti, can be looked at in terms of collisons of particles and solutions, but such reactions may be considered alternatively in terms of interfacial reactions. Cracks in metals and their spread can be viewed in terms of mechanics and stress, but there is an electrochemical aspect in the anodic dissolution at the tip of the crack. Are biological reactions collisonal reactions in solutions, or are they interfacial reactions occurring through charge transfer at interfaces?

5. In future towns, energy will come from atomic and solar sources; the medium for the energy will be electricity or hydrogen; the town is likely to be self-contained (materials recycled) except for its energy supply. At first CO_2 will be rejected into the atmosphere but if energy costs can be sufficiently reduced, it may be advantageous to recycle the CO_2. The processes which run the recycling must be non-polluting and many are therefore likely to be electrochemical. Transportation is likely to use stored electricity or hydrogen-working fuel cells (the only alternative is methanol); many machining processes and recovery of materials seem likely to be electrochemical. Polluted liquids will be electrodically regenerated.

Sewage will be processed electrochemically and probably reversibly.   Medical electronics - the electronic-electrodic combination - will be highly developed.

6.   Peeking over the top of the century, two particular advances can be made out.   One, the relation of light to electricity and the general development of photo-electrochemical processes; the other, the relation between the electrochemical potentials developed in the brain and other mechanisms in the body, all of which are associated with electrochemical currents and potentials.   A zero growth, stable world must first of all have a zero population growth.   It must thereupon be made clean. The processes will be expected to last indefinitely and energy sources will be solar, perhaps fusional.   Electricity and the electrochemistry-favourable reactant, hydrogen, must then be a considerable part of the material future.

The Still Dying Branch of the Electrochemical Tree: Nernstics

There is no doubt that however one eulogizes on the social usefulness of electrochemical technology, there is a festering patch of the field which needs amputation, or at least a clean-up job. I refer to the debris of the Nernstian Hiatus.   Greatness dies with difficulty.   Nernst was great.   Even now some electrochemists are nernstic.   Their thinking is thermodynamic and simplistic.   Moreover, there exists a large and younger group who consider electrode processes principally in terms of concentration gradients near electrodes.   Their thinking assumes $i_o = \infty$.   They should be called the Nernstian Simplistics.   It is important to avoid teaching young students such a Tenacious Error.

A Mature Definition of Electrochemical Science

It is the part of interdisciplinary electrochemical science - electrodics - which must be given a mature definition;   indeed it was given one by a Russian scientist, Kyvstiakovsky (11) in 1910:
'Electrochemistry is the science which deals with conversion of
    matter to electricity;   and/or electricity to matter'.
The definition is comprehensive and has within it excitement and a suggestion of a heap of applications still to come.

Overpotential: A Core Concept in Future Technology

The difference between the pre-1950 (Nernstic) and the post-1950 (B-V-G)* electrochemistry is in the concept of overpotential. In the old electrochemistry, overpotential had largely a nuisance

* B-V-G = Butler, Volmer and Erdey Gruz

value. In the new electrochemistry, it is that which moves one off from equilibrium. In modern terms, overpotential is the shift of the Fermi level necessary to bring about a flow of electrons in one direction at a given rate. This is its rational aspect.

However, there is another aspect of overpotential, that of analogy. Thus, the concept has overtones of aspects of experience wider than Electrochemistry. It is analogous, for example, to inflation. Without some degree of inflation, modern economies do not "go". If the inflation is zero, there is no progress in the sense of increase in living standards (forward current). In deflation, economies go backwards, and in analogy, a negative overpotential causes a current reversal, a backward flow of the growth rate, a downturn; eventually a collapse.

Correspondingly, if overpotential is great, the reaction rate is great, the economy goes fast, but it must run into a limiting situation where the overpotential can be increased indefinitely (around the limiting current) without increasing the value of the current. This is the analogue of galloping inflation where one uses more and more money ($\equiv$ Overpotential) with increasingly little effect on obtaining something.

Correspondingly, there is an analogy between overpotential and *reality*. Thus, one has an ideal concept of a thing in itself, which we may compare to the thermodynamic evaluation of an electrode reaction. Reality is found to be always less good than expected. Something is lost, just as in electrochemical energy conversion, the overpotential is the numerical quantity which represents that which is lost.

Lastly, there seems something which connects the atmosphere of modern and ancient life in overpotential. In mediaeval life - the long life of the European Middle Ages where people watched the sheep drifting over the hill - one had equilibrium, zero overpotential, or the reversible region anyway. Modern life is life under high tension, high overpotential: everything comes and goes quickly. In ancient life, one was going to no place, waiting for death, which was to be savoured. In modern life, one is going somewhere quicker and quicker, though one can savour life less and less and one is increasingly unsure as to where one is going, or why.

Away from the analogies, overpotential is the core quantity which relates chemistry to electricity, chemical energy to electrical energy, the physics of the energy level in the electrode to the electrochemical quantum mechanics of the electronic states in solution. It is a bridge concept which joins the affluent economy powered by fossil fuels, to the solar-hydrogen-electrochemical infrastructure of the future, zero growth, clean, high energy economy of the post-industrial world.

PART III   ELECTROCHEMICAL ASPECTS OF THE NEW TECHNOLOGY

The Time

     Predictors suffer a rough passage through Science Straits.
Scylla is the one year perspective in which industrial companies
operate, giving them clear short sight, and a time of adjustment
and ability to change longer than the stopping time.   Charybdis is
Science Fiction.   We set our sights between these, some five
decades into the future.   Our time scale is similar to that of
Meadows's "Limits to Growth".

     One must measure the time at which one should try to estimate
things in comparison with the time it takes to develop new technologies.   Friborg (12) distinguishes between the number of years the
fundamental work has to go on before a commercialisation is obtained,
the number of years the applied work has to go on, and the number
of years the commercialisation stage has to go on.

     Tafel began electrodics in 1903.   If Friborg's Laws are true,
commercialisation of electrodic concepts should be widespread in the
1980's (Fig. 3).

Electrochemical Technology of the Present

     Some interesting conclusions were brought out by Wengloski (13)
in the only study of the Electrochemical Industry published (1966).
The United States electrochemical technology of 1963 was 15% of U.S.
manufacture and 35% of chemical industry.   These fractions are
high, compared to the number of Ph.D.'s produced in electrochemistry
in U.S. universities.   The number of Ph.D.'s produced per year in
Chemistry in the U.S. is c. 1,000, and in Electrochemistry c. 10,
or 0.1%.   Were the subject of the new researchers to reflect the
realities of electrochemical economics, one would expect several
hundred new Ph.D.'s per year in Electrochemistry.   The composition
of the electrochemical industry in the 60's is shown in Figure 4.

     The electrochemical industry of 1963 in the U.S. was old
electrochemistry and amounts to 35% of chemical industry.   It seems
likely indeed that the one-generation future will bring an electrochemical industry which will be more than half of chemical industry,
and indeed it seems likely to guess that, in a two-generation future
(with all energy sources giving heat or electricity) chemical
industry will become largely electrochemical.

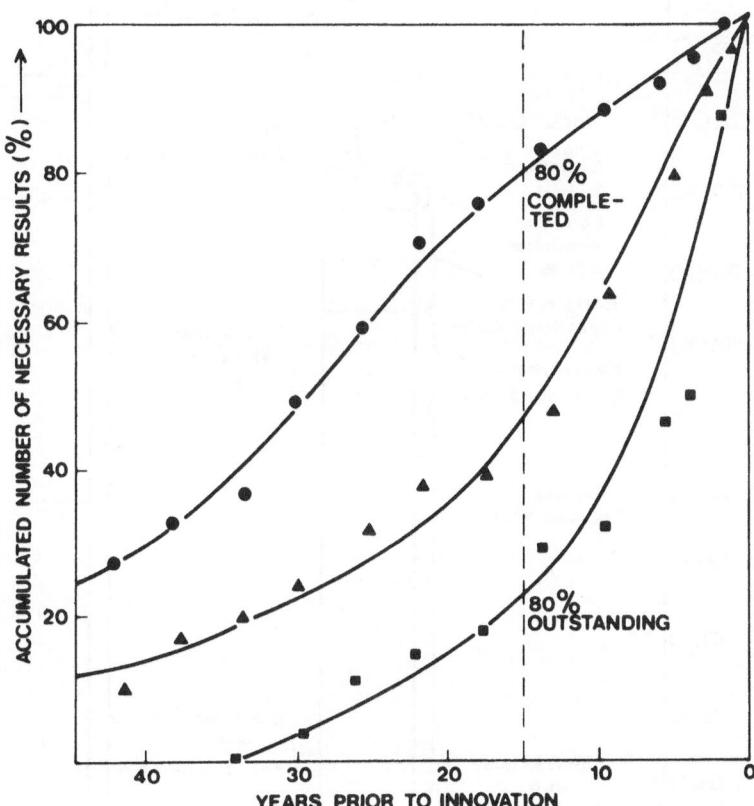

Fig. 3 Number of years of basic research (●), applied research (▲)
and development (■) necessary for the realisation of
innovations (Friborg (13)).

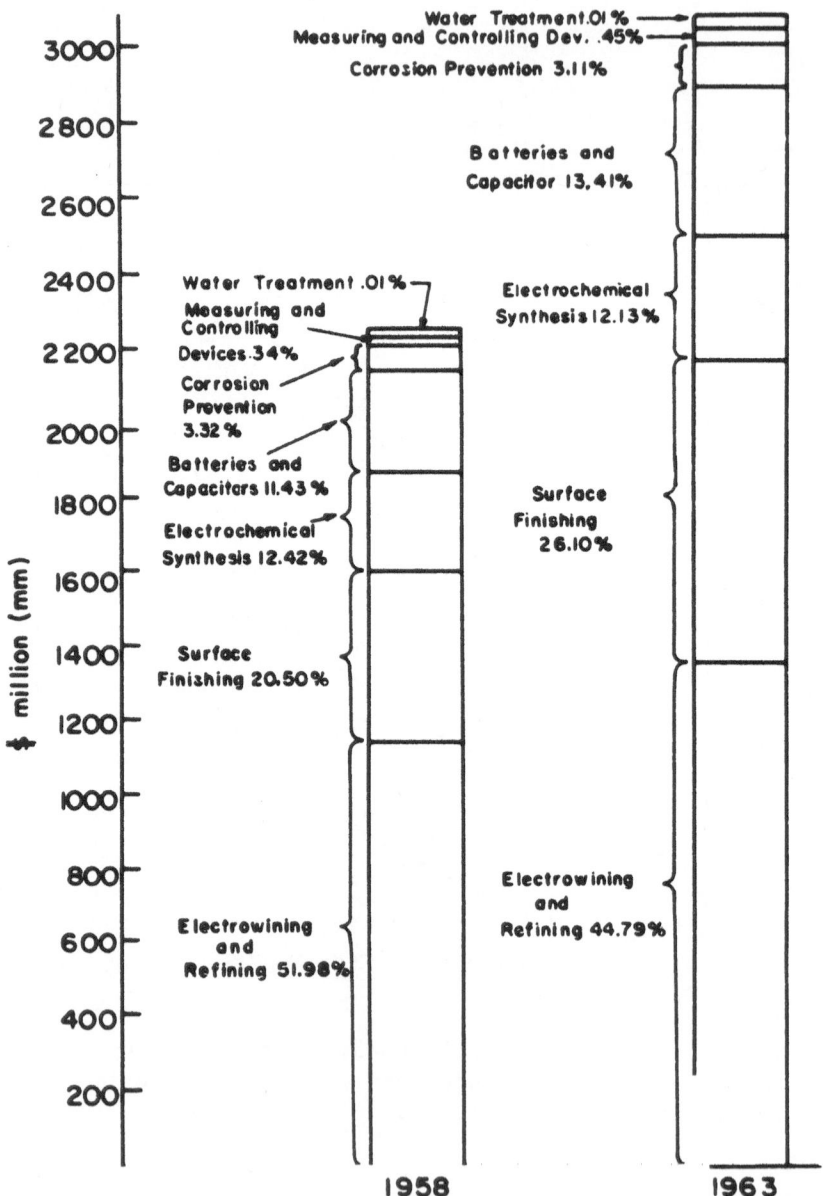

Fig. 4   Composition of the electrochemical industry by-product
         types; 1958, 1963.

Spontaneous Electrochemical Reactions

A spontaneous electrochemical product produces a fuel cell in which the objective is not to produce electricity but materials.

Suppose that a majority of our manufacture at present carried out by spontaneous chemical reactions, producing wasted heat energy, can be carried out in their electrochemical alternate technologies. We may exemplify by quoting dichlorethane made by spontaneous electrochemical reaction using ethylene at one electrode, chlorine dissolution at the other.   By-product heat is not worth collecting because of the difficulties of storage, mixing and transmission. This does not apply to by-product electrical energy recoverable from electrochemical reactors.   This could be collected, fed from the factory into a grid, and added to the energy supply.

The study of spontaneous electrochemical reactions is a neglected area.   Thus, the illusion exists that, if a product is to be electrochemically made, it has to be made with the use of electricity to drive the reactions.

The Electrochemical House

Likely aspects of electrochemical applications within the house are:

1.   Fuel (natural gas, but more likely hydrogen) is piped to the house and a basement fuel cell produces electricity.

2.   Solar hydrogen or roof-top energy is collected, converted to electricity, and stored in batteries ("Solar I", a house at the University of Delaware, has such a system).

3.   Roof-top collected electricity electrolises water and produces a hydrogen fuel supply for cars.

4.   In-house treatment of sewage: this will be mentioned below.

An Electrochemical Industry?

Here the one-generation future would look to a decreasing fossil fuel-driven industry, replaced by a combination of a hydrogen- or electrochemically-driven industry.

Where the accent goes here will depend upon whether we develop nuclear sources with the likely easy availability of electrical energy, where the (very large) reactor is less than 500 kilometers from the user area;   or whether we build a solar-hydrogen economy with long pipelines bringing electricity several thousand kilometers from highly insolated areas in the Southern Hemisphere.

TABLE III

Industrial Chemical Products and

Likely Orientation in a Post 2000 Industry

Group of Substances Produced	Orientation
Alkali and chlorine	Electrochemical
Industrial gases	Electrochemical
Dyes	Chemical
Organic Chemicals	Chemical (Hydrogen and Oxygen); and Electrochemical
Plastics, Resins	Chemical
Organic substances	Electrochemical and some Chemical

Another influence will be the ease of conversion of the present chemical processes to electrochemical ones or ones running with hydrogen. Table III above gives comments on the likelihood.

The availability of pure oxygen at cheap rates from a hydrogen economy is relevant to the evolution of processes. One of the important aspects in the balance between hydrogen and electrochemical processes arises from the spontaneous electrochemical reactor outlined above.

Electrochemical Transport

The lead acid battery would provide urban transportation, 60 kilometers per hour and 60 kilometers. However, lead resources are insufficient and this eliminates the prospect.

A more reasonable couple for urban transportation would be the nickel-iron battery, now in its commercialisation stage, with development by Westinghouse, or the sodium-sulphur cell. These batteries would give about two times and about five times the range of the lead acid battery, respectively*. They use cheap and abundant materials.

---

* One misunderstanding is that engineers associate batteries with the lead acid battery, the battery which has the poorest performance of all batteries in respect to its energy per unit weight.

Attention to battery-driven cars is overdone, and a rational electrochemical solution would be the use of fuel cells which would make the range of electrochemically-driven vehicles larger than that of gasoline-driven ones because of the greater efficiency. The lack of effectiveness of fuel cells in giving acceleration can be overcome by using hybrids with batteries. The most attractive suggestion (14) is a nickel-hydrogen battery used for acceleration and overtaking modes, charged by a fuel cell in which hydrogen is stored at high pressure within the tubular structure of the car. The electrochemical solution to transport - easily extended to trains and ships, though not to aeroplanes - will be in competition with the direct use of hydrogen.

There is an anti-electric car influence from manufacturers who would prefer to *modify* their mass production plant to build methanol-fuelled cars rather than to rebuild them, as they would have to with an electric car situation. The Japanese are affected by the same considerations.

Electrochemical Fuel Production: Hydrogen

Of the future energy media (electricity, hydrogen, methanol) the most likely - hydrogen - can be produced by the cyclical chemical method which involves a series of chemical reactions which would be driven by heat at 700-800°. It has disadvantages, including doubt that the degree of cyclicity will be complete. For this and other reasons, it seems unlikely that such a method would become economic.

The main way of making hydrogen seems certain to be a version of the electrochemical synthesis. However, the disparity between present industrial production and the possibilities shown in Fig. 5 promises much development work.

There seem to be three possibilities:-

1.   The high temperature way of producing hydrogen. Steam is electrolised at 1.3 volts.

2.   The devices which avoid electrochemical evolution of oxygen and involve, for example, oxidation of cuprous to cupric with subsequent thermal decomposition of the chloride.

3.   Photo-orientated methods in which light is introduced into a semiconductor which is a part of a cell and which introduces hydrogen in a photo-driven fuel cell.

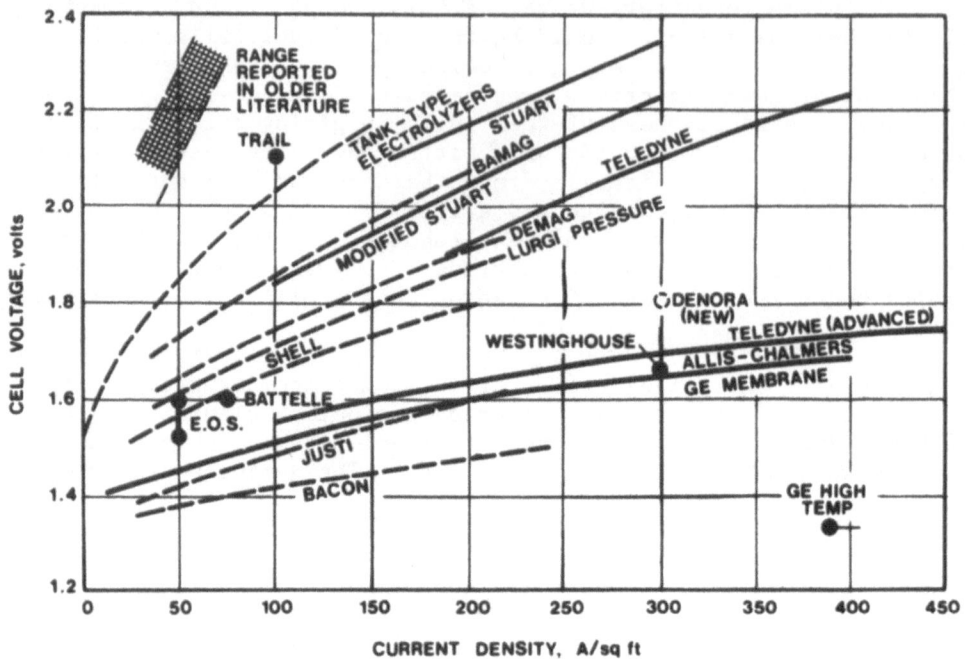

Fig. 5 Cell operating performances of various advanced electro-
        lysers (Gregory (15)).

Electrochemical Waste Disposal

The present electrochemical treatment of sewage introduces
the material into sea-water and produces hypochlorite by electrochemical means.

Other electrochemical possibilities involve spontaneous
electrochemical reactions in which high pressure oxygen is
introduced into a cell and would take part in a cathodic reaction,
the anodic reaction of which would be the oxidation of the cell-
ulosic and pyridonium-based sewage material to carbon dioxide.

The presence of in-house waste disposal would reduce the unit
cost per house.

Electrochemical Recycling

The attraction of electrochemical recycling is the availability of the extra variable of potential which allows one to pick out one substance at a time from the solution.

Reclaiming the constituents of scrapped cars by electrochemical means has received consideration and a table which shows the constituents of cars is given in Table IV.

Electrochemical Food Production

Perhaps one exaggerates by calling the proposal here "electrochemical", but one stage could be electrochemical, namely the reduction of atmospheric CO_2 to formaldehyde.   Further reactions would involve enzymatic synthesis of proteins and the latter stage has been achieved starting with the feed stock of hydrocarbons. Could other foodstuffs be synthesized?

TABLE IV

Breakdown of Metals in a Typical Car

Metal	Pounds	Price/lb	Value
Iron	3000	0.05/lb	$150.00
Copper	32	0.55/lb	17.60
Zinc	54	0.15/lb	8.10
Aluminum	51	(0.80/lb)	(5.10)
Lead	20	0.16/lb	3.20
Nickel	5	1.28/lb	6.40
Chromium	5	3.00/lb	15.00
		total	$200.30

Electrochemistry in Surgery

There are several applications of electrochemistry in surgery. For example, the use in Russia of electro-anaethesia, from which there are no chemical after-effects.

The control of blood clotting by using electrochemical circuitry as carried out by Sawyer and Srinivasan, is another example (16).

The Electrochemical Stability of Metals

The area here is wide and includes not only corrosion control but surface finishing and electrochemically-produced coatings. The era into which we are entering will have to make a careful control of the pollutional intention. A throw-away society will be replaced by a society where there is an attempt to make every material last to the longest possible degree before energy is to be spent in recycling it. Corrosion control will become mandatory.

A Hydrogen Economy in Electrochemistry

The large-scale adoption of a hydrogen economy would not contradict the views expressed in the present article. Many electrochemical concepts can only be carried out economically with hydrogen because of the high exchange current density associated with reactions involving hydrogen.

Methanol as a Fuel?

The third energy medium which may be becoming increasingly important is methanol. It could be produced from the remaining fossil fuels, including coal, so that we can envisage a few decades in which methanol could be a predominant fuel for applications in which gasoline is being replaced. It would have attractive aspects in being similar in appearance to gasoline. However, there are two counts against methanol from atomic or solar energy: (a) it would have to be produced from atmospheric carbon dioxide, and (b) hydrogen would have to come from the electrochemical decomposition of water. Hence, the manufacture of methanol would involve an extra stage and it is difficult to see that it would be cheaper per unit energy than the energy from which it came. How much would people pay for the greater ease of storage?

PART IV ACTION

Inferences from the Manhattan Project and the Nasa Moon Project

In considering the transformation of the fossil fuel technology to a solar-hydrogen-electrochemical economy, we are reminded of two great projects run by big government science. Both of these have been carried out under good conditions, one of war-time stimulus, where the entropy of commercial competition is eliminated; and the other with the stimulus of competition with the U.S.S.R., and no internecine strife. The main lesson to be learnt from these projects is that when a big government research effort is made, with enough money, the results can take us in a decade to a realization of happenings which are the Science Fiction of the previous generation. Thus, it is reasonable to suggest that two of the more remarkable achievements in the history of mankind have been the attainment of controlled atomic energy, and the moon landing. These were achieved respectively in three and ten years, both by spending sums on Research and Development in the region of billions of dollars per year. These achievements are relevant when we consider our Energy Predicament.

The lack of electrochemical education is the greatest barrier to progress. The field has poor definition among science professionals. The majority of chemists have little concept of electrodics. Of those educated in electrodics (a number in the region of a thousand) less than half have had training as electrochemical engineers.

Misunderstandings

Due to the small number of people who understand fundamental electrochemistry, the electrochemical case is sometimes greeted with sceptism. For example, one of the commonly given disadvantages of electrochemically-driven cars is that they will take a long time (i.e. eight hours) to recharge. Of course such a difficulty does not exist in the battery rental solutions. This is typical of the present uninformed state of the opposition to widespread electrochemical technology.

Is Fundamental Research Still Needed?

There are many areas where knowledge is poor. To name three:

1. The quantum mechanical theory of electron transfer; in particular the theory of the transmission coefficient.

2. The structural properties of the double layer at solid interfaces.

3. The nature of protective layers, and how they affect
    electrochemical processes occurring on them.

A Major Illusion: Comparing with the Present Price

One of the difficulties of appreciating the significance of
predicted advances is that engineers think of the present price of
the competition and the price of the under-developed alternative.
This is, of course, an irrational view.   Thus, one often hears as
an objection to solar energy converters: "They are not as yet
economical".   Such questions become chicken-and-egg ones: the
alternatives never will become economical unless they are researched
and developed.   The role of the fundamental scientist and he who
predicts the future is to estimate the probability of the fertility
of such research and development.

Applied Research: Still Fuel Cells

Fuel cells suffer from aging due to slow crystal growth which
reduces electrode area.

So much in the electrochemical picture depends upon the attainment of something like the theoretical predictions for fuel cells
that this difficulty should be the object of a concerted international effort.

There is an Opposition ....

The rate of realisation of the proposals implicit in this
article will not be governed alone by scientific considerations.
Capital has inertia.   But inertia springs from those who have
been trained in one field.   Electrochemical applications will
increase if it is made sufficiently attractive for young women and
men to become trained in electrochemistry at a graduate level.

The comparison of the predictions of Friborg (12) and some of
the time scales looked at in this article suggests that we are
faced with a serious situation with respect to the time at which we
can obtain our new energy sources and interface them with the
relevant electrochemical and hydrogen technologies.   Can we do
that in 25 years?   Do we want to make the effort?

REFERENCES

1. M. Sayigh, University of Petroleum and Minerals, Dharan, Saudi-Arabia, Private Communication, November 1975.
2. J.G. Winger, *Outlook for Energy in the U.S.A. till 1985*, Energy Economics Division, Chase Manhattan Bank, June 1972.
3. N.C. Turner and M.A. Elliott, presented at the A.C.S. Meeting Boston, 1972.
4. W.R. Linden, *Analysing World Energy Supply*, Institute of Gas Technology, 10/8/74.
5. J. Arthur, quoted by D.B. Thompson, Industrial Work, 26 November 1973, p.17.
6. V. Rothschild, Chairman, British White Paper on Energy Conservation, 1974.
7. K.J. Vetter, *Electrochemische Kinetik*, Springer, Berlin, 1961.
8. J.O'M. Bockris and B.E. Conway, Record of Chemical Progress, $\underline{25}$(1), (1964) 31.
9. J.O'M. Bockris and D. Drazic, *Electrochemical Science*, Taylor & Francis, London, 1972.
10. J.O'M. Bockris and A.K. Reddy, *Modern Electrochemistry Volume 1*, Plenum, New York, 1970, p.16.
11. L. Antropov, *Theoretical Electrochemistry*, Mir Publishers, Moscow, 1972.
12. G. Friborg, Forskning och Framsteg, Stockholm, $\underline{3}$, (1969) 2.
13. G. Wenglowski, *Modern Aspects of Electrochemistry*, Plenum, New York, 1966, Vol. 4, p.251.
14. J. Appleby, Laboratoires de Marcoussis, France, Private Communication, April 1975.
15. D. Gregory, *Blue Book in the Hydrogen Economy*, 1972.
16. L. Duic, P. Sawyer and S. Srinivasan, J. Electrochem. Soc., $\underline{120}$ (1973) 348.

F.T. BACON, O.B.E., M.A., F.R.S., M.I.Mech.E.

THE BRUNO BREYER MEMORIAL MEDALLIST 1976

F.T. BACON, O.B.E., M.A., F.R.S., M.I.Mech.E.

THE BRUNO BREYER MEMORIAL MEDALLIST 1976

E.C. Potter

Chairman

Electrochemistry Division

The Royal Australian Chemical Institute

Mr Conference Chairman, Delegates and Guests of The Fourth Australian Electrochemistry Conference:

We now have reached the most auspicious moment of the Conference. The Bruno Breyer Memorial Medal is an award of The Royal Australian Chemical Institute through its Electrochemistry Division. The bronze medal carries the likeness of Bruno Breyer on the obverse and the recipient's name on the reverse, and is awarded by the Division to a distinguished man or woman of electro-chemistry in commemoration of Australia's own most distinguished electrochemist, Bruno Breyer, who died in 1967. There have been two previous recipients of the medal, Professor Graham Hills in 1968 and Professor John Bockris in 1972. The third recipient is with us this evening and will shortly be receiving the medal from my hands. He is Mr. Francis Bacon, the originator and developer of the successful hydrogen/oxygen fuel cell that bears his name.

The name, Francis Bacon, may conjure up for you memories of a literary kind from over three centuries ago, and our Breyer Medallist can indeed trace his relationship back to that earlier man of letters. That alone, however, could hardly earn him the present award, and I hasten to explain further. Francis, or Tom Bacon, as so many know him, was educated at Eton and Cambridge where he took the B.A. Mechanical Sciences Tripos in 1925. Leaving Cambridge, he carried out development work for C.A. Parsons and Company until the outbreak of war in 1939. His first serious

work on the hydrogen/oxygen fuel cell began in 1940 at King's
College, London, but was interrupted by more pressing wartime anti-
submarine development work in Scotland.   In 1946 he returned to
Cambridge to begin a prolonged period of development of the hydrogen/
oxygen fuel cell.   This culminated, with the support of the
National Research Development Corporation, in the production of
the highly successful Bacon Fuel Cell, one of the very few fuel
cells that have ever achieved practical engineering application.
As is well known, the Appollo space missions of the late sixties
and early seventies relied upon the Bacon Fuel Cell.

Tom Bacon has received many honours and awards for his
remarkable and unique work.   He became an officer of the Order of
the British Empire in 1967 and was elected to Fellowship of The
Royal Society in 1973.   However, astonishing as it may appear,
the award he is about to receive is the first from a body devoted
exclusively to electrochemistry.   Tom Bacon has accepted the
invitation of the Electrochemistry Division of The Royal Australian
Chemical Institute to come to Australia to receive the Bruno Breyer
Memorial Medal and to deliver a memorial address which he has
entitled "The Development and Practical Application of Fuel Cells".

We are delighted that he has brought Mrs Bacon with him on
this most pleasing of occasions, and we look forward eagerly to
having them with us for the remainder of the Conference and for
some time afterwards.

It now gives me the greatest pleasure to ask Francis Thomas
Bacon to come forward to receive the Bruno Breyer Memorial Medal
and afterwards to deliver the memorial lecture.

BRUNO BREYER MEMORIAL LECTURE on

THE DEVELOPMENT AND PRACTICAL APPLICATION OF FUEL CELLS

F.T. Bacon

Westfield
Little Shelford, Cambridge

1. INTRODUCTION

First of all, I must express my deep thanks to you all, and
especially to Professor Bockris, for inviting me to come all the
way from England to deliver the Bruno Breyer Memorial Lecture on
my chosen topic of fuel cells. I appreciate this honour very much.

So far, fuel cells have only been used in space, and also in
a few military and other special applications, but I have always
believed that they would finally achieve widespread use in competition with existing electrical generators; this is proving to
be extremely difficult but great progress has been made in recent
years, and the prospects are much brighter than when I first
became interested in the subject about 44 years ago.

Next, I must explain that I am not an electrochemist; I was
trained basically as an engineer, but I have been blessed with a
number of good friends who have kept me right on the electrochemical aspects of the many problems which have had to be faced,
besides of course the problems concerning surface chemistry,
physics, metallurgy and so on. However, I am now retired, and
have no supporting staff to advise me, so I can only speak in
very general terms.

Finally, I must explain that I have not done any practical
work of any kind for many years, so in reality I have just been a
spectator; this is rather a grave disadvantage, as I have no
recent practical experience to draw on, but I have endeavoured to
keep abreast of the more important fuel cell developments in
recent years, and have tried to see how fuel cells might prove to

be of value in the rapidly changing conditions in the world energy
position with which we are now faced. I must admit that I find it
extremely difficult to keep up-to-date with all the literature on
the subject, and I am very ready to be corrected if what I say is
wrong.

Having said this, I think it is important to review very
briefly the practical applications which have so far been
achieved in this field, then discuss the various limitations in
present day fuel cells which have so far prevented widespread use,
and finally suggest ways in which the limitations could perhaps
be overcome, first by basic scientific work in the laboratory and
then by the laborious and expensive development work which is
always required before any novel device of this kind can be made
to compete on level terms with conventional electrical generators,
which have many years of careful development behind them.

Professor Bockris has told me to "try to make the lecture as
future orientated and speculative as possible;" I will endeavour
to do this, although I know it is highly dangerous to speculate
in any new field of this kind, and I shall no doubt be proved
wrong in some respects when the final tale is told.

2. PRESENT STATE OF THE ART

2.1 *Applications in Space*

It is well known that the first important application for fuel
cells was in two space missions, the Gemini and the Apollo; there
is no need for me to describe the units in detail, as this has
been done many times before. It is only necessary for me to say
that in both missions, pure hydrogen and pure oxygen were used as
reactants, both reactants being carried in spherical vacuum-
insulated vessels at slightly above the critical temperature, and
at moderate pressure. The Gemini design of the General Electric
Company (Liebhafsky & Cairns (1), Russell (2)) was based on the
use of a solid electrolyte which took the form of an acidic ion-
exchange membrane; at that time, the membranes had rather a
limited life, but since then a greatly improved membrane has been
evolved and this is said to be extremely durable. A cell temperature well below 100°C was used, and this meant that quite large
amounts of precious metal catalyst were necessary on both
electrodes.

The Apollo design of Pratt & Whitney Aircraft (Morrill (3);
Ching, Gillis & Plauche (4)) was based on the system which we
evolved in various departments of Cambridge University (England)
in the years 1946-1956. One of our main objectives in this

Cambridge work was avoidance of the use of precious metal catalysts, and in the Apollo design nickel, in the form of porous sintered metal, was employed for the hydrogen electrode or anode, and porous sintered nickel coated with a layer of lithiated nickel oxide for the oxygen electrode or cathode. The electrolyte was 75–85 per cent potassium hydroxide, and the operating temperature was up to $260^\circ C$. The average specific reactant consumption ($H_2 + O_2$) was 0.8 pound (0.36 kg.) per kWh. The Apollo design of fuel cell was subsequently used in the Skylab mission and in the Apollo section of the Apollo–Soyuz link–up mission.

Although it is understood that both designs fulfilled all requirements, they have now been superseded by improved units.

Both the Gemini and the Apollo designs suffered from certain disadvantages, the chief of which were limited life and very high cost; also in the case of the Apollo unit, a slow and rather elaborate heating–up process was required, the electrolyte being solid at room temperature. So a new design with a maximum power rating of 5kw has been adopted for the forthcoming Space Shuttle Program, the fuel cells providing the primary electrical power supply for the Orbiter vehicle. (Rice & Bell (5)). An aqueous alkaline electrolyte of potassium hydroxide is employed, and this is contained in a thin asbestos matrix; as the electrolyte volume is so small, it has been necessary to add a porous backup plate, adjacent to each hydrogen electrode, as an electrolyte reservoir, and this makes the control of electrolyte volume less critical, while still allowing close spacing of the electrodes. Catalyzed screen electrodes are used, the catalyst presumably being platinum black on the anode and a gold–platinum alloy on the cathode. The temperature of operation is believed to be about $80^\circ C$. As the cell temperature is so much lower than that of the Apollo units, the waste heat cannot all be removed by the circulating hydrogen, and a liquid coolant has to be fed through the cell stack between cells. At the average power rating of 3.7 kW., the specific reactant consumption ($H_2 + O_2$) is about 0.79 lb./kWh, or slightly less than that for the Apollo units. A goal of 5,000 hours endurance was specified, and it is under-stood that this has now been achieved. In spite of the fact that precious metal catalysts are used, the cost per kw is said to be substantially less than that of the Apollo units. Moreover, the specific weight has been reduced by a factor of 4, and the operating life has been improved by a factor of 5; also, these units are self–starting from cold and do not need an external source of power during the warming–up period prior to launch; this is of course due to the use of precious metal catalysts on the electrodes, instead of the nickel and nickel oxide used in the Apollo units.

The ion-exchange membrane (IEM) design of General Electric
has also been greatly improved, and a new design has been evolved
(Chapman (6)), based on a 10 mil thick membrane which is described
as a perfluoro linear polymer to which sulphonic acid radicals
(SO_3) have been chemically linked and immobilized. This new
technology is said to have demonstrated impressive advancements
in the areas of life, power density, weight and cost. The
electrodes consist of a thin film of catalyst embedded in an
electron-conducting metal screen which is pressed onto each face
of the membrane. A thin porous Teflon film is overlaid on the
surface of the cathode, to provide the necessary wetproofing
characteristic. A liquid coolant is used to remove waste heat.
The most striking feature of these more recent membrane cells is
invariant performance over really long periods of time (Nuttall
(7)); for example, a 4-cell stack has accumulated more than
34,000 hours of operation, mostly at an operating temperature of
82^oC, and with current densities cycling up to 260 amps/ft^2; the
performance has remained invariant over the entire period. Another
single cell module has been tested at a temperature of 104^oC and a
current density of 500 amps/ft^2 for 5,000 hours, again with no
measurable degradation in performance. What is the precise
scientific reason for this remarkable life capability, which is
believed to be unique in the fuel cell field?

Apart from the Gemini space flights, improved versions of the
IEM fuel cells have been used in the Biosatellite programme, and
in another high altitude Aerostat programme. And in addition to
these fuel cell applications, the same basic cell design is being
used in the water electrolysis mode to generate oxygen for space-
craft and submarine life support systems, and in an oxygen con-
centrator mode (which combines an air-breathing fuel cell cathode
with an electrolysis anode) to generate oxygen for the crew in
combat aircraft. (Titterington & Fickett (8); Nuttall &
Titterington (9)).

2.2 *Tentative Conclusions from Space Applications*

It is of course an immense advantage that these rather special
applications for fuel cells, and also for advanced designs of
electrolysers, have been achieved in the United States, and a
great deal of experience has been accumulated. What has all this
told us so far? I would suggest the following:-

1) Hydrogen/oxygen fuel cells can now be made which will last for
a long time without degradation in performance, even when operating at relatively high current densities. This is of course
of fundamental importance in almost all commercial applications.

2) All these specialized fuel cells will now start-up from cold, without the assistance of an external source of power, but this has necessitated the use of substantial amounts of precious metal catalyst.

3) It is essential to humidify the reactant gas, especially in the case of the IEM cell; this prevents localized drying out and the consequent stressing of the solid polymer electrolyte.

4) It has been found necessary to circulate a coolant, such as glycol, between adjacent cells or between every other cell, to remove waste heat; we used the circulating hydrogen to remove both the waste heat and the product water, and this method was also used in the Apollo units. But with the lower cell temperature and gas pressures, coupled with high current densities, now possible, a liquid coolant becomes essential.

5) The trend now, in most but not all fuel cell designs, is towards the use of a trapped electrolyte; this allows very close spacing of the electrodes, thus reducing electrolyte resistance, but prevents the electrodes from touching. In the arrangement which we used in Cambridge, we had a free electrolyte which was circulated through all the cells in parallel, and this often led to cell reversal when the battery was cold, owing to shunt currents between cells; this would be awkward in a practical unit.

6) Although the specific weight in these new designs has been reduced by a factor of 4, and operating life has been extended by a factor of 5, the cost per kw, though much less than that of the two early units, is still much too high for any ordinary commercial application. With regard to specific weight, it is perhaps important to point out that it is the overall specific weight of a fuel cell system, complete with its fuel supply, which is really relevant. This factor depends on the length of mission, the weight of the fuel cell itself being most important for brief missions, and the weight of the reactants most important for long missions.

7) The control of temperature, and of the rate of removal of the product water, do not seem to have presented any serious problems.

8) Much experience has now been obtained regarding the long-term corrosion of electrodes and other cell parts in both acid and alkaline environments; this will be invaluable in any future applications of fuel cells.

9) With regard to the use of pure oxygen in fuel cells, it has been stated that almost any material, not already in the oxidized state, will ignite and burn in a pure oxygen environment given a

sufficient energy release or an elevation to its auto-ignition
temperature; so great care must be exercised in the selection of
materials if pure oxygen is chosen as the oxidant, instead of air.
In this connection, I think it is correct to say that the only
serious accidents in the American space programme have been
associated with the use of oxygen rather than hydrogen; one
accident was concerned with the use of pure oxygen as the atmosphere in the capsule itself, and the other was thought to be due
to the ignition of electrical insulation in one of the liquid
oxygen containers.

2.3 Other Special Applications

What other special applications for fuel cells have been
achieved in recent years? The United States Army has made use of
very simple hydrazine/air units for the generation of electricity
for battery charging etc. in the field. Apart from this,
Engelhard Minerals and Chemicals Corporation has put two small
matrix type phosphoric acid fuel cell units on the market, one of
12 watts and the other of 750 watts, the cell temperature being
about 130°C; both can be fuelled with compressed hydrogen, and
the larger one can be supplied with an ammonia cracker, thus pre-
senting the great advantage, with respect to transport and storage,
of a liquid fuel. (Collins & Adlhart (10), Collins, Michalek &
Brink (11), and Adlhart (12)). These units are intended for use
as long-life power sources for telecommunication relays, offshore
navigational aids etc.; and in spite of the high initial cost, a
number of units are in operation and are proving their worth,
owing to reduced maintenance costs, when compared with small
engine-generators. An operating life of 15,000 hours is tenta-
tively projected, and this estimate is based on a great deal of
experience with single cells and cell stacks. I regard this as a
very important development for the following reasons, 1) this is
a genuine terrestrial application, 2) these units use atmospheric air instead of pure oxygen, and 3) they are hopefully
achieving a market, albeit a small one, in competition with conventional power sources. Let us hope they prove to be a commercial success.

As regards underwater applications for fuel cells, a 20 kw
hydrogen-oxygen unit has been built by Pratt & Whitney Aircraft,
now known as United Technologies, for the United States Navy.
(United Technologies (13)). This is for the propulsion of the
Deep Submergence Search Vehicle, and will make use of spherical
vessels for the storage of the two gases under pressure. An
endurance test on a 30 cell battery over a period of 2,345 hours
showed only a very small drop in performance An alkaline
electrolyte of KOH is employed, screen electrodes activated with

platinum catalysts, the electrodes being separated by an asbestos matrix. Cooling is achieved by a heat-exchange liquid, some of the waste heat being extracted through the wall of the pressure vessel surrounding the cell stack, and the rest from the condenser.

Other types of fuel cells intended for underwater use are being worked on, principally by Alsthom/Exxon, at Massy in France; (Warszawski et.al. (14), and Heath et.al. (15)). This is said to be the largest fuel cell effort outside the United States. Most of the work in the past has been concerned with soluble liquid fuels, hydrazine hydrate being dissolved in the anolyte and hydrogen peroxide in the catholyte, the electrolyte being potassium hydroxide; units as large as 100 kW. have been built, and power densities as great as 1 kW./dm^3 or 28 kW per cubic foot have been attained, a truly remarkable achievement; this is largely due to very clever engineering design; for example, twenty cells are incorporated in each centimetre of stack length, or fifty per inch. However, the cost of the reactants is high, and the life of the cells is at present rather short, though an operating life of 500 hours has now been demonstrated. So, although eminently suitable for a number of underwater applications, for example the propulsion of small submarines, it is difficult to see how wide-spread use of this kind of fuel cell could be achieved. Since 1970, most of the Alsthom/Exxon resources have been devoted to methanol/air systems, and although nothing has been published about this work, it must be presumed that this has required a major evolution of the original concept; I have been assured that great progress has been made towards the development of a commercial design of their methanol/air fuel cell, especially with regard to the use of low cost catalysts.

And apart from this, it is believed that they are now working on hydrogen/air systems. I look forward very much to the time when results can be published on these further developments, particularly with respect to hydrogen/air cells.

As regards possible applications, they are looking at both stationary and mobile uses - specialized uses at first, but ultimately vehicle propulsion and even large scale power generation.

A rather different approach to the problem of building a fuel cell unit operating on hydrazine and hydrogen peroxide has been made by the Siemens Research Laboratories in Germany (Cnobloch et al.(16)); they have built a 7 kw. unit in which the hydrogen peroxide is decomposed outside the cell stack, the gaseous oxygen being fed into the cells in the usual way. In Germany, much progress has taken place on the use of Raney metals and alloys

as electrocatalysts, and this design incorporates a Raney nickel
alloy on the anodes and Raney silver on the cathodes; when the
unit is shut down, the Raney nickel is protected against oxygen
by filling the battery with nitrogen. Although this design, including the control gear etc., has been worked out in great detail,
it is difficult to see how a fuel cell using hydrazine and hydrogen
peroxide could achieve widespread use.

Further interesting work has been carried out by the Varta
Batteries AG in Germany on fuel cells fed with a hydrogen/nitrogen
mixture obtained either from ammonia or hydrazine (Sprengel (17));
air is used as oxidant. The use of ammonia is attractive on
account of its relatively low cost, when compared with hydrazine.
Once again, Raney alloys are used as electrocatalysts, in conjunction with an alkaline electrolyte.

2.4 *Fuel Cells for Dispersed Power Generation*

Much the greatest effort to convert the fuel cell into a
device suitable for ordinary applications, in competition with
existing electrical generators, is of course being made by United
Technologies, formerly Pratt & Whitney Aircraft. They have decided that fuel cells must be made to accept a commercially
available hydrocarbon fuel, and air as oxidant. This is a far
more difficult proposition, and we never attempted it in our work
at Cambridge for this reason; we knew that it was beyond our
powers, and this is why we adhered to the use of pure hydrogen
and pure oxygen; fortunately a totally unexpected use for fuel
cells of this type arose in the American space programme.

There are certainly many attractions in the idea of generating
electricity locally, instead of in large central power stations;
these attractions have been widely discussed, and there is no need
for me to detail them now; (Orlofsky (18), Lueckel et al.(19),
King et al.(20), Fickett (21)); but one obvious advantage would
be that the waste heat, which is inevitable in any energy conversion process, could almost certainly be used for space and
water heating; this is usually difficult with conventional fossil-
fueled power stations which are mostly sited some distance from
city centres, and this applies even more force in the case of
nuclear stations.

The first big effort of United Technologies was to produce
relatively small hydrocarbon/air units, the fuel being natural
gas; this is known as the TARGET programme. (Pratt & Whitney
Aircraft (22)). The units consist of a steam reformer, complete
with shift converter, the fuel cell power section, and a D.C. to

A.C. Inverter. The initial units had a power output of 12.5 kw., but the present program is oriented towards the development of a 40 kw. unit, with an overall efficiency of 35 per cent; it is hoped that finally an efficiency of 40 per cent will be attained.

The second big effort, which is being financed by United Technologies, the Edison Electric Institute, and also by nine electric utilities, has as its objective the construction of 26 MW units; (Fickett (21), Iammartino (23)); these will be fuelled either with natural gas or liquid distillates. It is expected that the overall efficiency will be 37 per cent, though the potential exists to improve this to 40 or even 45 per cent. (United Technologies (24)).

In both cases, the improved efficiency of 40 or 45 per cent is to be attained by a different design of cell; although very little technical information has been published, the present designs are, it is believed, based on the use of very strong, e.g. 98 per cent, phosphoric acid as electrolyte, trapped in a porous or fibrous matrix of a plastic material; many of the cell parts are of carbon or graphite, and the electrodes are of carbon paper or felt, activated with platinum or platinum alloy catalysts, the temperature of operation being somewhat below 200°C. However, considerable progress has been made recently with the molten carbonate cell, the electrolyte being a mixture of alkali metal carbonates in a ceramic matrix, the operating temperature being about 650°C. The electrodes are porous nickel structures, no noble metal catalysts being required; moreover there is no need for a separate shift converter. The improved overall efficiencies of up to 45 per cent which I have just mentioned refer to this design. In the past it has been extremely difficult to demonstrate constant performance using molten carbonate cells, but now it has been shown that a stable performance can be achieved over a period of more than 7,000 hours. I do not know how this has been done, but it is said to be due to an improved design of anode; and the aim now is to demonstrate a 40,000 hour stability, using low cost manufacturing processes.

It is believed that the first generation 26 MW powerplant, based on the phosphoric acid cell, is planned for commercial service by the end of this decade; and the second generation plant, based on a molten carbonate electrolyte, is aimed at commercial application in 1985. In view of the magnitude of this program, United Technologies are now asking for Government help.

2.5 *Tentative Conclusions from other Applications*

I have tried to outline what I believe to be the major efforts

in the fuel cell field at the present time. What can we learn
from all this painstaking and devoted work? And can we detect
any overall trends? I would suggest the following:-

1) As regards fuel, compressed hydrogen is not favoured except
for very small units and for underwater use; liquid hydrogen
is judged to be impracticable and too expensive, except for
space use, though some people would not agree with this conclusion; hydrazine works well, but is too expensive for general use; cracked ammonia also works well, and is much cheaper
than hydrazine; as a result of a massive effort in the United
States, both liquid and gaseous hydrocarbon fuels have been
shown to be practical, but they must be steam-reformed before
entry into the cells; work is proceeding on the direct use of
methanol, but very little has been published about this.

2) As regards oxidants, it has been shown that air can be used
successfully instead of pure oxygen, without sacrificing too
much performance; but if an alkaline electrolyte is employed,
a CO_2 scrubber becomes essential which adds to maintenance
costs. Hydrogen peroxide may be used either dissolved in the
electrolyte, or it may be decomposed externally and the oxygen
fed into the cells in the usual way. But in any case, it is
probably too expensive except for special applications and
particularly underwater uses.

3) As regards electrolytes, potassium hydroxide solution is
still favoured where pure hydrogen, or a hydrogen-nitrogen
mixture is used as fuel gas, and where pure oxygen is chosen
as oxidant, the main exception being the acidic-ion exchange
membrane used by G.E.; the main reasons for preferring alkali
are, I think, better performance than with acid, and also a
greater freedom in the choice of materials of construction.
When air is used as oxidant, I do not know of any really practical way of overcoming the problem of gradual carbonation of
an alkaline electrolyte from the CO_2 content in air. I will
return to this again later. A great deal of experience has now
been obtained with phosphoric acid as electrolyte, using either
steam-reformed hydrocarbons, cracked ammonia or pure hydrogen
as fuel, and air as oxidant; I have already referred to the
long life claimed for the Engelhard design, the cell temperature
being 130°C. I do not know the operating temperature of the
United Technologies phosphoric acid cells, and I have not seen
any definite figures for life, but it is obvious that they, and
also the gas and electric utilities, have great confidence in
this system; and it was stated in 1973 that the economic operating period was approximately 16,000 hours. (Martin (25)).
Next, great progress has been made recently with fuel cells
employing molten carbonate electrolytes, this work having been

intensively studied at the Sondes Place Research Institute in the U.K., and more recently at the Institute of Gas Technology in Chicago. Molten carbonates have always been the natural choice, if mixtures of hydrogen and carbon dioxide are used on the fuel side, and now that much longer life of the cells has been demonstrated, their use as electrolyte in batteries fed with steam-reformed hydrocarbon fuels looks much more hopeful.

In most recent fuel cell developments, the electrolyte is held in a matrix of some porous or fibrous non-conducting material, and is not circulated, the exceptions being those cells which depend on soluble reactants. And lastly, the new ion-exchange membranes appear to have some unique properties when used as electrolytes; however, the cost of the catalysts and of certain cell parts is still rather high, though there is said to be hope that these costs can be reduced substantially in the future.

4) As regards electrode design, there seems to be general acceptance in most low temperature cells of the so-called screen electrode, the basis of which is often a woven wire gauze onto which is deposited a mixture of catalyst — often platinum black with a large surface area — and Teflon. In phosphoric acid cells, carbon paper or felt is used successfully, instead of wire gauze, thus reducing cost. These electrodes have unique properties in establishing an extensive triple contact of gas, electrolyte and electrochemically active electrical conductor, which is so essential if high current densities are to be achieved.

However, in spite of the hydrophobic properties of Teflon, it is sometimes difficult to prevent the electrodes from becoming flooded with electrolyte, so it is usual now to trap the electrolyte in a porous or fibrous matrix of some insulating material which has hydrophilic properties. Thus, the degree of flooding of the electrodes can be controlled by the total volume of electrolyte, the pore size of the matrix material being in general smaller than that in the electrodes. In the case of the 98 per cent phosphoric acid cells, the total volume of the electrolyte can be controlled fairly easily because the vapour pressure varies quite rapidly with concentration; so the rate of evaporation of the product water into the air stream is controlled automatically, to some extent, under all conditions of load. The same kind of principle was used in the Apollo cells, which used 75–85 per cent potassium hydroxide solution, although in this case the electrodes were of biporous design and the electrolyte was free.

It is very satisfactory that so many different approaches
are still being made to the difficult problem of converting the
fuel cell into a commercial proposition for ordinary use, and
it is not easy to decide at present which will be most rewarding in the long term.

5) Finally, problems concerning temperature control, rate of
water removal, instant response to rapid changes in load, etc.,
seem to have been solved satisfactorily; and no special difficulties seem to have arisen when individual cells are built up
into a battery. The vital factor must always be the design of
the cells themselves. However, it should be mentioned here
that the cost of the so-called auxiliaries, such as pumps,
blowers, controls etc., is still a cause for anxiety, especially in the more complex fuel cell systems.

3.   PROBABLE MAIN REASONS FOR DROP IN CELL
PERFORMANCE WITH TIME

I have always believed that durability is a vital factor,
especially of course in any commercial applications of fuel cells.
This does not seem to have been discussed very much in published
papers, so I think it is worthwhile to make a tentative list of
the main reasons for a fall in performance with time:-

1) Poisoning of the catalysts, especially in the case of the
fuel electrode, or anode, although air cathodes are not always
immune from this trouble. This refers mainly to low and medium
temperature cells. Sulphur in any form in the fuel gas is
usually detrimental; and carbon monoxide tends to poison
platinum catalysts below a temperature of about 135°C. Poisoning can also arise from impurities coming from corrosion products from cell parts, such as diaphragms, cell frames or
gaskets.

2) Loss of surface area of the catalysts, for example platinum
black, owing, for instance, to solution and reprecipitation of
platinum crystallites (Tseung & Dhara (26)), or to sintering
of other metals in the case of high temperature cells.

3) Increase in electrical resistance of any oxide layer on the
cathodes.

4) Corrosion of metal parts, particularly cathodes, and deposition of these metals in the diaphragms or on the anodes.

5) Gradual oxidation of porous metals, if used, leading to a decrease in pore size and thus flooding of the micropores.

6) Gradual flooding of pores in hydrophobic electrodes.

7) Accumulation of inert gases inside the electrodes.

8) Gradual carbonation of alkaline electrolytes, due to the CO_2 content of air, or to the CO and CO_2 in fuel gases.

9) In molten carbonate cells, there is the danger of cracks developing in the paste electrolyte, due to thermal stresses in temperature cycling.

This is only a tentative and probably incomplete list of the troubles which may be encountered in producing a really long-life fuel cell. Corrosion has of course always been our greatest enemy, but it has been shown that careful choice of materials and meticulous attention to cell design can lead to really long life. Rates of corrosion increase rapidly as the temperature is increased, though on the other hand cell performance is improved. So if a long life is required, it is best on the whole to be conservative with respect to temperature of operation.

I would like to see all these problems brought into the open and discussed freely by the people who have real practical experience; only in this way shall we achieve real progress, and I feel sure it would also increase confidence in the whole fuel cell idea. In fact, I believe that in the future it will be found that long life is one of the fuel cell's main attractions.

However, engine-generators do not last for ever, so there is no reason to be despondent if fuel cells are not quite perfect in this respect either, especially in the early stages of development.

4.   STORAGE OF ELECTRICAL ENERGY

I have always had a conviction that fuel cells would provide an answer to the difficult problem of storing electrical energy both efficiently and economically. This problem has sometimes been referred to as the greatest unsolved problem in technology; and it is beginning to look as if at last there is some hope of fulfilment.

I originally put forward the hydrogen/oxygen cell basically as a storage device, the two gases being produced electrolytically in electrolysers, and then stored at high pressure in steel gas vessels. (Bacon (27)). I have often been laughed at for

suggesting that this would ever be accepted in practice, and the
fact still remains that nobody likes to use compressed gases if
they can avoid it; this applies particularly to mobile applications. It has now been shown that atmospheric air can be used
quite effectively as oxidant in fuel cells, but the difficulty of
storing hydrogen really effectively remains. It could be argued
that we shall have to wait until we have piped hydrogen generally
available, as may happen in the future if the Hydrogen Economy
comes to fruition; the hydrogen would be stored in the high
pressure pipelines, and could then be used in stationary applications of fuel cells. But could some other form of hydrogen
storage be adopted before this comes about?

In the United States, a private car has quite recently been
driven on hydrogen, the gas being used as fuel for a conventional
internal combustion engine; the hydrogen was stored as a liquid,
using the latest developments in cryogenics developed for the
space programme. But the general opinion seems to be that liquid
hydrogen will not be accepted as a fuel for general use for many
years to come, owing to problems of safety in the storage of large
volumes of the liquid, and the difficulties involved in its
transfer to a vehicle.

I have already referred to the use of liquid ammonia as a fuel
for fuel cells, and this could be regarded as the most obvious and
practical way of storing hydrogen, though it has limitations,
especially for traction. The use of hydrazine hydrate also has
limitations. Would it be possible to complex ammonia with some
other material, with the object of reducing the vapour pressure
substantially?

What about the use of metal hydrides as a very direct way of
storing hydrogen? Many of us have been interested in this possibility for a long time. Recently some first class basic work on
the subject has been initiated. The major part of this work, has,
I think, been done at the Brookhaven National Laboratory, Upton,
New York; (many authors (28-33)); but some other excellent work
has been reported from Battelle, Geneva (Jonville *et al.* (34)),
and from the Philips Research Laboratories in Holland (van Mal,
(35)).

Workers at Brookhaven have carried out some important
engineering development work on an iron-titanium alloy, but many
other metals and alloys have been suggested, and it remains to be
seen what the final outcome is. The iron-titanium alloy is
probably rather heavy for the majority of mobile applications, but
it has other favourable qualities and it is relatively cheap; as
a result of this it has been decided that initially this hydride
storage should be tried on a practical basis as a load-levelling

device by an electric utility. The plan is to use a commercially available low pressure electrolyser, a gas compressor, a series of tubular storage vessels filled with the alloy powder and incorporating means for introducing or removing heat, and a fuel cell stack, plus the usual control gear. In spite of the complication involved in this arrangement, it is believed that it can be made to pay for itself. It would have to compete with alternative peaking devices, such as gas turbines, pumped water storage and batteries. In order to get an overall picture of this scheme, it is of course necessary to refer to the published literature.

One criticism which is always levelled against storage devices of this kind is the poor overall efficiency which is usually shown; this is due, to a large extent, to the inefficiency of the oxygen electrodes, in both the electrolyser and fuel cell modes, and I would like to refer to this again later. Another criticism is life, especially of the fuel cell; but now I think it can be fairly said that long life has been demonstrated, certainly in some designs.

This, to me, is much the most interesting development in the whole field of energy storage, especially as some real engineering work is being done, as well as the basic studies on metal hydrides. There is no doubt that a novel storage device is very badly needed. Automotive applications are also being considered at Brookhaven, the hydrogen being used in an engine; this could be a very useful stepping-stone towards the eventual use of hydrogen/air fuel cells in mobile applications, the inducement being the higher efficiency which should be attained with fuel cells, thus leading to a lower weight of the hydrogen storage device. Present-day low temperature hydrogen/air fuel cells have been shown to have a thermal efficiency of about 50 per cent; and the higher temperature hydrogen/oxygen units used in the Apollo programme showed an overall efficiency, during the whole Apollo 11 mission, of 72 per cent based on the total heat of the reaction and 87 per cent based on the free energy. How do these figures compare with small engines? I am assured that a petrol engine, on the test bed, on full throttle, and on constant speed and load, may give an efficiency as high as 30 per cent, based on the lower calorific value of the fuel; however, on a standard driving cycle, as defined by the Environmental Protection Agency in the United States, it may be as low as 10 or even 8 per cent. Diesel engines are of course better, and I am told that the corresponding efficiency figures are of the order of 40 per cent on the test bed, and 15 per cent on a standard driving cycle. It should also be borne in mind, when making these comparisons, that the efficiency of a fuel cell as a general rule improves on light loads, whereas the reverse is true of an engine; and in practice, on a standard driving cycle, the power unit is working for much of the time, on relatively light loads.

It must be admitted that, looked at superficially, the metal hydride method of storing hydrogen, appears unfavourable on a weight basis; it has been stated that in practice the composition limits may be expected to range from $FeTiH_{0.10}$ to $FeTiH_{1.85}$, to give a total of 1.67 wt. per cent available hydrogen. However, in stationary applications, it is volume rather than weight which is the critical parameter, and this is the reason why Brookhaven is initially concentrating on stationary rather than mobile uses. It is worthwhile pointing out that on a volume basis, the hydride contains more hydrogen than does liquid hydrogen.

Other metal hydrides, particularly magnesium hydride, would show a better weight ratio than the iron-titanium alloy, but in other respects would be less favourable, especially with regard to heat of formation and decomposition temperature. Also certain alloys of the formula AB_5, where A is a rare earth metal and B is Fe,Co,Ni or Cu can absorb up to seven hydrogen atoms per AB_5 unit, and many of them have high dissociation pressures at room temperature. I, for one, intend to watch these developments with the greatest interest.

5.   PROPOSALS FOR FUTURE WORK ON FUEL CELLS

I have often wondered what I would do, in the light of present knowledge, if I had the opportunity to start work on fuel cells all over again, assuming that I had the facilities and the finance, and of course the youth and strength.

I would first try to decide the difficult question of what application one should have in mind. Sir Winston Churchill once observed that looking ahead was good but that on the whole it was useful only to look ahead as far as one could see. How far ahead can we see at the present time? I can see fuel cells coming into use as long-life power sources for telecommunication relays etc., using either compressed hydrogen or cracked ammonia, but the market is probably not very large, and units have already been developed for this purpose. Likewise, the market for underwater power sources is probably not very large and demand for them has been strangely slow in coming. The idea of generating electricity locally by means of fuel cells using a hydrocarbon fuel and air is certainly very attractive, but he would be a bold man indeed who tried now to compete with large and powerful organizations such as United Technologies, who have vast experience to draw on. Mobile uses for fuel cells would seem to be too difficult at the present time, though I have always hoped that this would be achieved in time, especially when naturally occurring hydrocarbon fuels really become scarce and expensive; also, at present, it would be very difficult to compete, on the basis of capital cost, with mass produced engines. Dissolved fuels, such as methanol and hydrazine, are certainly interesting, but as an engineer I am waiting to see results published which show really long life and reasonable cost for methanol fuel cells, and moreover in the case of hydrazine there seems very little prospect that the cost of the fuel will ever become competitive for ordinary purposes.

So once again I am driven back to energy storage. I see a great need in the world for a novel and improved method of energy storage, and so far this has beaten us all, although great efforts are being made in the sphere of high energy batteries. In my view, the main difficulty with respect to the use of fuel cells for this purpose has always been the lack of a really effective method for storing hydrogen; the other difficulties are mainly concerned with life and cost of the fuel cells themselves, and I personally believe that both will be overcome in time. I have already referred to the excellent work at Brookhaven and elsewhere on metal hydrides; the iron-titanium alloy developed at Brookhaven looks very hopeful for stationary use, and the programme has now been broadened to cover not only electric utility storage, but also the storage of solar and wind energy. Will some other metal hydride, or more likely metal alloy hydride, be discovered which is lighter than iron-titanium? This would broaden the field of application even further. And moreover, it should be remembered that hydrogen itself is a unique fuel as regards weight, especially if it can be used really efficiently in a fuel cell; it has always seemed remarkable to me that the small four-seater private car, that was converted to fuel cell drive by Union Carbide in the United States in about 1969, used less than 2 kg of hydrogen for a range of over 200 miles; (Kordesch (36 & 37)); the disadvantage was that the total weight of the six lightweight gas cylinders to contain the hydrogen was 82 kg.

So, if I was given the chance, I would look very closely into novel methods for storing hydrogen, and especially hydrides. I feel very sure that Brookhaven is working on the right lines.

I have not referred so far to the possible use of self-contained regenerative hydrogen/oxygen, or hydrogen/nickel oxide fuel cells (Crowe (38)). A great deal of work has been done on these systems, almost entirely in the United States, the objective being to develop a storage device for space stations, in conjunction with solar cells. It was hoped that a system could be developed which was lighter than the usual nickel-cadmium cells. I had great hopes of this at one time, but it seems now that it is going to be difficult to beat our old friend nickel-cadmium, on the grounds of either energy density, efficiency, life or reliability. This is disappointing.

The question of cost is a thorny subject, as fuel cells have
the reputation of being very expensive. Great efforts are being
made to reduce costs, and United Technologies have made great
progress with regard to the use of carbon and graphite parts in
the phosphoric acid cell, though, as far as we know, small amounts
of precious metal catalysts are still required. In the Union
Carbide alkaline cells, much use was made of carbon and plastics,
and a light loading of noble metal catalyst was used on the anodes;
the cathodes could be made to operate with carbon alone, but I
think it was usual to activate them with silver or preferably with
a cobalt-aluminium spinel. Personally, I have always been drawn
towards the use of an alkaline electrolyte, partly because of the
better performance which is often attained, but mainly because it
is more likely that cheaper catalysts can be employed and also
because corrosion problems in general are less severe; there is
also the hope of the evolution of a more nearly reversible oxygen
electrode for use in alkali. The developers of molten carbonate
cells have always argued that they avoid the use of costly
materials, especially of course catalysts.

The whole question of electrocatalysis is of course of the
greatest importance, and it is very gratifying that there is now
so much interest in the subject; I am sure there are many people
here who know much more about the subject than I do.

As I have already mentioned, one of the main criticisms
levelled against the use of hydrogen/oxygen systems for energy
storage is the rather low overall efficiency; this is mainly
due to the inefficiency of the oxygen electrodes in both modes,
namely gas evolution and gas absorption. Some very interesting
work has been going on at City University in London for some years
with the objective of evolving a more nearly reversible oxygen
electrode. (Tseung & Bevan (39 & 40)). I am sure that electrochemists are familiar with this work; it is only necessary for
me to say that in an alkaline electrolyte fuel cell, not only are
hydroxyl ions (OH^-) formed at the oxygen electrode, but also
perhydroxyl ions (HO_2^-); this results in a mixed potential which
is significantly lower than the theoretical value of 1.23V at
25°C and 1 atm. pressure, which should be obtained if hydroxyl
ions only are formed; thus the open-circuit voltage of the
hydrogen/oxygen cell is normally 1.00-1.10V. The best results
at City University have been obtained by the use of complex
oxides having a perovskite structure as electrocatalysts for
oxygen ionization in alkali, and in particular strontium doped
lanthanum cobaltite ($La Sr Co_2 O_8$), the cost of which, incidentally, would be far less than that of platinum. The low
temperature performance was poor, but at 150°C and above the
performance was superior to similar electrodes catalysed with
platinum black. Moreover, the reversible oxygen potential was

attained at both atmospheric and at higher temperatures, and the so-called peroxide step in the voltage-current density character-istic was eliminated. Not being an electrochemist, I do not fully understand the theory on which this work is based, but the ob-jective is to help to break the oxygen-oxygen bond in the O_2 molecule.

Unfortunately, it is extremely difficult to get any support for fundamental research of this kind in the U.K. now. I should mention here that some valuable work has also been going on at City University, London, on gas evolution in Teflon-bonded porous electrodes. The factors influencing the strength and stability of this type of electrode have been discussed in detail, and it was concluded that if the electrocatalysts were prepared by chemical precipitation or thermal decomposition methods, the bonds linking the individual particles together would be strong enough to resist disruption by the pressure built up in the pores during the gas evolution process.

Moreover, results showed that these Teflon-bonded porous electrodes can give a much higher performance than ordinary porous electrodes or foil electrodes for gas evolution reactions. This should be of considerable interest in the development of more efficient electrolysers, besides rechargeable metal-air batteries and chlor-alkali plants. All this information either has been, or soon will be, published in *Electrochimica Acta* (Vassie & Tseung (41)).

One other subject which I would like to see investigated more thoroughly is the use of atmospheric air in hydrogen/air cells with an alkaline electrolyte. Is this a hopeless quest? My own experience is almost entirely with alkali, so it may be that I am prejudiced; but there do seem to be a number of solid advantages over the use of acid.

A number of years ago, a method was described by Pratt & Whitney (now United Technologies) for the electrochemical de-carbonation of a potassium hydroxide electrolyte; (Handley & Meyer (42)); carbonate ions in the circulating electrolyte were converted to carbon dioxide at the anode of a special cell op-erating at a high current density; this special cell was in-corporated in the stack and it contained a diffusion barrier; this led to a build-up of concentration polarization until the electrolyte adjacent to the hydrogen electrode approached neutrality and finally started to evolve CO_2 into the hydrogen stream; this was then vented to atmosphere. This scheme never got as far as practical application, as far as I know.

My question is this; — could the same principle be used for decarbonating the alkaline solution in a CO_2 scrubber? It is of course only a vague idea. And I must admit that if all the claims for the acidic ion-exchange membrane cell, with respect to performance, life and future capital cost, are borne out in practice, my arguments for alkaline electrolytes would probably fall to the ground. On the other hand, it may be asked whether there is any chance of developing a nearly reversible oxygen electrode in acid, as opposed to alkali; I have never heard of this possibility.

There is one other field which I have often suggested should be examined really scientifically, namely gasket materials and design. This may seem a very lowly field for scientific research, but gaskets are undoubtedly of the greatest importance in both fuel cells and electrolysers, especially as they must also act as electrical insulators. At present, most of the information on the subject appears to be in the hands of the manufacturers, who are naturally loath to part with it. What makes a really good leakproof gasket? What materials do they contain, and what kinds of contamination should one expect in say strong acid or alkali? I may be wrong, but I do not know of a single standard work on the subject, to which one could refer for the fundamentals of sealing. Is this a subject suitable for investigation by a University Department?

Recapitulating, I would like to see a really determined effort to solve the age-old problem of energy storage, both on a small scale for the storage of solar and wind energy, and also on a larger scale for peak-shaving etc. by electric utilities; this would involve work on improved forms of electrolyser, on hydrogen storage, and on fuel cells designed specially for this kind of application. It may be that we are so far behind the United States in all these fields, that we could not make a useful contribution, but I would like to think that we could.

After achieving some practical experience in this field, I would like to return once more to the even more difficult task involving the use of fuel cells in mobile applications.

6. CONCLUSIONS

What can one say in conclusion? Quoting the great Francis Bacon of Elizabeth the First's time, he once said "It were good that men in their Innovations would follow the example of Time itself, which, indeed, innovateth greatly, but quietly, and by degrees scarce to be perceived". In our particular case, I take this to mean that we should not be too ambitious to start with;

bringing the fuel cell down to earth, where it must compete with other highly developed power sources in some special application for which it has special advantages, is one huge step. The use of atmospheric air instead of pure oxygen is another big step; and the widespread use of fuel cells for power production or storage is yet another. The first two steps, involving the use of fuel cells in some ground application, and the use of air instead of pure oxygen, are, I think, on the verge of being achieved in a genuine commercial application. This will give us extremely useful experience, and we can then build on this for the future.

On the other hand, Francis Bacon also said "He that will not apply new remedies must expect new evils; for time is the greatest innovator". In our case, I would interpret this as a warning that we should start preparing now for the time when fossil fuels become really scarce and expensive. Novel devices such as fuel cells take a long time to develop, and we have only just started. I would expect that our early efforts to produce practical fuel cells will be laughed at, in the same way that we now tend to laugh at the early motor cars; development cannot apparently be hurried.

I have not talked at all about the eventual exhaustion of fossil fuels, about the possibility of changes in climate caused by the discharge of vast quantities of CO_2 into the atmosphere or indeed by the ever increasing quantities of waste heat, about the Hydrogen Economy, or about synthetic fuels in general, or about alternative sources of energy such as nuclear, solar, wind or wave energy; all these have been dealt with in many excellent publications. However, these alternatives all really need the addition of some really practical and economic storage device before they can achieve their full potential. Should we not concentrate our thoughts and actions on this important, though elusive, objective? One other subject that I have not discussed in detail is electrolyser design, using all the experience which has now been obtained with fuel cells. I feel sure that this is also a very important subject for research and development; I would like to see electrolyser and fuel cell voltages gradually brought closer together; only then can we make out a really good case for this method of energy storage.

I am acutely aware of the fact that I have not really said anything new; it is the young who have the new ideas, and you must not expect too much of an old man of 71. All I can hope for is that I may perhaps have provided some material for further discussion and no doubt argument. Quoting the great Francis Bacon yet again, he once said "They are ill discoverers that think there is no land when they can see nothing but sea."

We, in the fuel cell world, have had many ups and downs; my own work has in fact been closed down four times. Nevertheless, I believe there are more fuel cell enthusiasts than ever, at the present time, and I feel sure we shall be proved right in the end. Professor Bockris has been an enthusiast since the very early days; and once again I must thank you all, and John Bockris in particular, for this wonderful opportunity to give my own personal views on this fascinating subject.

ACKNOWLEDGEMENT

The author would like to express his deep thanks to Mr. T.M. Fry, of Associated Nuclear Services, for help in preparation of this lecture.

REFERENCES

1. H.A. Liebhafsky and E.J. Cairns, *Fuel Cells and Fuel Batteries*, Wiley, New York, 1968, p. 587.

2. J.H. Russell, *19th Power Sources Conference*, PSC Publications Committee, New Jersey, 1965, p. 35.

3. C.C. Morrill, *19th Power Sources Conference*, PSC Publications Committee, New Jersey, 1965, p. 38.

4. A.C. Ching, A.P. Gillis and F.M. Plauche, *Proceedings of the 7th Intersociety Energy Conversion Engineering Conference,* 1972, Paper No. 729064, p. 368.

5. W.E. Rice and D. Bell, *Proceedings of the 7th Intersociety Energy Conversion Engineering Conference,* 1972, Paper No. 729067, p. 390.

6. L.E. Chapman, *Proceedings of the 7th Intersociety Energy Conversion Engineering Conference*, 1972, Paper No. 729076, p. 466.

7. L.J. Nuttall, *Solid Polymer Electrolysis Fuel Cell Status Report*, General Electric, Aircraft Equipment Division, Wilmington, Mass.

8. W.A. Titterington and A.P. Fickett, *Proceedings of the 8th Intersociety Energy Conversion Engineering Conference,* 1973, Paper No. 739020, p. 574.

9. L.J. Nuttall and W.A. Titterington, *Conference on the Electrolytic Production of Hydrogen*, City University, London, England, February, 1975.

10. M.F. Collins and O.J. Adlhart, *25th Power Sources Symposium*, PSC Publications Committee, New Jersey, 1972.

11. M.F. Collins, R. Michalek and W. Brink, *Proceedings of the 7th Intersociety Energy Conversion Engineering Conference*, 1972, Paper No. 729006, p. 32.

12. O.J. Adlhart, *Proceedings of the 7th Intersociety Energy Conversion Engineering Conference*, 1972, Paper No. 729163, p. 1097.

13. United Technologies, Private Communication, 1975.

14. B. Warszawski, B. Verger and J.C. Dumas, Marine Tech. Soc. J., 5(1) (1971) 28.

15. C.E. Heath, B. Verger, C. Hespel and P. Fauvel, *Proceedings of the 9th Intersociety Energy Conversion Engineering Conference*, 1974, Paper No. 749069,, p. 646.

16. H. Cnobloch, H. Kohlmuller and D. Kuhl, Energy Conversion, 14 (1975) 75.

17. D. Sprengel, Energy Conversion, 14 (1975) 123.

18. S. Orlofsky, *Proceedings of the Eighth World Energy Conference*, Bucharest, 1971, Paper 2, p. 5.

19. W.J. Lueckel, L.G. Eklund and S.H. Law, Trans. I.E.E.E., New York, 1972, Paper No. T72, p. 235.

20. J.M. King and S.H. Folstad, *Proceedings of the 8th Intersociety Energy Conversion Engineering Conference*, 1973, Paper No. 739056, p. 111.

21. A.P. Fickett, *American Power Conference*, April 1975.

22. Pratt and Whitney Aircraft, *A Review of the TARGET Program*, 1972, PWA, East Hartford, Conn.

23. N.R. Iammartino, *Fuel Cells: Fact or Fiction*, Chemical Engineering, May 27, 1974, p. 62.

24. United Technologies, *A National Fuel Cell Program*, April 30, 1975.

25. C. Martin, Aviation Week & Space Technology, Jan. 1, 1973, p. 56.

26. A.C.C. Tseung and S.C. Dhara, Electrochim. Acta, 20 (1975) 681.

27. F.T. Bacon, BEAMA Journal, 61(199) (1954) 6.

28. R.H. Wiswall and J.J. Reilly, *Proceedings of the 7th Intersociety Energy Conversion Engineering Conference,* 1972, Paper No. 729210, p. 1342.

29. J.J. Reilly and R.H. Wiswall, Inorganic Chemistry 13 (1974) 218.

30. J.J. Reilly, K.C. Hoffman, G. Strickland and R.H. Wiswall, *26th Annual Power Sources Conference,* May 1974.

31. J.M. Burger, P.A. Lewis, R.J. Isler, F.J. Salzano and J.M. King, *Proceedings of the 9th Intersociety Energy Conversion Engineering Conference,* 1974, Paper No. 749034, p. 428.

32. C.H. Waide, J.J. Reilly and R.H. Wiswall, *THEME Conference,* Miami, 1974.

33. F.J. Salzano, E.A. Cherniavsky, R.J. Isler and K.C. Hoffman, *THEME Conference,* Miami, 1974.

34. P. Jonville, H. Stohr, R. Funk and M. Kornmann, *THEME Conference,* Miami, 1974.

35. H.H. van Mal, *THEME Conference,* Miami, 1974.

36. K.V. Kordesch, J. Electrochem. Soc., 118 (1971) 812.

37. K.V. Kordesch, *Proceedings of the 6th Intersociety Energy Conversion Engineering Conference,* 1971, Paper No. 719015, p. 103.

38. B.J. Crowe, *Fuel Cells: a Survey,* NASA SP-5115.   Technology Utilization Office, NASA, Washington, D.C. 1973.

39. A.C.C. Tseung and H.L. Bevan, U.K. Patent Application No. 19292/70, 1970.

40. A.C.C. Tseung and H.L. Bevan, J. Electroanal. Chem., 45 (1973) 429.

41. P.R. Vassie and A.C.C. Tseung, Electrochim. Acta 20 (1975) 759, 763.

42. L.M. Handley and A.P. Meyer, *24 Power Sources Symposium,* PSC. Publications Committee, New Jersey, 1972, p. 188.

ELECTROCHEMISTRY IN THE SOLAR ECONOMY

A. J. Appleby

Laboratoire d'Electrolyse du C.N.R.S.

92-Bellevue, France

INTRODUCTION

Since the leadtime for the widespread application of any new
technology is of the order of thirty years, valid discussion of the
above topic is restricted to the 21st century.   Any predictions
concerning energy utilization, or the state of society, at that
time are necessarily hazardous, especially in view of the recent
Ford Foundation Report (1) on the energy future of the United States
to the end of the present century.   This report, which represents
an extrapolation of current energy technology and conventional
economic analysis, envisages a wide range of scenarios for what
remains of the present fossil-fuel economy, involving vastly different predictions concerning energy consumption and G.N.P.   The
dangers of such analyses are made immediately apparent by the fact
that the postulated figures for the baseline year (1975) are fairly
wide of the mark.   However, in spite of the above comments, I
believe that some reasonable assumptions may be made concerning the
energy economy of the O.E.C.D. countries (and of other nations with
similar development) into the 21st century and beyond.   Currently,
about 35% of the primary energy input in these countries is represented by imported oil.   For the E.E.C. countries and Japan,
whose energy requirements are (1973) about 35% of the O.E.C.D.
total (2), imported oil fuels are approximately 67% of the overall
economy.   It is clear that this remarkably low-cost supply of
primary energy has allowed a new, twentieth-century, industrial
revolution in these countries.   The question, therefore, is
whether this revolution may be perpetuated into the post-fossil-
fuel age.

51

   While it is currently not evident that the end of the 20th
century should correspond to that of the age of fossil fuel, in
view of the enormous reserves of coal in certain O.E.C.D. countries
(in particular, the U.S.), economics give a reasonable insight into
why this will be so.   Coal was displaced from its predominant
position as a primary energy source by oil and natural gas because
it required too much energy (in the form of work) for its extraction
and transportation.   Work (i.e. electricity) costs based on nuclear
power are now lower than those based on both oil and coal, in
particular when the environmental consequences of utilization of
coal are considered.   In addition, recent evaluations of the
problems of coal extraction show that its use as a substitute
energy source for oil or gas will be limited (3).   As indicated
previously, environmental, labor and transportation problems show
that coal supplies will not be able to keep up with the historical
projection of future energy requirements in the U.S. (4), and we
may assume that the same conclusion will be even more appropriate
in the case of other O.E.C.D. countries lacking coal reserves.
Based on historic growth rates, oil production will be unable to
keep up with demand after the end of this century (5).   The
importance of oil conservation as chemical feedstock is underlined
by the fact that currently almost 10% of all oil used in the O.E.C.D.
countries (11.3% in North America) is for non-energy purposes (2).
The use of petroleum-derived plastic materials is certain to grow
in the future, for purely economic reasons: their energy contents
are lower than corresponding metal or glass products, and are in
addition recoverable;   their lower densities mean less energy consumption in use.   Conservation of petroleum resources to provide
several centuries of recyclable raw material is therefore vital.

   We conclude that if society is going to exist in anything
resembling its present form into the 21st century and beyond, then
the economy will be largely fueled by non-fossil energy.   I cannot
however accept the view that the supply of this energy will be
demand-limited, and that society will then be even more energy-
intensive than it is today, consuming twice as much energy per ˙
capita in the year 2000 as today (6).   The reason for this is that
future non-fossil energy will be very capital-intensive, hence very
costly.   Energy is needed by society both in the form of heat and
work.   Primary energy to provide the latter (for electrical generation and transportation) corresponded to over 52% of the total
primary energy requirement in the U.S. in 1973 (2).   The remainder
of the energy requirements consisted of the direct use of fuel for
heating purposes (about 30% for industrial process heat, the
remainder for use in homes and commercial premises).   We will
briefly examine the effects of providing this energy in a non-fossil-
fuel form by considering nuclear fission power, for which costing
is well established.

Currently, the electric utilities in the U.S. consume about 12% of the total available capital for industrial investment (7). Using the latter figure, which largely results from adding over 7% of system capacity per annum, we can make some approximate guesses concerning the capital requirements of a nuclear energy industry providing all sectors of society, expanding at the same rate as the G.N.P., with replacement every 25 years. Similarly, using the recognized figures of $3.60/Gj (1970 dollars) (8) for nuclear electricity, and 90c/Gj for nuclear (HTR) heat (9), we can estimate total energy cost as a percentage of the G.N.P., according to a number of different scenarios. For simplicity, the energy consumption pattern in the different sectors is considered to be the same as that today. Energy breakdown data are from Ref. 10.

Scenario 1: All electricity provided by local nuclear plants, all heat input provided by non-fossil hydrogen produced by electrolysis at remote power plants. Hydrogen cost (delivered) $5/Gj (1970 dollars) (4,11). Capital requirement: 145% of maximum available; Energy cost: 21% of G.N.P.

Scenario 2: As above, but with all residential and commercial energy use provided by electricity with higher utilization than at present. Capital requirement: 125% of maximum available; Energy cost (without heat pumps) 17.6% of G.N.P., (with heat pumps) *ca.* 16% of G.N.P.

Scenario 3: As above, with heat pumps with a coefficient of performance of 3 installed in all residential and commercial premises, and with all process heat provided by nuclear heat. Capital requirements: 82% of maximum available; Energy cost: 10.9% of G.N.P.

Scenario 4: As above, but with all synthetic fuel IC engined private automobiles replaced by electric vehicles. Capital requirement: 65% of maximum available; Energy cost: 8.6% of G.N.P.

While the above estimated costs may change somewhat according to the technology used (e.g. thermochemical hydrogen, which will cost $3.75-$5.00 in 1970 dollars, calculated for practically attainable efficiencies (4)), the general picture will not be affected. In fact, the energy cost per unit of G.N.P. may be much worse than the above figures if capital has to be raised on the free market, since low (7-8%) interest rates are assumed. These would not apply to future society, in which demand for raw materials of increasing energy content as resources become more dilute, accompanied by increased energy costs, will cause a highly inflationary economy. It may of course be argued that a breakthrough in technical progress may occur, e.g. for solar energy utilization or fusion power, that will lower costs or capital requirements. I feel that this is improbable, on the grounds that the increasing

difficulty of application of a technology is always accompanied by
an increasing capital-intensiveness.   This is even more apparent
in the case where energy prices rise more rapidly than those of
other goods and services, since the special materials required by
these new technologies are generally scarce and possess high energy
contents.   Costs of new installations will therefore rise faster
than the growth of the economy as a whole as energy prices rise,
and vice versa, in a feedback effect.   The overall situation does
not change if we examine the use of some of the 'softer' technologies:
eolian, wave or ocean thermal power (12,13), or solar thermal
electricity (14).   For the first two energy sources, we can expect
investment costs of the order of $500 per peak kw, written off over
a limited number of hours per year, to which an energy storage
potential (e.g. via electrolytic hydrogen) must be added.   As will
be shown below, total costs for such processes will be higher than
those for hydrogen production from nuclear heat.   An exception
may perhaps be made for the case of electrical energy derived from
thermal gradients in the tropical oceans (13).   Such installations,
based on optimistic investment costs of $250/kw, may conceivably be
cheap at source in the form of bus-bar electricity (perhaps just
over half the cost of conventional BWR or PWR power – of the order
of 7 mils per kwh, or about $2.10/Gj), but will require long-distance
energy transport to be economically useful.   Such energy transport,
in the form of hydrogen carried by long-distance pipeline (3000
miles), will add $1.00/Gj to the minimum production cost of $3.00/Gj,
based on very optimistic technology (see Ref. 11).   Such plants
will only be appropriate for the case of Scenarios 1 and 2 above.
Similarly, we will see below that other types of solar power systems
have capital requirements and energy costs that are at least as
high as those for nuclear power, when their low utilization per
year and storage requirements are considered, the only exception
being low-temperature solar heating.

      In view of the above, it is clear that Scenarios 1 and 2
(direct substitution of non-fossil energy) will be unattainable in
the future, and that primary emphasis will have to be placed on a
policy of conservation.   This policy will in itself be capital-
intensive, but will be so for the consumer, rather than the producer
(e.g. investment in heat-pumps or improved insulation).   Its overall
effect will be much healthier than a policy based on direct substitution: it may be argued that such a policy will be possible
if energy capital is raised outside the money market, e.g. as a tax
corresponding to a certain percentage of the G.N.P.   However, such
a program would not only have dangerous political, economic and
environmental consequences, but would only serve to delay (and
render more difficult) the introduction of the inevitable conservationist society.

      The energy conservation policy will apply to all countries,
not only to those with the highest per capita energy use (North

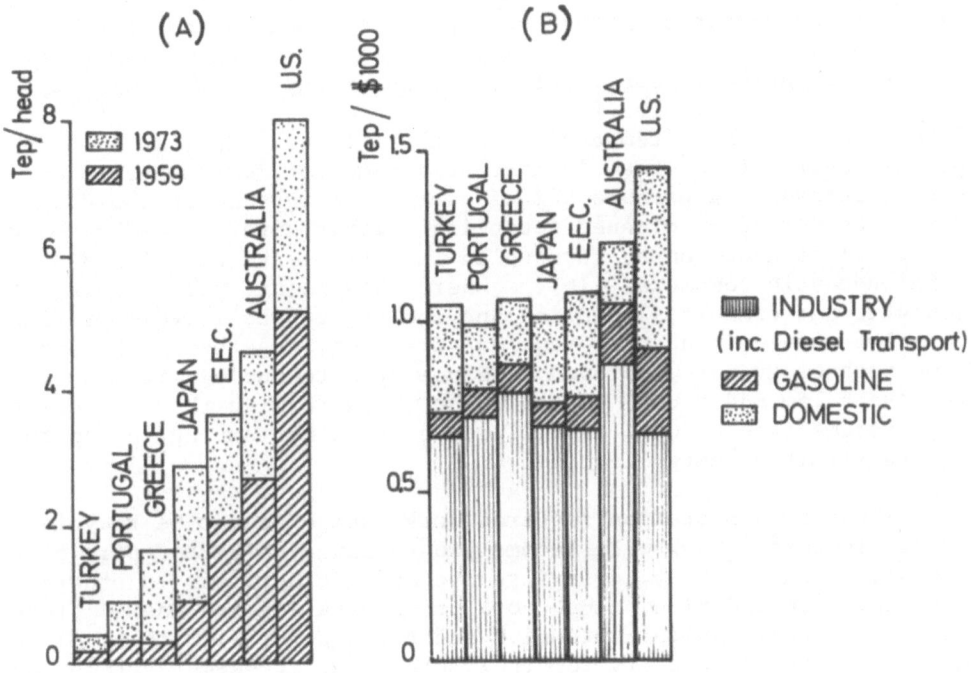

Fig. 1   Energy Consumption for some O.E.C.D. Countries: (A) per
Inhabitant for 1959 and 1973, and (B) per Unit of Gross
National Product for 1971.   Tep = Tonne Equivalent
Petroleum.

America and Australia).   This is clearly shown by the plots in
Fig. 1 above, which give energy data per inhabitant, and per G.N.P.
unit, in a range of O.E.C.D. countries for recent years (2).
Despite a fairly wide variation of total energy use per $1000 of
G.N.P., it can be seen that the useful part of this energy (i.e.
total energy minus local heating and electricity use, together with
road gasoline use), representing a good approximation to the capital-
producing part of the G.N.P., is independent of national income*.
The ability of different countries to raise capital to construct
new, non-fossil, energy sources will therefore be independent of
national income and energy use, and all will be constrained to follow
the same conservationist course.   Evidently, those countries which
have not yet developed the habit of high personal energy use will

---

* The case of Australia seems to be somewhat anomalous.   This is
perhaps a result of the very high percentage of coal use in its
economy.   Other exceptions are Norway and Luxembourg (15).

have fewer problems in following such a policy.   This point empha-
sizes why conservation should be applied as soon as possible.

Conservation in some sectors of the energy economy will be
comparatively difficult (for example, those requiring energy in the
form of work or high-temperature heat).   In contrast, much of the
primary energy input that is currently used for low-temperature
heat (representing perhaps 50% of current energy use in the U.S.)
may be conserved by planned recycling, either directly or by the use
of efficient heat-pumps, in coupled industrial-industrial and indus-
trial-domestic complexes (16).   Where this is not practicable,
domestic and commercial heating and cooling can be largely provided
by solar panels if heat storage and transfer, together with insulation technology, are improved.   Since such technology is expected
to have a low additional capital cost compared with that for existing systems (see below), it will represent a net saving in the future
capital-limited energy economy.

Since transportation requires work, which cannot be recycled,
energy in this sector will become progressively a larger part of the
overall energy economy as savings are made in heat use in other
sectors.   In addition, since both transportation and energy production are highly capital-intensive and require (or will require) very
energy-intensive industry for their creation, it seems certain that
future society will not be able to afford the luxury of both.
Currently, industry is essentially transportation-based, and is using
valuable mineral resources, as well as valuable oil, at an alarming
rate.   To avoid a future 'minerals crisis' (17), as well as to allow
the realization of a non-fossil energy economy in terms of capital
and industrial possibilities, transportation (i.e. the private auto-
mobile) will have to be seriously pruned.   This will cut down the
overall energy requirement of society, in particular in North America,
and allow the possibility of a reorientation of industry to give a
stable, low-growth, self-perpetuating economy based on replacement
and renewal of non-fossil energy production facilities, with emphasis
on low waste and recycling of mineral resources.

As we have seen above, solar energy, in the form of locally
supplied low-grade heat, will make an important contribution to the
realization of this economy.   The object of this paper is to examine
ways in which electrochemistry, associated with solar primary energy,
will be used to supply the higher-grade heat and work input in such
a society, and to provide a valid alternative to the dangers of a
plutonium economy.

THE ENERGY INDUSTRIES IN THE SOLAR ECONOMY

Unlike nuclear power, solar-based systems will be able to
provide high-grade energy (in the form of electricity or chemical

fuel, i.e. hydrogen) over a range of plant scale varying from a few kilowatts to many gigawatts.   While the optimum type of plant, from the viewpoint of capital-effectiveness, will depend on geographical conditions (energy received per square meter per annum, distance between energy producer and user), there is overwhelming evidence that plants can be constructed at almost any latitude. Such plants will require an optimum of both direct and indirect solar technologies (for example: in the tropics, direct radiation plus ocean thermal gradients; at high latitudes, direct solar radiation in summer, wind and wave energy in winter).   This optimum will act in such a way as to maintain a yearly mean energy supply from (or for) a particular area.

At distances greater than about 1000 miles from the major energy production areas, energy transfer to the regions of industrial and domestic consumption will be better made in the form of non-fossil chemical fuel (hydrogen (11)), rather than delivered electricity.   Such a system will provide a storage capacity to smooth out daily and seasonal peaking, but has the disadvantage (compared with electricity) of requiring high-efficiency end-use technology.   In contrast, local systems will require only a simplified storage concept, on the accumulator pattern, since long distance transport will not be necessary.   Such storage may well refer to heat as well as to electricity.   The overall economy will require both hydrogen and electricity, depending on the type of output required.

Hydrogen Production

Non-fossil hydrogen can be manufactured from heat either directly (via cyclic thermochemical processes (18,19,9)) or by electrolysis, using thermally derived electricity (11).   Combinations of both processes may in some cases be energetically advantageous. The availability of radiation allows other possibilities (photosynthesis, radiolysis).   Below, we will briefly examine these concepts in the context of a solar economy.

Thermochemical hydrogen production.   Recent work, established on the basis of reversible thermodynamics (20), implies that in some chemically favorable cases overall thermal efficiency may be as much as 62% of that for the equivalent Carnot cycle (based on data in Ref. 20, using a calculated efficiency based on the free energy of hydrogen produced rather than on its high heating value, and assuming a maximum cycle temperature of $925^{\circ}C$: corresponding values for simpler two-step cycles at more elevated temperatures may theoretically be somewhat higher).   Actual conversion efficiencies for electricity production as a percentage of the limiting Carnot value are about 67% at lower temperatures ($550^{\circ}K$), which as a result of materials limitations falls to about 60% at $1125^{\circ}K$.

The factors that will therefore determine the ultimate production
method depend on the relative capital costs apart from that of the
heat source, the practical (as distinct from the theoretical)
efficiency for thermochemical cycles, and the efficiency for the
conversion of electricity into hydrogen by electrolysis.   Capital
costs for generation equipment, inverters and electrolyzers may be
expected to be about $200/kw (11) electrical input.   If we suppose
100% conversion of electrical energy into heat at the high heating
value of hydrogen in such electrolyzers (see below), then the value
added cost of this equipment corresponds to about $1.30/Gj of hydrogen
produced.   Since we have shown earlier that the cost of the chemical
facility (in value-added terms) will probably be of the same order
of magnitude as this figure (4), the main parameter controlling the
economic production of hydrogen will be the conversion factor for
heat into work in both cases, so that the value-added figure for
the non-fossil heat source is reduced to a minimum.

An electrochemical analog of the thermal processes is useful
to obtain an idea of the irreversible factors involved in practical
chemical reactions.   If we neglect diffusion-controlled limitations,
we can use the thermal equivalent of the Butler-Volmer equation (21),
i.e. the Marcelin-De Donder equation (22), for reactions with real
rates*:

$$k = k_o \left(\exp \overrightarrow{A}/RT - \exp \overleftarrow{A}/RT \right) \qquad (1)$$

in which k is the forward net rate, k_o is the equilibrium exchange
rate, and $\overrightarrow{A}$, $\overleftarrow{A}$ are the forward and backward affinities.   If for
simplicity we consider a gas-solid process (ideal for a separation
step, with forward rate independent of reactant composition), a
simple calculation shows that the overall rate will be a useful
fraction of the rate determined in the absence of the back reaction
for affinity values of about 1.3RT (23).   Since such rates are
normally used for process scaling, the corresponding affinity values
should represent a reasonable approximation to those that will occur
in practice.   These irreversible thermodynamic losses must be summed
over all chemical reaction steps, and probably also over all diffusion and separation steps, in the overall process (23).

It may of course be argued that Equation 1 will not apply to
the higher temperature reactions in thermochemical processes, since
the Arrhenius law will indicate rates that are sufficiently high so
as to be essentially reversible under these conditions.   Fig. 2
indicates why this should not be so.   In any process that proceeds
at constant heat flux (as from a nuclear reactor) under steady-state
conditions, the heat-transfer surface will increase exponentially
as the temperature difference between the hot point in the system
and the site of reaction increases.   A simple calculation shows

---

* Note typographical error in Ref. 23.

that for reactions of similar energy of activation, the effect of
the available reaction surface will compensate the Arrhenius term.
This is illustrated in Fig. 2.   In consequence, the affinity effect
will apply to all reactions at all temperatures.

   A further important point is the effect of the affinity terms
on the total heat input into the system.   Classical thermodynamics
indicates that any affinity input should be equated to a direct
free energy input, which should in consequence be substituted for
part of the $T\Delta S$ term for heat input in each step (e.g. overpotential
in an electrolyzer).   This will not necessarily be true if the
affinity input occurs outside the reacting system, e.g. as pumping
work (analogous to IR drop in the leads of an electrolyzer), though
it is of course possible that some of this can be recovered as heat
in a subsequent lower-temperature step.   The effect of these
additional work terms, combined with pumping inefficiencies, will
be to substantially increase system heat requirements.   Limitations
are also placed on the types of reactions: gas-solid or gas-liquid
processes will be preferable to avoid recombination and separation,
but such reactions will be irreversible (diffusion-limited); similarly, homogeneous rapid gas-phase processes are unusable because
of the problem of recombination.   The effect of all these phenomena
will be to degrade efficiencies by about 30-40% for 4-step processes.
Overall, it would seem that thermochemical processes involving 3-5
reaction steps, operating at maximum temperatures less than $900^{\circ}C$,
will be less efficient than direct electrolytic processes (23).
Optimum efficiency will require preferably simple two-step processes,
with the higher-temperature step operating above $900^{\circ}-1000^{\circ}C$.   The
small number of steps involved will reduce the overall affinity
summation: even though the low-temperature step may have to be
driven electrochemically, the overall energy required will normally
be less than that for a purely thermal process requiring separation

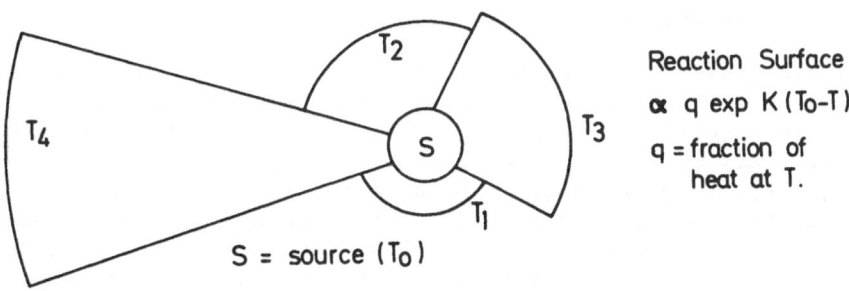

Fig. 2   4-Step Thermochemical Process (Schematic).   k will be
         constant for reactions of similar activation energy.

and pumping work.   A possible high temperature step involves dissociation of a metal oxide (24), for example (20):

$$2CdO \rightarrow 2Cd + O_2 \quad (1225^{\circ}C) \tag{2}$$

followed by the electrochemically-driven dissolution of the metal
in water, with the evolution of hydrogen.   While such high-temperature dissociation processes are possible on paper, the recombination
problem on cooling the emerging gases remains to be resolved.   To
avoid this, steps that are only reversible when catalyzed will be
necessary.   By far the most attractive reaction from the viewpoints
of economics and known engineering problems is the inverse contact
process (26):

$$H_2SO_4 \rightarrow H_2O + SO_2 + \tfrac{1}{2}O_2 \tag{3}$$

operating at about $900^{\circ}C$.   Such temperatures will be relatively
easily realized using solar collectors of the linear array type.
The second, electrochemical, step of the above process (low-temperature SO_2-depolarized electrolysis of H_2SO_4) has not so far been
discussed in detail.   An analysis is given in the following section.
Overall, the process looks very favorable, and may permit a real
efficiency (hydrogen high heating value) of between 50 and 55%, for
a relatively small investment in plant ($150 - $200/kwT of hydrogen
produced).   If we assume a real cost of about $1.50/GjT for solar
heat (annual mean for clear weather climate) (27), an overall
hydrogen cost of about $4.00/Gj will be possible.   This figure
compares favorably with the lowest estimates for either electrolytic
hydrogen (11) or that derived by multi-stage thermochemical cycles
from nuclear heat (4).   It must be pointed out that the use of
higher-temperature dissociation processes, which are interesting
from the viewpoint of theoretical yield, will not necessarily result
in a lower cost product: collector efficiency decreases rapidly
at higher temperatures, so that higher areas with greater optical
perfection are required (cf. the Odeillo furnace at $2000/peak kwT
(28) - about $48/GjT annual mean cost).

   Electrolysis.   While the Allis-Chalmers system has been cited
as an answer to the problem of increasing the efficiency of low-
temperature electrolysis (see Ref. 11), it is often forgotten that
it differs only marginally from current (Zdansky-Lonza (29)) pressure
electrolyzers, in particular in respect to its use of higher surface
electrodes and thinner separators.   Since recent work has shown
that nickel is a more effective catalyst than any of the platinum
group metals for oxygen evolution in alkaline solution at temperatures
greater than $60^{\circ}C$ (30), future gains in performance will result from
such areas as hydrogen electrode catalysis and higher temperature
operation.   The latter will require new, thinner, separators of
high chemical resistance.   Thin sheets of Teflon-bonded potassium

titanate appear to have the properties needed for work at temperatures up to $ca.$ 150°C (25). Non-noble hydrogen electrode catalysts for use in alkaline solution are currently under development. The most promising materials are high-surface-area cobalt molybdate (30), and molybdenum dioxide. Both compounds should give a virtually reversible hydrogen electrode performance, equivalent to that for platinum black electrodes. Overall, it is not unreasonable to expect a maximum potential of 1.6V using catalyzed sintered-nickel electrodes, at a current density of $0.4A/cm^2$ and at a pressure of 30 bar using standard electrolyzer hardware operating at 120°C (30). Based on currently quoted prices (31) for industrial equipment operating at lower current densities ($ca.$ $0.2A/cm^2$), an ultimate cost of less than \$100/kw seems possible for the electrolyzer and its associated equipment.

Acid electrolyzers offer an alternative possibility. At present, the most promising concept involves the use of stable cationic exchange membranes (polymerized fluorinated sulfonic acids) as electrolytes. Such materials may be stable to temperatures up to 150°C. The advantages of these systems are considerable: excellent corrosion resistance (since the supply circuit to the cell will contain only distilled water), simple technology and high rates (32). Such advantages are unfortunately offset by high cost, due to the necessity of noble metal catalysts and a (at least currently) high investment cost for the acidic ion-exchange membrane ($cf.$ Ref. 32). Even so, costs of under \$100/kw at a working potential of 1.6V are predicted (32). As we have indicated earlier (33), the development of a stable anionic exchange membrane, perhaps using fluorinated quaternary ammonium hydroxides, would considerably reduce the overall capital costs of such systems, since noble-metal catalysts would no longer be needed.

A preliminary conclusion of the figures given above for electrolyzer investments show that a value-added cost for hydrogen of the order of \$0.7/Gj, resulting from the electrolyzer and its additional equipment, seems reasonable based on extrapolations of current technology and accountancy. Overall hydrogen costs (based on high-heating-value) will be about 8% over electricity prices plus \$0.7/Gj. For example, for bus-bar solar electricity costs of 12 mils/kwh - comparable to nuclear costs for PWR and BWR reactors (8) - we can expect hydrogen prices of \$4.70/Gj. Such electricity costs may be attained using solar systems requiring investments less than about \$600/kw on an overall annual basis. As we have indicated above, such costs may be reasonable for electricity based on eolian (34) or wave (12) power in certain favorable locations, where the mean power density is close to the annual maximum (e.g. wave power in the Outer Hebrides, at 77kw/m (12)). In addition, such electricity costs seem well within the estimated possibilities of plants deriving heat from ocean thermal gradients (see introduction). All these plants will necessarily be remote from centers of energy use, and

will only be justified by the production of electrolytic hydrogen.
Combined thermal-electrochemical plants can only be situated in
clear sky regions at low latitudes (Southern U.S., Australia, N.
Africa; or as solar islands (35) in tropical oceans).

Thermal-electrochemical plants may use a cycle of the SO_2 type
referred to above or alternatively may use direct high-temperature
electrolysis.   As we have pointed out earlier, such systems are
not attractive unless heat is available at temperatures above 900°
(33), so that the electrolyzer can operate close to the reversible
potential for the mean gas mixture used, the $T\Delta S$ term being provided
from the environment.   An extrapolation of current predicted
performance indicates largely ohmic polarizations of about 175 mV
at $1A/cm^2$ (33,36) for typical systems.   If this figure can be
attained under practical conditions at reasonable cost, we can
anticipate electrolysis at 1.15V under high-pressure conditions,
with the heat of evaporation of water and the high-temperature $T\Delta S$
term provided by solar heat.   An approximate calculation based on
the above values, using data in Ref. 33, indicates an overall high-
heating-value efficiency (for hydrogen at 1 atm.) of about 45%,
assuming a solar heat - electricity conversion of 40%.   The SO_2
system is even more attractive in overall efficiency terms (see
above), provided that a low-polarization hydrogen-evolution/SO_2
oxidation electrolysis unit can be developed.   The overall efficiency quoted above assumed 200 mV overall loss (at both electrodes
plus IR drop), which implies development of a high-efficiency SO_2-
oxidation electrode.   Such an electrolysis unit does not at present
exist.   In alkaline solution, SO_3^{2-} oxidation is relatively rapid,
but cells will require not only high-resistance ion-exchange membranes to avoid neutralization of the electrolyte, but will need an
energy-efficient procedure for converting alkali sulfates into
sulfuric acid.   In acid solution, the SO_2 oxidation step is very
irreversible (37), and it is well known that preferential production
of H_2S and S, instead of hydrogen, occurs at the cathode at higher
temperatures (38).   In addition, the preferred reaction product is
dithiate ion (38).   These difficulties may perhaps be avoided by
the use of a membrane (fluorinated sulfonic acid) electrolyzer of
the type proposed for water electrolysis (32), for which temperatures
up to 160^oC should be attainable, resulting in lower polarizations
at the anode.   No SO_3 or SO_4^{2-} transfer through the membrane is
expected, so that pure hydrogen will be evolved at the cathode.
Such an approach presents a great electrochemical challenge from
the viewpoint of catalytic studies.   In addition, study of certain
molten systems (e.g. bisulfate) may prove interesting for this
process.

All the above solar energy systems (focussed clear-sky solar
heat, heat in tropical oceans, high-energy wind- and wave-power)
will be complementary in the energy economy, depending on their
meteorological location.   The common element among them will be

the necessity for production of hydrogen as an energy storage and
transmission medium, to improve overall system cost up to the
consumer.   This point will be much more important for remote solar,
as distinct from remote nuclear, systems, due to the problem of
daily and seasonal energy storage.   Hydrogen production will be
entirely electrochemical, or thermal-electrochemical, since economics
and irreversible thermodynamics argue against direct thermochemical
procedures.   Because solar cells of photovoltaic or photochemical
type are not so dependent on focussed concentration of solar energy
as are heat-dependent systems, since they can use diffuse light,
they may be more appropriate for local energy use, alongside solar
heat collectors.   Their possibilities are discussed in the next
section.

Local Energy Production

   Photovoltaic Systems.   Recent developments in the Edge-Defined-
Film-Growth method (39) for growing continuous monocrystalline
ribbons of silicon and other materials have given some hope for the
eventual production of low-cost silicon photocells.   Prices on the
order of $380/peak kw (40,41) have been quoted.   For CdS/Cu_2S cells
of polycrystalline type costs may be eventually even lower, despite
an efficiency only about half that of monocrystalline Si cells (42).
It would seem that the above prices are close to the floor level -
in the case of silicon cells, they would be $38/m^2, compared with
$30/m^2 for steerable mirror heliostats (27), and about $20/m^2 for
the cheapest possible solar heat panel (43).   The latter represents
the probable lower limit for self-supporting panels and simple
technology, suggesting a lowest-possible cost of *ca.* $250/peak kw
at 8% efficiency for the simplest technology CdS cells.   Averaged
over the year in a temperate climate, such cells should allow
electricity production at about 20 mils/kwh.   This cost is quite
reasonable, since it corresponds closely to current industrial
electricity costs in the U.S., and will be close to delivered
nuclear electricity costs based on current nuclear fuel prices (12
mils/kw for bus-bar electricity (8), perhaps 18 mils/kw delivered).
To be of use, however, such systems require energy storage.   This
could be either via electrolytic hydrogen or electrochemical
batteries.

   Hydrogen production using solar cells does not seem to be very
attractive, due to the high cost of the solar photoelectricity.
If we assume reasonable costs of $50/kw and 90% efficiency for the
low-temperature electrolyzer and its auxiliaries, hydrogen costs
will be about $9.00/Gj - much higher than delivered hydrogen costs
from remote direct- or indirect-solar plants.   Investment costs
and conversion losses in fuel cells would give prices for electricity
in hours of darkness at least 1.7 times greater than the above (i.e.
50 mils/kwh).   These very high costs partly result from the necessity

of amortizing the electrolyzer only over the effective yearly daylight hours.   A battery system is clearly a better solution: at
70% overall efficiency, with an investment cost of $100/kw and a
suitable lifetime, a delivered annual mean electricity cost of about
38 mils/kwh should be attainable.   This compares well with electricity costs for fuel-cell conversion of hydrogen from remote solar
plants, delivered locally at $6.00/Gj (20 mils/kwhT).   At present,
such batteries do not exist.   However, an intensive effort is
currently being directed towards their development, in connection
with the program for off-peak storage of electrical energy in a
nuclear economy (44).   It would seem that the chances of success
are high, and that electrochemical storage on the local level will
be the most important factor in rendering a solar economy possible.
We will briefly examine some of the candidate battery systems below.

Electrochemical Photosystems.   Energy systems using photo-
electrochemical phenomena have been recently reviewed (45).   Such
reactions can be separated into two broad classes: those in which
photon energy homogeneously dissociates certain compounds, e.g.
NOCl (46) whose products can then be separated and stored, and those
in which the reaction occurs on separated sites, in which automatic
component separation occurs.   The former suffer from the same
fundamental problems as those in thermochemical cycles, in particular,
those of recombination and component separation (see, for example,
Ref. 47), which considerably reduce their high theoretical quantum
efficiency.   The second type of process may involve the production
of useful yields of chemical compounds at separated electrodic sites,
or, as an alternative, production of an electric current.   Both
varieties of the second type of process are electrochemical.

The most interesting of the first type of process may be NOCl
dissociation, followed by recombination in a fuel cell (46).   So
far no costs for this system have been worked out, but it is reasonable to suppose minimum fuel cell costs $ca.$ 4 times higher than those
for the $H_2 - O_2$ reaction (due to the lower cell potential).   Based
on minimum collector costs (for solar energy) of $50/peak kw, at
80% efficiency, we may expect about 5 mils/kwhT for the fuel plus
oxidant.   Overall electricity costs, after fuel cell conversion,
may be about 40 mils/kwh, which will be competitive with those given
above for solid-state photovoltaic devices with battery storage.

For processes of the second type, the most attractive, at
least from the theoretical viewpoint, are those that lead to storage
of separated chemical products during periods of insolation, similar
to the NOCl system discussed above.   There is no question that
dissociation of water to yield hydrogen and oxygen would be the
most exciting of such reactions.   However, for spontaneous dissociation of the H-O bond photons of about 5eV energy would be
required, in the case of both homogeneous and heterogeneous processes.

For the latter, the adsorption of reaction products may lead to lower energies of activation for the dissociation process, so that lower limiting photon energies are possible. This factor, combined with the advantage of the automatic separation of the reaction products, has encouraged a good deal of recent research.

Thus far, cells containing semiconductors with gaps of the order of 3eV, combined with reversible counter electrodes (TiO_2) (48), or in other arrangements (anthracene (49)), have been shown to be able to decompose water on illumination. Such systems are of no practical interest, since light capable of being adsorbed by the semiconductor (i.e. that of wavelength shorter than 3eV) accounts for only a small part of the solar spectrum. Efficient photosensitive collectors with a longer wavelength threshold imply not only a smaller band gap, but also a stronger adsorption of OH radicals. The overall result is a photocorrosion process, rather than water decomposition (45,50). It therefore appears improbable that efficient one-step photosystems for water decomposition using the majority of the visible solar spectrum will ever be produced. However, there is no fundamental reason why a light-rechargeable battery using a lower-energy redox couple or couples cannot be constructed. Practically, such a system is difficult to conceive: unless the products of the photocouple are gases, an efficient membrane will be needed, introducing high resistance losses. In addition, the photoelectrode will almost certainly have to be illuminated from the rear, posing great problems of construction of transparent thin-film electrodes and current collection. In principle, two such visible light systems, one producing oxygen and the reduced form of a redox couple, which is recycled to the second system, where the oxidized form of the redox couple and hydrogen are produced, can be used for water decomposition with automatic separation of hydrogen and oxygen. The process would be an exact analog of biological photosynthesis (33,51,55). However, from the engineering viewpoint such a system would be much more difficult to realize than a photovoltaic cell coupled to an electrolyzer.

We should perhaps mention in passing that homogeneous chemically-catalyzed photolysis of water (with reaction intermediates stabilized by bond formation) (52), which is at first sight attractive, will almost certainly prove impractical due to the enormous difficulties involved in handling and separating stoichiometric hydrogen and oxygen mixtures. We cannot, however, exclude the possibility of the eventual use of biological systems for water dissociation, since these will involve automatic component separation, together with the possibility of a regenerable structure that is perfectly adapted to the task without the necessity of high design, research, and development costs.

The greatest hope for the application of electrochemical systems to high-efficiency photovoltaic energy conversion seems to lie in the use of light-activated electrodes in contact with redox systems, the overall cell containing a reversible counter-electrode so that a constant composition of the redox couple is maintained, with the production of an electric current.   Two limiting cases of such photocells may be considered, namely (i) a stable n- or p-type semiconducting electrode, whose gap is less than the energy of the light to be adsorbed, in contact with the redox system (53), and (ii) a soluble photosensitive dye in contact with a stable metallic electrode and a redox system (54).   Physically, the limiting cases represent band and exiton models respectively.   In practice, the use of a soluble photoactive compound results in low efficiency since transfer to the electrode is diffusion-controlled and quenching effects are considerable.   Conversely, the efficiency of a system using a suitable semiconducting electrode may be high (equivalent to that of a corresponding solid-state photovoltaic cell), but again the problem involves finding semiconductors of suitable band-gap that are stable in the presence of water under highly oxidizing conditions.   Inorganic semiconductors with band-gaps in the visible energy range are normally unstable under oxidizing conditions.   However, a wide range of insoluble photosensitive dyes exists (porphyrins, phthalocyanines).   These compounds show remarkable stability with respect to oxygen.   They are exact analogs of the active molecules involved in the oxygen evolution reaction in natural photosynthesis (51), and may serve to give yields equivalent to those of conventional photocells when applied in thin layers on stable metallic electrodes (*cf*. Ref. 55), for example in the presence of an Fe^{2+}/Fe^{3+} couple and a stable counter-electrode.   More research is required in this important area.   An important final point concerning such systems is their enormous economic potential from the viewpoint of local energy production and storage.   As we will see later, a possible low-cost battery storage system is that involving redox reactions.   The couple used in the photocell could be also used in the battery.   The fact that redox systems will normally require a very large volume of stored solution will be no disadvantage, since the photocell itself can be used as a solar heat collector, and the battery for heat storage.   The expected energy uptake (*ca*. 10% as electricity, 90% as heat) is reasonable for many domestic requirements, and combined capital costs should be relatively low.   This interesting area requires further examination.   Some solar cell concepts (33) are shown in Fig. 3.

Local Storage of Solar Energy

Conventional Batteries.   Currently, an important research effort is being directed towards the development of storage batteries with costs on the order of $20/kwh, $100/kw for peakshaving applications (44).   Maintenance-free lifetimes of 5 - 10 years are expected. Such batteries will be very suitable for economic storage of solar electricity.

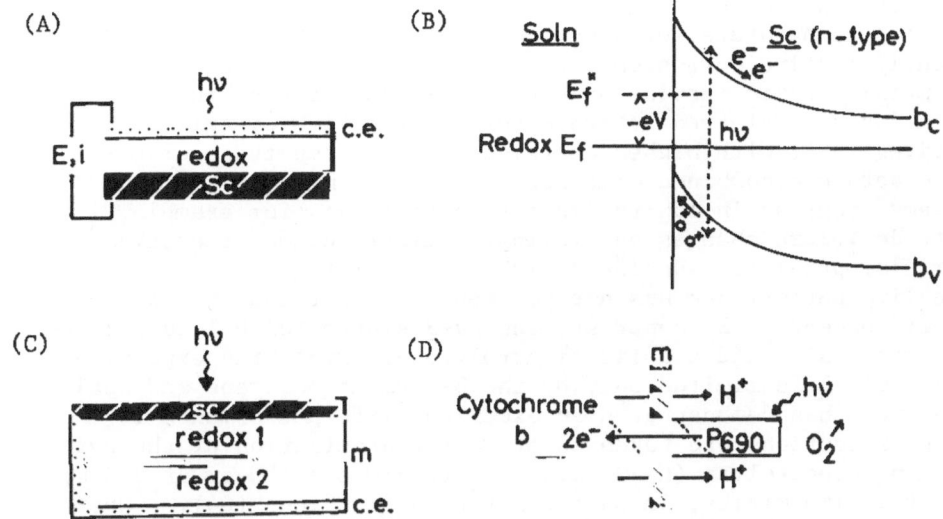

Fig. 3   Electrochemical Solar Cell Concepts.   (A) Photovoltaic
System.   c.e. - transparent counter electrode;   Sc -
semiconducting light-sensitive electrode.   (B) Energy
diagram for cell (A).   (C) Storage cell.   Electrodes
are shorted, and the two redox couples are recharged in
the two compartments separated by membrane m.   Two cells
of Type (A) with shorted electrodes, with different (n-
and p-type) semiconductors and different redox couples
(one charged positively, the other negatively) are
equivalent to a single cell of Type (C).   The solutions
would be circulated, stored, and recombined in a fuel cell.
(D) Photosynthesis Analog of Cell B.   Redox couples are
H_2O-O_2 and Cytochrome B.   Electronic conduction from P690
center is via a semiconducting macromolecule.

        Candidate systems at present being examined are sodium-sulfur-
β alumina (56) and its variants (57), lithium alloy - metal sulfide
(58,59), zinc-chlorine (60) and fuel cell-type concepts using cheap
dissolved redox couples (61).   These seem to be the only systems
whose materials costs are sufficiently low for consideration (*ca.*
$2/kwh for sodium-sulfur, $10/kwh for lead-acid), and which possess
sufficiently high charge-discharge efficiencies (60-75% is required:
other potentially cheap systems, e.g. iron-air (62), have efficiencies that are too low - 25% in the latter case).   The above battery
systems have been briefly reviewed in a previous article (33).

Both high temperature concepts (Na-S and Li-sulfide, with operating temperatures of about 300°-350°C and 400°C respectively) currently involve corrosion problems and the development of specialized metal-ceramic insulating seals.   In the case of the Li-sulfide system (whose preferred active materials are lithium-silicon alloy negatives (59), with higher energy densities than those of Li-Al (58) - both are combined with FeS_2 positives) many of the same problems occur as in low temperature batteries: for example, electrode volume changes on cycling, redistribution of active materials, problems associated with separator function.   Long cycle-life performance has not yet been demonstrated for this battery concept.   In contrast, the Na-S system (with liquid electrode materials, and a solid electrolyte allowing free expansion) has a cycle life limited only by the β-alumina membrane and seals. Recent work has demonstrated an excellent life performance if simple precautions are taken to prevent contamination of the pure β-alumina electrolyte (Na_2O. $8.5Al_2O_3$ (63); $\rho = 15\Omega$-cm at 330°C) by certain impurities, in particular potassium ions (64).   Over 6,000 3-hour cycles have been obtained using laboratory material, representing 1,200 Ah charge passed (64).   This is equivalent to 1,800 complete daily cycles for an optimized sulfur cathode structure, or five years operation under solar conditions.   Thermal shock resistance appears not to be a major problem.   In addition, seals and metal parts with adequate corrosion resistance are now being successfully developed.   Other important advantages of the sodium-sulfur battery are its zero self-discharge rate, high charge-discharge efficiency, and its inherent ability to be designed to incorporate over-charge and over-discharge protection.   These factors imply few problems on the incorporation of single cells into batteries.   At present, only the problems of its mass production remain to be solved.   There is considerable confidence that the system will be capable of fulfilling its promise.

Among the low temperature systems mentioned, zinc-chlorine has yet to exhibit good cycle life, since it suffers from important materials and mechanical problems.   The highly corrosive chloride electrolyte, containing dissolved chlorine, requires the use of positive electrodes of dimensionally-stable-anode type (ruthenium dioxide coatings on titanium (65)), titanium heat exchangers, and elaborate pumping equipment.   In addition, refrigeration is required for chlorine storage as hydrate.   The redox battery systems require further developments in membrane technology to improve component separation and reduce ohmic drop.   In addition, they will probably need the use of complex ions (preferably organic iron complexes) for high performance at low cost.   Their energy densities will be very low, due to the amounts of solvent required: however, as we have seen above, this property can be turned to great advantage in the solar economy.

Other Energy Storage Concepts.   Considerable interest existed
in thermally regenerative battery systems during the 1960's (*cf.*
Refs. 66, 67).   The most interesting of these was the sodium-
bismuth molten salt battery system (O.C. voltage of intermetallic
compound about 0.5V) (67).   This was abandoned due to self-discharge
problems associated with the molten salt electrolyte.   Use of the
Na-β alumina technology, as in the sodium-sulfur battery, could
eliminate this difficulty.   If the Na-S system can be produced for
$100/kw, a cost of $500/kw is possible for the sodium-bismuth system,
including alkali-metal heat-pipe solar collector.   Year-round
electricity costs may be about 20 mils/kwh.   A scheme of such a
system is shown in Fig. 4.   The great advantage of such systems is
their storage capacity and their potentially high efficiency (in
this case about 20% overall: the battery is at 300°C, the evaporator
at 900°C).   Other, more efficient, concepts involving a high-temperature thermochemical process with recombination in a fuel-cell at
potentials lower than those for $H_2 + \frac{1}{2}O_2$ require examination.   An
example that can immediately be cited is hydrogen iodide decomposition, whose kinetics are well known (68).   If the problems of recombination and separation can be solved, this and similar systems
will prove very attractive for energy storage and production.   An
overall efficiency (system plus solar collector) of 25% is a reasonable goal.

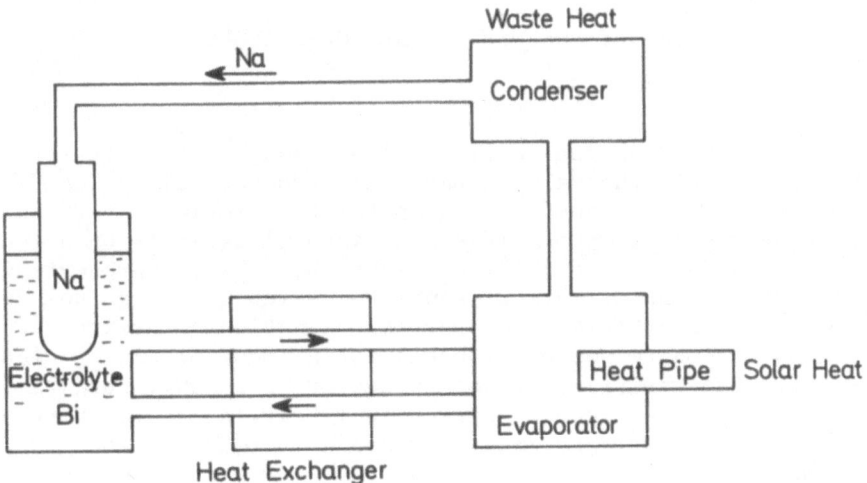

Fig. 4 Thermally Regenerative Cell Concept.

While the present article is not concerned with the thermal
storage of solar energy, a comment in passing is perhaps appropriate
in this respect.   Thermal storage will, in the future, be an
extremely important element in reducing the energy load on the
future economy.   It may be obtained directly, on a local basis,
from solar heat, or alternatively from indirect solar energy, e.g.
eolian work, using frictional storage.   In general, only storage
via latent heat of fusion has been proposed.   While the advantages
of this procedure for high temperature heat are obvious, we should
not forget that about half of the total energy input into society
is required for heat at low temperature.   Low cost solar panels
could supply an important part of this heat requirement, but even
so an alternative heat source will be needed to fill in during the
winter period.   Traditional solar building concepts call for this
energy input in the form of electricity or natural gas.   Conventional heat storage is not practical, due to insulation problems
and to quantities of liquid required.   The use of the heat of
hydration of a number of cheap chemical compounds may provide a
solution to the problem of domestic energy storage from summer to
winter.   While this may be considered to be outside the scope of
this article, the physical phenomena involved can be considered to
be electrochemical, at least in the ionic sense.   In particular,
we can suggest the use of anhydrous calcium chloride, regenerated
by evaporation in summer, as a suitable heat storage medium (150
kcal/kg, $ca.$ 175 whT/kg, with no necessity for heat insulation).

ENERGY END USE IN THE SOLAR ECONOMY

Fuel Cells

Various types of fuel cells will be needed, not only for some
of the more exotic thermal- or photo-electrochemical cycles discussed
above in the solar economy, but also for the simple conversion of
hydrogen produced at remote locations into electricity on a local
scale.   Such fuel cells will, at least initially, represent only
minor modifications of current concepts, which will use fuel derived
from either steam-reforming or partial oxidation of hydrocarbons.
The molten carbonate system probably is the most promising of the
latter from the viewpoint of cost-effectiveness (69), but its use
in a purely hydrogen economy appears unlikely, since for both
electrolyte stability and kinetic reasons it requires continuous
supply of CO_2.

Since the subject of fuel cells is dealt with elsewhere in
this symposium (70), it suffices to add only a few comments.
Electrochemical hydrogen oxidation in the absence of CO and CO_2
from reformates will still involve the problem of atmospheric CO_2.
In this case, the continued use of an acid electrolyte seems
appropriate.   For these systems, new types of electrolyte will

replace phosphoric acid, which has relatively poor conductivity and oxygen electrode performance. The use of organic fluorosulfonic acids, either as high polymer membranes (as in the electrolyzer system described in Ref. 32) or as lower polymers in matrix-type cells (71), seems to be the most likely choice for hydrogen oxidation. At least on the local sub-station level, hydrogen fuel will be available at high pressure (perhaps 70 atm (11)). This pressure difference represents almost 7% of the room-temperature high-heating value (assuming oxygen is available at the same pressure). Part of this energy would be worth recovering. This may be achieved by using the fuel cell in a combined cycle with a gas turbine, whose first stage compresses the cathode gases, in which excess hydrogen from the anode exhaust is burnt on exit from the fuel cell to drive the turbine. Such a system will decrease the relative cost of the fuel-cell component (which will always be several times higher per kw than that of the turbine) by increasing overall efficiency and/ or power density. Such a concept would be best applied to a high-temperature fuel cell (e.g. doped zirconia, or molten carbonate using processed photosynthetic fuel or garbage) since waste heat can be recovered at 40% efficiency by a steam turbine, allowing a global efficiency of 70%. A further important practical advantage of a pressurized fuel-cell system is the lower gas volume throughput, thus reducing loss of electrolyte by evaporation and consequent reduction of cell life. This will be a particularly important consideration for hydrogen cells based on fluorosulfonic acid matrix systems (71) and in the molten carbonate system (69).

Transportation

In the solar economy, much personal transportation (e.g. commuting to work) will be replaced by communications networks and computer links, which are less energy-intensive than social movement (72). Urban transportation will be entirely electric. The same will not necessarily be true in country districts, in agriculture, or in developing nations: this is made clear by the fact that the yearly energy requirements of a 1 tonne vehicle (15,000 km/ year, 0.2 wh/tonne-km at motor driveshaft) with a 30% efficient Stirling engine are only equivalent to the clippings of a medium-sized lawn (400 m^2), assuming a modest photosynthetic efficiency value of 1% (73,74), or the leaves from three large trees.

Where pure hydrogen produced in remote locations is available, the most logical vehicle power source will be the hydrogen fuel cell. We have previously given a conception of such a vehicle (33), for which peaking power would be provided by nickel-hydrogen batteries (75) stored in the compressed hydrogen reservoir of the vehicle. If the high-pressure storage system proposed should prove to represent a possible safety hazard, hydride storage can be considered (76), though at some weight penalty (33). With an

estimated hardware weight equal to 2.3 times that of the zinc-air
battery discussed below, a hardware cost on a high volume basis of
about $100/peak kw, or $40/nominal kw, seems possible for an electric vehicle.   This figure corresponds to $4/lb of fuel cell
structure, which will consist essentially of mild steel and plastic
in an alkaline system.   Detroit-style mass production may eventually
reduce this cost by a factor of two.   This figure assumes the use
of low-cost, stable electrode materials.   The latter problem may
be considered to be solved for the air electrode, since active
carbons of adequate activity give a reasonable life (3000 h (78)),
but the anode still requires platinum, albeit in low loadings ($ca.$
0.7 mg/cm^2, or perhaps $50/kw).   A prime goal of electrocatalytic
research must be the elimination of this high cost material in such
systems.   Currently, catalysts giving equivalent performance to
that of platinum hydrogen electrodes under equilibrium conditions
exist (30), but they possess poor stability when polarized at low
anodic overpotential.   Eventually, the same catalysts will be used
in nickel-hydrogen peaking batteries, which will accept regenerative
braking and which will be contained in the compressed hydrogen
reservoir (33).

Where locally produced electricity is the predominant energy
medium, the most important factor will be the balance between cheap
'bus-bar' solar electricity, and its utilization in transportation.
Typically, vehicle use will be mainly during the hours of maximum
insolation.   In such cases, it will not be economical to recharge
the vehicle storage batteries during the night from a battery
system that has previously been charged with solar electricity
during daytime, since the effective cost differential between the
equivalent rates will be of the order of 1:2 in the best possible
cases.   A possible solution to this problem will be the use of
exchange batteries, stored in service stations, and charged directly
from solar electricity.   However, a rapid look at the economics of
such a concept shows immediately that its costs with respect to the
consumer would be at least of the same order of magnitude as those
for overnight charging from stored solar electricity, due to the
increased capital inventory per vehicle required in the form of
extra batteries.   A much more satisfactory concept, requiring a
low overall capital outlay, is via the use of a fuel cell whose
oxidized products can be electrochemically recharged in a fixed
location during the day.   Such a fuel cell would necessarily be of
metal-air type.   The 'refuelability' aspects of vehicles using
this type of power source, equivalent to those of current gasoline-
powered cars, will be a great advantage from the consumer viewpoint.

Refuelable metal-air cells may be divided into two classes:
those in which solid electrodes of flat-plate type are necessary,
due to the highly reactive nature of the metal used, and those in
which it is possible to add a less reactive metal in powder form
to the electrolyte.   The former are best characterized by Li (77)

air and Al-air systems, the latter by Zn-air (78).   The reactive
metal systems will involve storage, handling and loading problems,
together with a complex molten-salt extraction technique which
cannot be carried out at the local service-station level.   In
addition, reprocessing losses may be quite high (10% estimated for
lithium (80)), and overall efficiency of the couples is quite low
(about 30% or less) since the metal can never develop its thermodynamic potential in aqueous solution.

By far the most promising system is the zinc-air powder fuel
cell (78,79).   While its energy density is quite modest compared
with that potentially available for Li-air (ca. 110 wh/kg (79),
compared with perhaps 200-250 wh/kg), it is still sufficient to
give a practical vehicle performance (110 km/h maximum speed;
150 km range at 50 km/h for a 1 tonne urban vehicle (79)).   Regen-
eration of the spent electrolyte (KOH plus colloidal zincates)
would be carried out by electrolysis in a service-station facility
during daylight hours, using simple, low-temperature electrolysis
methods.   Refueling with fresh zinc-powder electrolyte slurry
would be a simple operation, analogous to filling up with gasoline
(though about 5 times the volume would be needed).   This will
result in great customer convenience, and similar logistics to
those currently used for I.C. engined vehicles.   The primary fuel
cell system is now out of the laboratory stage, and is predicted
to cost $35/kw, when produced on a 100,000 25kw units/year scale
(79).   Since its energy density increases as relative power
requirements (kw/tonne of vehicle), hence hardware weights, fall,
it is particularly suitable for heavy vehicles (79).   Efficiency
is 40% overall.   For a further description, see Ref. 33.

Other Energy Uses

Energy in the solar economy will be delivered in three different
forms: namely, direct or indirect heat, direct or indirect electricity, and hydrogen.   In addition, we may add the possibility of
using renewable photosynthetic fuel in smaller quantities (perhaps
methanol or ethanol), together with that of simple inorganic
hydrogen carriers, in particular liquid ammonia.   The latter is
likely to find use in I.C. engines in some trains and long-distance
trucks (81).   As we have shown above, the domestic and commercial
sector will use direct solar heat and its ancillary storage for
heating and cooling, heat transfer being provided by heat pumps.
The electrical energy for the latter will be derived from either
solid-state or electrochemical photocells, or via thermally-regen-
erative systems.   A rather similar pattern will be seen in the
industrial sector, at least from the viewpoint of low temperature
heat utilization.   Process heat will be provided by hydrogen, and
reaction free energies in the form of electricity.   A review of

the important electrochemical contribution to such an economy has
been given in Ref. 33.

CONCLUSIONS

This brief review has attempted to predict a possible future.
Obviously, such a task is uncertain, since it is becoming increasingly difficult to measure the future in terms of extrapolation of
the recent past.   However, it is clear that we are approaching a
turning point of the same magnitude as that of the period 1789-1848,
a point of real change in Western society, in which simultaneous
economic, energy, food and population crises will occur.   Organ-
ization will be required to preserve society in something resembling
its present form.

As we run out of fossil fuel, we must change from a society
whose capital base is invested in the utilization of energy produced,
to one in which available capital is used for the production of
that energy.   This conclusion is inevitable, since the possibility
for any type of society to raise capital is limited: low capital
requirements for energy subsidize rapid industrial growth, and high
capital for energy must result in a low overall industrial growth
rate.   The economic basis of the future society will be the employment of people to expand and replace the industrial infrastructure
for non-fossil-fuel energy, rather than that for transportation,
which is the case at present.   Industry in such a society will be
much more materials-conservative than at present, so that the possibility of a future minerals crisis will be reduced.   A further
important advantage, from the social viewpoint, is that general
employment will follow the economic plan dictated by replacement
of existing energy equipment over long leadtimes, so that an intrinsic buffering with regard to short-term economic changes will occur.

It will be observed that many of the concepts suggested above
are equally applicable to a nuclear economy, which may be capable
(based on present estimates) of carrying out the same task at
lower cost.   This line of thinking must be resisted, not only on
the grounds of the fiscal and social costs that will be bequeathed
to future generations by a full development of nuclear energy, but
also by the fact that the economics of extraction and processing
of nuclear fuel are such that constant dollar costs of bus-bar
nuclear power may be expected to at least double (*cf*. uranium costs
in Ref. 82).   This situation will not drastically change with the
development of either the breeder reactor or nuclear fusion.   In
the near future, the solar economy will therefore be not only soc-
ially, but economically, desirable.   Its generalized application
will depend on the use of electrochemical concepts, both for energy
storage and end-use.

REFERENCES

1. S.D. Freeman *et al.*, *A Time To Choose*, Energy Policy Project of The Ford Foundation, Ballinger, Cambridge, Mass. (1975).
2. *Statistics Of Energy, 1959-1973*, O.E.C.D., Paris, 1974.
3. H. Douglas, Energy Research Reports,1(5) (1975) 1.
4. A.J. Appleby, *Energy, Electrochemical Symposium, Imperial College*, London, April 1975; A.J. Appleby, to be published in *Energy Policy*.
5. M.A. Elliot and N.C. Turner, Presented at *A.C.S. Symposium on Non-Fossil Fuels*, Boston, Mass., April 1972.   See also H.W. Menard and G. Sharman, Science, 190 (1975) 337.
6. Ref. 1, p.366; F. Felix, Electrical World, November 1, 1975, p.64.
7. Ref. 1, p.255.
8. S. Law, *Proceedings of the Cornell International Symposium and Workshop on the Hydrogen Economy*, Cornell University Press (1975) p.212.
9. R.H. Wentorf and R.E. Hanneman, Science, 185 (1974) 311.
10. Stanford Research Institute, *Patterns of Energy Consumption in the U.S.*, Menlo Park, Calif., November 1971.   Office of Science and Technology, Washington, D.C.,January 1972.
11. D. Gregory, *A Hydrogen Energy System*, American Gas Association, Cat. No. L21173, 1973.
12. S.H. Salter, Nature, 249 (1974) 720.
13. C. Zener, Physics Today, 26 (1973) 48.
14. A.B. Meinel and P.M. Meinel, *Ibid.*, 25 (1972) 44.
15. J.D. Parent, *Institute for Energy Analysis Workshop on the Effect of Energy Consumption on the Economy*, National Academy of Sciences, Washington, D.C., May 1974.
16. Oak Ridge National Laboratory, *Nuclear Energy Centers, Industrial and Agro-Industrial Complexes*, O.R.N.L. 4290, U.S. A.E.C., Washington, D.C., November 1968.
17. D.H. Meadows, D.L. Meadows, J. Randers and W.W. Behrens, *The Limits to Growth*, Potomac Associates, Washington, D.C., 1972.
18. G. deBeni and C. Marchetti, Euro Spectra, 9 (1970) 46.
19. J.E. Funk and R.M. Reinstrom, Ind. Eng. Chem. Proc. Des. Develop., 5 (1966) 336; J.E. Funk, *Proceedings of the Symposium on Non-Fossil Fuels*, Boston, Mass., April 1972, A.C.S., Washington, D.C., p.79.
20. D. Gregory and J.B. Pangborn, *Proceedings of the 9th I.E.C.E.C. Conference*, 749014, San Francisco, Calif., 1974.
21. T. Erdey-Gruz and M. Volmer, Z. Physik. Chem., 150A (1931) 203; J.A.V. Butler, Proc. Roy. Soc. (London), A157 (1936) 423.
22. P. Van Rysselberghe, J. Chem. Phys., 29 (1958) 640.
23. A.J. Appleby, Nature, 253 (1975) 257.
24. A.R. Miller and H. Jaffe, U.S. Pat. 3,490,871, January 20, 1970; *cf.* D. Souriau, Ger. Pat. 2,221,509, November 16, 1972.

25.   G. Kissel, P.W.T. Lu, M.H. Miles and S. Srinivasan, *Proceed-ings 10th I.E.C.E.C. Conference*, 759177, Newark, Del., August 1975.

26.   G.H. Farbman and L.E. Brecher, *Ibid.*, 759178.

27.   A.F. Hildebrandt and L.L. Vant-Hull in T.N. Veziroglu (Ed.) *Hydrogen Energy, Part A*, Plenum, New York, 1975, p.35.

28.   J.O'M. Bockris, *Ibid.*, p.9.

29.   D.H. Smith in A.T. Kuhn (Ed.), *Industrial Electrolytic Processes*, Elsevier, Amsterdam, 1971, Chap. 4.

30.   A.J. Appleby and J. Jacquelin, to be published in Revue Generale de l'Electricite. *Cf.* also Ref. 25.

31.   Lurgi, A.G., Technical Information.

32.   L.J. Nuttall and W.A. Titterington in T.N. Veziroglu (Ed.), *Hydrogen Energy, Part B*, Plenum, New York, 1975, p.441.

33.   A.J. Appleby, presented at *Electrochemical Symposium, Imper-ial College*, London, April 1975.

34.   R.L. Savage *et al.*, *A Hydrogen Energy Carrier-Systems Analysis* NASA–ASEE, University of Houston, Texas, 1973, p.48.

35.   W.J.D. Escher in T.N. Veziroglu (Ed.), *Hydrogen Energy, Part A*, Plenum, New York, 1975, p.209.

36.   Ref. 11, p. III 36.

37.   I.P. Voroshilov, N.N. Nechiporenko and E.P. Voroshilova, Elektrokhimiya, 10 (1974) 1378 and references quoted therein.

38.   P. Pascal, *Nouveau Traité de Chimie Minerale, Volume 13b*, Masson, Paris, 1956, p.1242.

39.   T.F. Ciszek and G.H. Schwuttke, *Proceedings of the Inter-national Conference on Photoelectric Power Generation*, September 1974, Hamburg, p.159.

40.   A. Kran, *Ibid.*, p.177.

41.   M. Wolf, Energy Conversion, 14 (1975) 49.

42.   J.J. Jordan, *Proceedings of the International Conference on Photoelectric Power Generation*, September, 1974, Hamburg, p.221; R.J. Mytton, K.W. Boer, *Ibid.*, pp. 205, 627; L.D. Partain and M.M. Sayad, in Ref. 27, p.45.

43.   J.A. Day, A.F. Clark, W.C. Dickinson and A. Iantuono, *Proc-eedings of the 10th I.E.C.E.C. Conference*, 759112, Newark, Del. 1975.

44.   N.P. Yao and J.R. Birk, *Proceedings of the 10th I.E.C.E.C. Conference*, 759166, Newark, Del., 1975.

45.   M.D. Archer, J. Appl. Electrochem., 5 (1975) 17.

46.   W.E. McKee, E. Findl, J.D. Margarum and W.B. Lee, *Proceedings of the 14th Annual Power Sources Conference*, May 1960, p.68.

47.   O.S. Neuwirth, J. Phys. Chem., 63 (1959) 17.

48.   A. Fujishima and K. Honda, Nature, 238 (1972) 37; A. Fujishima, K. Kohayakawa and K. Honda, Bull. Chem. Soc. Japan, 45 (1975) 1041; J. Electrochem. Soc., 122 (1975) 1487.

49.   H. Kallmann and M. Pope, J. Chem. Phys., 30 (1959) 585; Nature, 188 (1960) 935.

50.   D.R. Dixon and T.W. Healy, Aus. J. Chem., 24 (1971) 1193;
        H.R. Schöppel and H. Gerischer, Ber. Bunsenges. Phys. Chem.,
        75 (1971) 1237.
51.   R. Hill and F. Bendall, Nature, 186 (1960) 136;   R.K. Clayton,
        Proc. Nat. Acad. Sci. USA, 69 (1972) 44.
52.   V. Balzani, L. Moggi, M.F. Manfrin, F. Bolleta and M. Gleria,
        Science, 189 (1975) 852.
53.   H. Gerischer, J. Electroanal. Chem., 58 (1975) 263.
54.   E. Rabinowitch, J. Chem. Phys., 8 (1940) 551, 560;   C.G.
        Hatchard and C.A. Parker, Trans. Farad. Soc., 57 (1961) 1041.
55.   J.H. Wang, Proc. Nat. Acad. Sci. USA, 62 (1969) 653.
56.   N. Weber and J.T. Kummer, *Proceedings of the 21st Annual
        Power Sources Conference*, 1967, p.37;   S. Gratch, J.V. Petro-
        celli, R.P. Tischer, R.W. Minch and T.J. Walden, *Proceedings
        of the 7th I.E.C.E.C. Conference*, 729008, San Diego, Calif.,
        1972.
57.   J. Werth, U.S. Pat. 3,847,667, November 12, 1974.
58.   E.C. Gay, W.W. Schertz, F.J. Marino and K.F. Anderson,
        Proceedings of the 9th I.E.C.E.C. Conference, 749122, San
        Francisco, Calif., 1974;   E.C. Gay, F.J. Martino and Z.
        Tomczuk, *Proceedings of the 10th I.E.C.E.C. Conference*,
        759097, Newark, Del., 1975.
59.   S. Sudar, L.A. McCoy and L.A. Heredy, *Ibid.*, 759099.
60.   P.C. Symons, U.S. Pat. 3,713,888, January 30, 1973;   *Inter-
        national Conference on Electrolytes for Power Sources*,
        Brighton, England, December 1973.
61.   L.H. Taller, *Proceedings of the 9th I.E.C.E.C. Conference*,
        749142, San Francisco, Calif., 1974;   M. Warshay and L.O.
        Wright, NASA Technical Memorandum TMX-3192, February 1975.
62.   H. Cnobloch, D. Gröppel, D. Kuhl, W. Nippe and G. Siemsen
        in D.H. Collins (Ed.), *Power Sources 5*, Academic Press, New
        York, 1975, p.261;   O. Lindström, *Ibid.*, p.283.
63.   Y. Le Cars, J. Thery and R. Collongues, C.R. Acad. Sci.
        Paris, 274 (1972) 4.
64.   Y. Lazennec, Presented at *I.S.E. Conference on Primary and
        Secondary Batteries*, Marcoussis, France, 1975.
65.   A.T. Kuhn and C.J. Mortimer, J. Electrochem. Soc., 120 (1973)
        231.
66.   H.A. Liebhafsky, *Ibid.*, 106 (1959) 1068.
67.   H. Shimotake and E.J. Cairns, *Proceedings of the 2nd I.E.C.E.C.
        Conference*, 679095, Miami Beach, Fla., 1967.
68.   M. Bodenstein, Z. Physik. Chem., 13, 56 (1894);   *Ibid.*, 22
        (1897) 1;   *Ibid.*, 29 (1898) 295.
69.   J.M. King, *Proceedings of the 10th I.E.C.E.C. Conference*,
        759038, Newark, Del., 1975.
70.   F. Bacon, this volume, Chap. II.
71.   R.N. Camp and B.S. Baker, Contract No. DAAK02-73-C-0084,
        U.S.A.M.E.R.D.C., Fort Belvoir, Va., 1973;   NTIS, Springfield,
        Va., AD 766313/1.

72.   E. Dickson and R. Powers, *The Video Telephone, A New Era in Communications, A. Preliminary Technology Assessment,* Cornell University Program on Science and Technology, 1973.

73.   D.F. Westlake, Biological Rev., 38 (1963) 385.

74.   T.R. Schneider, Energy Conversion, 13 (1973) 77.

75.   M. Klein, *Proceedings of the 7th I.E.C.E.C. Conference,* 729017, San Diego, Calif., 1972;   J.F. Stockel, G. Van Ommering, L. Swette and L. Gaines, *Ibid.,* 729019.

76.   R.H. Wiswall and J.J. Reilly, *Ibid.,* 729142.

77.   H.J. Halberstadt, *Proceedings of the 8th I.E.C.E.C. Conference,* 739008, Philadelphia, Pa., 1973.

78.   A.J. Appleby, J.P. Pompon and M. Jacquier, *Proceedings of the 10th I.E.C.E.C. Conference,* 759121, Newark, Del., 1975.

79.   A.J. Appleby and M. Jacquier, J. Power Sources, 1 (1976) 17.

80.   R.B. Ormsby, in *Hearings before the Committee on Commerce,* U.S. Senate, Oct. 7-10, 1975.

81.   R.L. Graves, J.W. Hodgson and J.S. Tennant in T.N. Veziroglu (Ed.), *Hydrogen Energy, Part B,* Plenum, New York, 1975, p.755.

82.   R.E. Lapp, Fortune, October, 1975, p.151.

The above list of references is not exhaustive. Others may be found in reviews on specialized topics, e.g. solar devices (Ref. 28), solar electrochemistry (Ref. 45), electrochemical storage (Ref. 44), wind power (W.E. Heronemus, *8th Annual Conference, Marine Technology Society,* Washington, D.C., 1972), and power from ocean thermal gradients (A.M. Strauss, *Proceedings of the 10th I.E.C.E.C. Conference,* 759117, Newark, Del., 1975). Many useful references to hydrogen production and use are given in Refs. 11 and 27. Other useful papers are to be found in recent *International Conference on Photoelectric Power Generation* and *I.E.C.E.C.* proceedings, as well as in the Journal, *Solar Energy.* The recently published book *The Solar Hydrogen Alternative* by J.O'M. Bockris (Australia and New Zealand Pub. Co.), which was not available at the time of writing, contains valuable additional information.

COMPETING PATHWAYS TOWARDS LARGE SCALE HYDROGEN PRODUCTION

J.O'M. Bockris

The Institute of Energy Studies
The Flinders University of South Australia

INTRODUCTION

Surveys of methods of producing hydrogen on a large scale have been published in recent times (1-3). Such articles will be taken as read. This article shows prospects visible in mid-1976.

SOURCE OF HYDROGEN

The only source will be water. The direct photolysis of this substance under solar light is not a prospect (4). The photons required to decompose water directly to hydrogen and oxygen are more energetic (*ca*. 6eV) than those available from the sun (2 eV). However, 6 eV photons might be producible on a large scale by injecting metal powders into plasma (5). When fusion has been maintained and built on a large scale, we shall have many decades before we shall know if fusion reactors will ever be competitive with massive solar and wind, with their simple engineering require-ments. A Hydrogen Economy will be needed before commercial fusion.

Catalysis at room temperature could produce hydrogen but only by involving some other substance, for example CO_2 (6).

Sea water is often supposed to be admissible as the feed stock for electrolysers: it is a highly conducting electrolyte. However, the electrolysis of sea water gives a partial current density for chlorine at the anode. A speculative analysis of the potential-dependence of the ratio i_{Cl_2}/i_{O_2} has been made by Bockris (7). If the cell potential (excluding IR) is less than 1.6 V, $i_{Cl_2} \to 0$. If the economics of low-potential electrolysers

79

are not favourable, one could use $Cl_2 + H_2O \rightarrow 2HCl + \frac{1}{2} O_2$ and reject excess HCl to the deep sea. However, the reaction has to be driven, thus requiring the cost of the extra heat. Under conditions availalbe in Australia, Saudi Arabia, etc., sea water could be solar de-salinated on a very large scale before feeding to the electrochemical reactors.

MAIN TRENDS OF THE PRESENT APPROACH

Research on large scale hydrogen production is occurring in many laboratories at this time but virtually all the work is on the chemical decomposition of water (8-10) (*v.i.*). Some work has begun on its photo-electrochemical decomposition (11).

Although other approaches have been described, particularly those on the photo-synthetic approach (12,13), there is no broad attack upon what must be regarded as one of the Greater International Research and Development Problems of the time, i.e. how to produce hydrogen from water at a price cheaper than that of natural gas in, say,the year 2000*.

Research on the electrochemical method of water splitting, particularly in the United States programme, is conspicuous by its absence, corresponding to the lack of contact with Reality in that country connected with the development of Modern Electrochemical Technology. For this reason, U.S. industry continues to use electrolysers designed many decades ago and working at twice the potential shown possible in electrochemical research reactors (thus almost doubling the price of the product).

ON LARGE SCALE THERMAL METHODS OF SPLITTING WATER

The present approach towards the thermal decomposition of water is associated with the names of Marchetti and de Beni**.

The electricity used in the electrochemical method of splitting water is not the energy source for the production of hydrogen but comes from, e.g. coal, via mechanical energy. The heat-to-mechanical energy requires a Carnot factor. This amounts to 0.3-0.4, so that the 75% efficiency of an electrolyser is reduced to a 25% efficient producer of hydrogen. This is an intrinsic difficulty in the electrochemical approach and Marchetti

---

* A massive injection of research funding in the late '70s might attain examples of a Hydrogen Economy by 2000 (54).
** The running of electrolytic processes at more than 100% efficiency of the electric energy used threatens no thermodynamic laws: it does require external heat (14).

suggested it could be overcome by using thermal decomposition.

Thus, the chemical (cyclical) process of water splitting involves the use of heat *directly* from an energy source (nuclear?), no detour via mechanical work would be needed to produce the intermediate fuel of electricity, and, therefore, (it was thought) no Carnot (in) efficiency factor for conversion to mechanical energy would reduce the net efficiency of the conversion of the primary energy source to hydrogen. *Therefore,* there could be (under conditions of ideally low intrinsic energy loss) a hypothetical 100% utilization of the free energy for the production of hydrogen and oxygen from water. Further, the standard free energy change for the overall decomposition at a given temperature could be reduced from that at, say, 2000°C, by application of knowledge of the sign of the entropy change in the partial reactions of a sequence of reactions leading to water splitting. Reactions with a negative entropy change would be run at high temperatures, those with a positive entropy change at low temperatures. The net $T\Delta S$ contribution would make the ΔG^{O} less positive, and less energy would be needed to attain hydrogen than the energy of direct thermal decomposition. The temperature of the heat needed would be reduced and enter the range (800-900°C) at which nuclear heat is available.

At first, the authors published (15) a large number of hypothetical cycles. Their approach was to find suitable stoichiometry, note appropriate entropy signs, and advise estimates of temperatures for each partial reaction, based on a knowledge of ΔH and ΔS (but neglecting accounting for the - mostly unknown - rate constants). The taking up of this approach by so many schools all over the world, and the omnipresence of computers to increase the number of hypothetical sequences, has lead to a proliferation of researches in this field.

The disproof of the basis of the approach is based upon knowledge not published at the onset of endeavours on thermal water splitting in 1970. It depends upon two concepts (16-19).

(i)   The thermodynamically calculable energy for driving an electrochemical process is that which causes the process to occur at zero rate (1.23 V for water decomposition at 25°C).

If we want the process to occur outside the reversible region, a greater amount of energy (i.e. a larger applied potential than 1.23 V) is needed and this extra potential (over the reversible thermodynamic value) is termed the overpotential for the rate of hydrogen evolution desired.

A similar factor occurs in thermal reactions, though has hitherto been unnoticed (see Appleby (18)). The energy

usually used in calculations to make them occur at equilibrium
(at zero rate) under isobaric and isothermal conditions is
given by the thermodynamic quantity used in the analyses.
If, now, the product is to be *produced at an appreciable rate,*
i.e. the reaction driven out of the equilibrium position to
a forward rate with significant driving force, appreciable
"overheat" will be required and this does not seem to have
been understood by those who advocated sequential thermal
decomposition as a way to avoid the Carnot difficulty.

(ii) The essential concept in the production of hydrogen by a
cyclical method is that the partial reactions are to be run
at *different temperatures and pressures* respectively, on a
cyclical basis. When one has to take a mass of reactant
and convert it to another pressure, or temperature, cyclically,
one has, of course, to do work, or inject energy, by compress-
ing and expanding or changing the working temperature of the
reactant and product masses. These energy use-ups were not
taken into account in the discussions of the economics of the
cyclical chemical method for hydrogen, except by Funk *et al.*
(16), who have given a detailed analysis of such changes.

When these aspects of the loss and gain in the economo-energy
analysis are made, the energy needed is as large as, and often
greater than, that needed in the electrochemical method with
its burden of admitted Carnot Loss. The cyclical thermal
method could be energetically preferable to electrolysis when
the sum of the overheats (required to reach a certain forward
velocity) in the partial reactions and the Carnot-like work
losses required to bring the systems to their various
necessary T's and P's, is less than the Carnot loss in
manufacturing electricity plus the overpotential losses in
the electrochemical reactors.

These remarks remove the essence of the argument for the
production of hydrogen by driving a series of partial reactions
at appropriate T's and P's*.

There are other reasons why the thermal water splitting
alternative has a lesser prospect than other methods (although few
of these are being researched) (20).

(a) Reactions only follow the free energy-preferred pathways at
sufficiently high temperature, e.g. 1000°C or more. At lower

---

* One refers to the *pure* thermal cyclical method, i.e. one in
which water and heat are the only net entities used. Thermal
cycling of an undesired anode product (e.g. I_2 back to I^-) could
play an interesting part in an electrochemical or photo-oriented
process.

temperatures, a reaction follows the optimal *kinetic* pathway,
and this by no means coincides with the largest free energy
driving force. Thus, when one sees one of the chemical cycles
written on paper, there is little likelihood that the path
indicated, and duly thermodynamically calculated, has a
relation to what would happen when the reactants are brought
into contact at the assigned T and P. Of course, this remark
applies more strongly to those paths which, for reasons of
their negative entropy changes, must be operated at low
temperature: the heat of activation for the fastest kinetic
pathway, not the thermodynamic free energy of reaction,
determines the product*.

(b) The degree of cyclicity would have to be exceedingly high
     (e.g. 99.99%), for if an undesignated reaction occurred to a
     significant extent, there would be a pile-up of the products,
     which could not be cycled, and this would worsen the economics
     and demand extra capital for machinery to remove the left-
     overs. The water-splitting plants will be big and left-over
     solids could be at the level of many tons per day.

(c) The complex multi-process plant required by the cyclical
     chemical method (pumps, flowtrains, heating elements, reactors)
     would be more expensive to build and maintain (high temperature,
     corrosion rate and materials cost) than the simple one step
     electrochemical reactor, where 100% Faradaic efficiency is
     reached without effort, and where discussion of side products,
     recycling, corrosion, pumps, heat losses, etc., could have
     no meaning.

(d) The cyclical chemical method implies availability of atomic
     heat at 800 to 900°C. The future of atomic reactors as the
     basis of a hydrogen economy after fission reactors run out of
     U^{235} in the era of 2000-2010 (when a Hydrogen Economy could
     be attainable) is parlous. If solar reactors are to supply
     massive amounts of energy in the future and to underlie a
     Solar-Hydrogen Economy, it will be difficult to obtain temp-
     eratures above 400°C on a large scale**.

---

*   Another example of the error in judgment to which attention is
    here brought is the Pourbaix diagram where thermodynamic know-
    ledge is used to predict the kinetic happenings of room temper-
    ature corrosion (55).
** Meinel (21) has been an advocate of extended piping using selected
    coatings which might bring a circulated fluid reaching 400°. The
    necessity of extreme high vacua on a very large scale obviously
    makes his suggestions uneconomic. The use of solar energy to
    reach high temperatures (e.g. 3000°) is established but a large
    area of concentrators are needed to heat a very small volume.

THE ELECTROCHEMICAL APPROACH AVOIDING ANODIC OXYGEN EVOLUTION

On the standard hydrogen scale, hydrogen evolves with a reversible thermodynamic potential of zero whilst oxygen is at 1.23 V on this scale. If one arranged for an alternative anodic reaction, e.g. $2I^- \rightarrow I_2 + 2e$, to occur in conjunction with hydrogen evolution, the cell potential for the production of hydrogen (hence, energy and cost) would be reduced. As the electricity costs (proportional to the cell potential) are the dominant ones in the electrochemical hydrogen production, the price of hydrogen would be reduced correspondingly. For example, 1.23 V and 0.52 V are the standard potentials at N.T.P. for the evolution of oxygen and iodine respectively and one could expect a reduction in the electricity cost for producing hydrogen by more than half.

Another advantage would come with this approach: the oxygen evolution reaction is the most sluggish of all electrochemical reactions. Its exchange current density at 25°C is the order of 10^{-10} A cm^{-2}. As the overvoltage (η) is related to the exchange current density (i_0) for evolution of hydrogen by the expression:

$$\eta = \frac{2RT}{F} \ln \frac{i}{i_0} \tag{1}$$

the evolution of oxygen is accompanied by a higher overpotential than that of other processes. The price of hydrogen by the direct electrochemical decomposition of water will be large. Suppose $(i_0)_{O_2} = 10^{-10}$ A cm^{-2}; and $(i_0)_{I_2}$ is 10^{-3} A cm^{-2}, the overpotential will be about $2.303 \times 2RT/F[\log (10^{-10}/10^{-3})] = 0.7$ V less in the Tafel region for I_2 deposition than O_2 evolution – the total cell potential for hydrogen evolution would be reduced from ~ 2.0 V to ~ 0.5 V, i.e. the electricity costs reduced to one quarter of the amount for hydrogen evolution accompanied by oxygen evolution.

This discussion is over-optimistic because it neglects the heat which would be necessary in the auxiliary thermal process to drive the I_2 to O_2 ($I_2 + H_2O \rightarrow 2HI + \frac{1}{2} O_2$), and return the I^- to solution. This endothermic reaction needs heat though its single T and P conditions will mean that the heat needed is the thermodynamically indicated (hence, cheap) heat and the overheat to make it occur at a suitable rate. Calculations have been published by Bockris (22). The net position is a reduction of the price by 40%.

These concepts can be advanced by an aspect first used by Juda and Moulton (23). Massive injection of a waste product is used and this undergoes an anodic reaction alternative to that of oxygen evolution. The overpotential for oxygen evolution is removed from the cost of producing hydrogen, and the cell potential (hence electricity costs) fall from those for the electrochemical decomposition of water to H_2 and O_2 to those for hydrogen evolution and the anodic reaction concerned.

There will be, of course, an overvoltage for the anodic
alternate reaction but the likelihood is that this will be less
than that of oxygen evolution because the exchange current density
of this process is so small. Moreover, a useful product (e.g.
H_2SO_4) is liable to result from the recycling of the waste product,
and the sale of this can be subtracted from the price of the
hydrogen*.   Half to one-third the thermodynamical minimum cost
should be easily reachable.

Research is needed in these directions. The prospect of
producing hydrogen at less than the cost of gasoline (per energy
unit), taking into account the sale of the anode product, seems
a realizable one. Much organic oxidative manufacture would be
possible.

TRENDS IN ELECTROCHEMICAL REACTOR RESEARCH

An illusion to which considerations of large scale hydrogen
production is subject, particularly in the United States, is that
few new contributions to electrolysis reaction engineering have
been made for several decades. This impression can be speedily
dispersed by understanding Fig. 1: over the last twenty years, the
potential for straight electrolysis, without the injection of
alternate reaction pathways, has fallen from > 2.4 V to 1.3 V.
Recent research has therefore potentially reduced the cost of
electrolytic hydrogen by a factor of 1.8.

The minimal use of electrical energy in the electrolytic
production of hydrogen would be in the high temperature electrolysis
of steam. The overpotential for the evolution of oxygen falls to
a negligible value at > 400°C. The reversible potential for the
decomposition of water decreases with increase of temperature.
This was brought out by Gregory (24), who introduced the concept
of a thermoneutral potential. At potentials below it, the process
of water splitting needs both electricity and externally applied
heat (i.e. below the thermoneutral potential, in the absence of
applied heat, the system falls in temperature if water is decomposed
at the thermoneutral potential by the application of a potential).
Above the thermoneutral potential, the process is using electricity,
but giving out waste heat.

---

* The fact that, eventually, the amount of the product would exceed
  the market does not devalue the concept of the sale of products
  in earlier stages of a Hydrogen Economy. Not only one product
  (e.g. H_2SO_4) need be made. The anodes in hydrogen reactors could
  be used as large scale oxidation sites for many kinds of industrial reactions, where at present chemical oxidation is used.

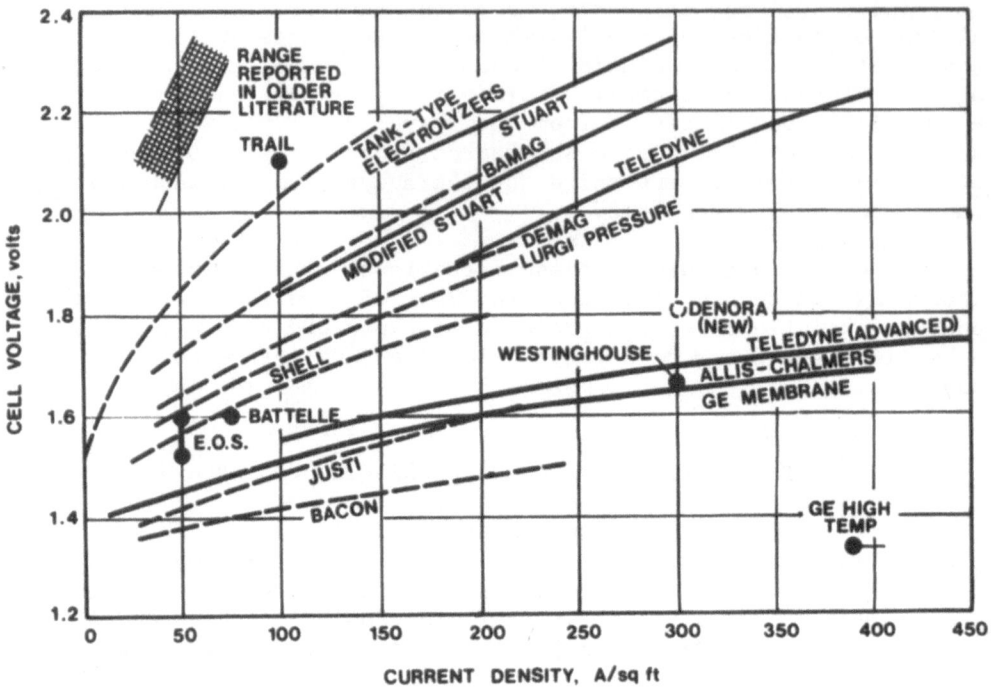

Fig. 1   Various more recent electrochemical hydrogen reactor
         results (Gregory (24)).   The contrast with classical
         electrolysers - upon which most calculations have been
         made - is startling.   Conversely, the newer results are
         from laboratory or pilot plant equipment.

        Now, ideally, we would like to have an electrode-oriented
process which utilizes "no externally generated electricity", i.e.
uses all heat.   Electrochemical processes avoid the Carnot cycle
efficiency factor, and if the overvoltage is reduced towards zero
by the use of high temperature, it is possible to conceive a
process which runs electrochemically at a very small applied
potential, whilst using heat to maintain the cell temperature, so
that the heat is effectively converted electrochemically to hydrogen.
This conversion of heat to hydrogen would occur at more than twice
the efficiency which would be needed if the decomposition ran via
electricity applied from without (because of the Carnot Cycle), or
if decomposition took place by the cyclical-thermal method which,
as we have now seen, involves "overheat" and a kind of pseudo
Carnot-cycle inefficiency, because of the work done in moving the
reactants round a P-T sequence.

The highest practical temperature at which we could assay the decomposition of steam by the electrochemical method would be some $1500^{\circ}C$ or $1723^{\circ}K$, at which temperature the reversible cell potential for water splitting to H_2 and O_2 at normal pressures is about half of that necessary at room temperature. As the overvoltage will tend to a negligible value, the necessary electrical energy (that which must be obtained after wasting 2/3 of the chemical energy of the basic fuel, e.g. coal, through a Carnot cycle loss) is now reduced to about one-third of that at room temperature (one-third and not one-half because there is now no overvoltage) and the rest of the energy is supplied by "pure heat", not suffering the Carnot fate. Reasoning in this fashion is new (25) but the General Electric Company reported the successful electrolysis of steam in 1966 (26).

Recently, Bevan, Singh and Bockris (27) have been investigating urania-yttria mixtures which show high electronic *and* O^{2-} ion conductivity at temperatures in the $1000^{\circ}C$ range, and thermodynamic stability in H_2 at > $2000^{\circ}C$. This material could act as the electrodes in electrolysers with a zirconia-yttria electrolyte.

Thus, a difficulty with the General Electric Cell is the metal collectors which would oxidize in O_2 and steam. This is overcome by the use of U_2O_3 as an electronic conductor (Fig. 2). Another difficulty is in getting sufficiently thin ZrO_2-Y_2O_3 sheets. This is overcome by vapor spraying them on the U_3O_8-Y_2O_3, the ionic conductivity of which is high enough so that a thickness of 1mm can be attained.

The proposed B-S-B cell contains a 1mm U_3O_8-Y_2O_3 layer, a 0.1mm ZrO_2-Y_2O_3 layer and a 0.1mm porous U_3O_8-Y_2O_3 layer. The thicker U_3O_8-Y_2O_3 layer would be the cathode. Possible reactions are shown in Fig. 2. O^{2-} diffuses through the U_3O_8-Y_2O_3 and the ZrO_2 (which has a negligible *electronic* conductivity). H_2 evolves on the outside of the U_3O_8-Y_2O_3. Does O_2 evolve at the U_3O_8-Y_2O_3-ZrO_2-Y_2O_3 boundary? If it does, it emits through the porous structure of the anode.

Results of this concept as a working cell have not yet been attained. Its possibilities for very high temperature (> $1000^{\circ}C$) electrolysis appear good.

PHOTOLYTIC METHOD

The photolytic approach to hydrogen evolution is one which can be carried out only indirectly because the gas phase or liquid photolysis of water requires photons bigger than those available in light from the sun. Will it be practical to inject plasma with elements which, at the heat concerned, would give out 6 eV photons (5)?

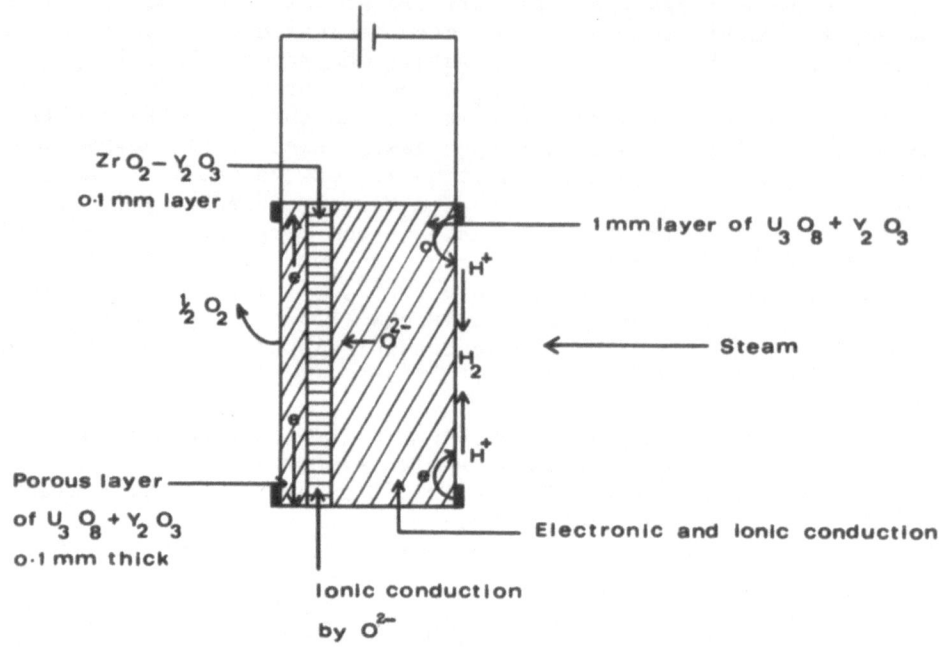

Fig. 2 A hypothetical high temperature hydrogen-oxygen product,
        using U_3O_8-Y_2O_3 as an electronic conductor. (Bevan, Singh
        and Bockris unpublished).

     Laser-induced photolytic methods have not been tried.
Commercial lasers hitherto produce sub-solar photons whereas 6 eV
photons would be needed.

PHOTOSYNTHESIS

Photosynthesis among Natural Products

     The use of the growth of plants - spontaneous photosynthesis -
to produce crops which could then be used as a fuel is considered
in Australia (28). Indeed, simple forest growth is sometimes
proposed. A plant cited is sugar cane and the activity has been
recommended by an Australian committee considering solar energy
(28). The approach has the advantage of lack of need for research.
The plants could be burned directly, forming steam; but they could
be heated in the absence of oxygen, to form fuels such as methane
and hydrogen; or fermented to alcohol.

The method of spontaneous photosynthesis does, however, own
to a disadvantage: it is only about 1% efficient (28). The best
laboratory-borne photosynthesis is 9% efficient (28). The amounts
of ground necessary for the collection of solar energy if the
efficiency were 9% would increase proportionately. It is already
so large that only large desert expanses can be considered.

Research on photosynthetic mechanisms and efficiencies are
needed; as is also development work on the fermentation of
alcohol (29).

Photosynthetic Decomposition of Water

The idea that water could be decomposed by the use of naturally
occurring enzymes was put forward by Rubin and Gaffron (30). The
enzyme could be obtained cheaply from bacteria. At first, one
imagines utilizing sections of the sea, baffled off into sea-solar
farms, into which the enzyme could be introduced on a large scale.
Collection of hydrogen from the production areas, particularly the
separation from oxygen, would involve the use of membranes on a
very large scale, and might be cheap*.

However, the reality of the possibilities of photosynthetic
production needs much more than this (31). It is necessary to have
an inbuilt light-absorbing apparatus. Thus, the process of the
photosynthetic production of hydrogen involves not only the
presence of a suitable enzyme but also the existence of two entities
known as, respectively, photosystem 1 and photosystem 2. These
substances are complex mixtures of photosensitive organic compounds,
interconnected by a chain of redox couples. Photosystem 2 receives
a photon and carries out the photochemical reaction which splits
water and releases oxygen. The residual hydrogen is then attached
to a hydrogen carrier within the photosynthetic apparatus (i.e. the
plant, the biological compound) and is released after excitation
of photosystem 1 under the action of an enzyme, such as hydrogenase
or nitrogenase.

These photon-collecting systems are not present in an enzyme
molecule. It is the enzyme which helps the release of the hydrogen,
but the presence of the enzyme alone in the water would not give
rise to the evolution of hydrogen by the action of light. It is
necessary to include in the system, the carrier, and some method
for acceptance of light.

---

* The estimate given (41), which is $0.3 per 10^6 BTU (1974 currency),
  depends upon an estimated reaction rate and this is taken in the
  middle range for photosynthetic reactions — there was little to
  go on at the time the estimate was made.

One may use an alga. Thus, the alga itself, if suitable, and living off water by the production of hydrogen in a photosynthetic reaction, contains photosystem 1 and photosystem 2, and the hydrogen carrier, as also enzymes. Thus, one grows the alga in the presence of carbon-dioxide and light.

When the carbon dioxide is switched off, the alga lives on water, producing hydrogen, so long as other substances upon which it can live (for example, nitrogen) are removed from the system. In light, the alga will decompose the water to produce hydrogen (it produces more oxygen than hydrogen) and the aim outlined above is qualitatively attainable.

The enzyme could be obtained from a blue/green alga, e.g. *Anabaena cylindrica* (32). The alga is grown in a nutrient solution. It contains nitrogenase. After the growth of the enzymes in bottles, irradiated and containing CO_2, the CO_2 is removed, argon introduced, nitrogen excluded, and the system decomposes water.

The rate of production is nanomoles per sec per mg. of protein. The efficiency of production of hydrogen from this enzyme was 0.3% (Fig. 3).

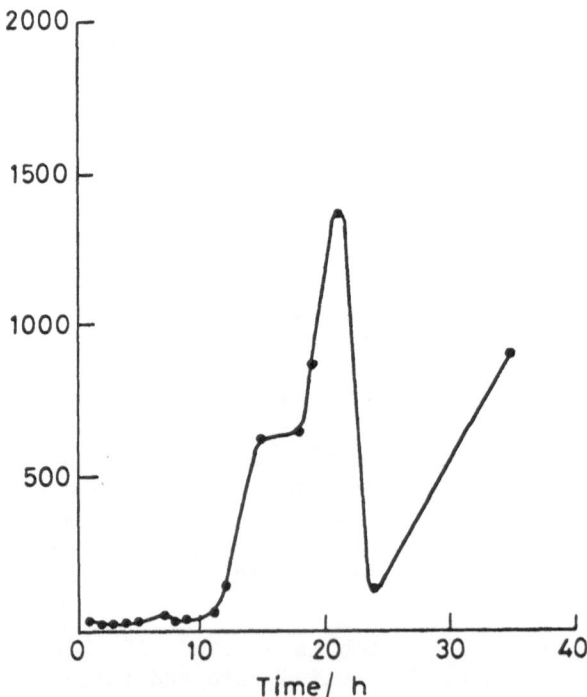

Fig. 3 H_2 in namomoles over an aqueous solution containing *Anabaena cylindrica*, irradiated (Neil *et al.* (33)).

The evolution of hydrogen from water by naturally growing
enzymes has been continued for 24 hours, and there are complex
processes occurring in the solution, because there is a period in
which the hydrogen over the solution is decreased, i.e. some
hydrogen is consumed, perhaps by decomposition products of the
alga.

It is necessary to develop stable systems which live for a
much greater time than those which have already been used. In
research studies, an increase in the rate of hydrogen evolution
by about 10 times is needed in order to make the hydrogen production
method worth considering for developmental attention.

The method is wide open for research. Firstly, many other
hydrogen-producing algae apart from *Anabaena cylindrica* are known
(32); the effect of pH, salt, temperature, have not been examined.
The concept of feeding the cultures with CO_2, then exposing them
to water in the absence of CO_2, then reintroducing CO_2 to revive
the culture, etc., has not been tried.

The difficulty with the photosynthetic approach to hydrogen
is that the theory of photosynthesis is still not detailed.
Attempting to relate the constitution of enzymes and the surrounding
systems to their efforts in producing hydrogen, is more difficult
than considering the solid state systems involved, say, in the
photo-electrochemical method.

THE PHOTO-ELECTROCHEMICAL APPROACH

A Combined Photovoltaic-Electrolyser Concept

One approach to the photo-electrochemical situation would be
to utilize photovoltaic couples, and then drive water electrolysers.
The efficiency of these is about 75%, and concepts exist for
raising it to 90%. The overall efficiency of light to hydrogen
conversion would be, then, about 6%.

Both the electrolyser and the photovoltaic couples need
research work, e.g. the lifetime of cheap cadmium-sulphide couples.

DIRECT PHOTO-ELECTROCHEMICAL APPROACH

General

In the photovoltaic situation, the electrons have to live
long enough to reach metal collectors, whereas in the photo-electrochemical approach the junction is that between the solid and the
solution, and the path length of the photon-activated electron is

less than that in photovoltaic couples. The need for single
crystals of high purification is reduced.

Early Work

Honda and Fujishima (11) were the first to report the photo-
electrochemical production of H_2. They used TiO_2 single crystals
and reported, using a cell of uniform pH, the evolution of oxygen
on the titanium oxide, and that of hydrogen on a platinum electrode
in the same cell. The reaction occurring at the platinum electrode
was the production of oxygen but later they showed (34) that
hydrogen could be evolved on platinum with light insolation of
TiO_2 so long as there was a pH gradient in the cell (a 0–14 pH
gradient would reduce the necessary cell potential 0.8V). This
method of utilizing a pH gradient cannot be utilized upon a large
scale because the pH would tend to become uniform with use.

The Ideal System for Photo-Electrochemical Use

A proper system for photo-electrochemical effect must have a
uniform pH throughout. This means that there must be irradiation
in both the cathode and the anode.

The basic relationship which must hold between the energy
gaps of two electrodes is

$$\Delta G^o = - \, (nFE_{rev} + h\nu) \tag{2}$$

$$\text{and: } (h\nu)_{cathode} > (E_G)_{cathode} \tag{3}$$

$$(h\nu)_{anode} > (E_G)_{anode} \tag{4}$$

Thus, the light would be directed firstly at the electrode
with the larger energy gap, and light not absorbed there would be
directed further to the other electrode; alternatively, a stream
of light could be directed at each electrode.

Only semiconductors can be considered for electrode systems
(35). Electron-photon interactions in metal systems are greater
than those in semiconductors so that electrons which could be
emitted as a result of photons do not reach the surface but are
deactivated. One is left with semiconductors which have an energy
gap of less than 2 eV. Combinations of such semiconductors (a p-
and n-type) should give rise to the evolution of hydrogen and
oxygen on irradiation. The kinetics of the evolution has, however,
to be examined (see below).

A Principal Difficulty of the Photo-Electrochemical Method

A difficulty of the electrochemical method is that the substrate itself may take part in the electrochemical interfacial
reaction. This happens at the anode because of the more positive
potential of the oxygen. Thus, the anodic electrode reaction for
the dissolution of the semiconductor surface may have a rate
comparable with that of the evolution of oxygen, whereupon the
electrode will be photo-electrochemically consumed.

This difficulty greatly reduces the number of electrodes
which can be used for a photo-electrochemical hydrogen/oxygen cell.
One solution is cladding - the substrate is covered with a protective layer of a substance which is known not to undergo competing
dissolution. Thus, a gold layer has been suggested (36). The
requirement is that the covering be transparent to light which
limits the thickness to 100 Å. For metals, such a thickness is
adequate for protection (passive layers are *ca.* 50Å), although the
danger of pits may be present.

Bockris and Uosaki (37) first showed that cladding could be
effective for CdS electrodes with thin TiO_2 films. Without the
cladding, they obtained an efficiency of 0.6% and with cladding an
efficiency of 2.4% (Fig. 4).

The Theory of the Hydrogen/Oxygen Cell

The qualitative picture of the evolution of oxygen at a photo-
anode electrode is as follows. A photon enters the electrode and
if it has an energy greater than that of the energy gap, is
absorbed by the semiconductor, activating an electron from the
valence band to the conduction band. Thereafter, the charge
carriers undergo a series of photon-electron energy vesting
collisions in the semiconductor, and a fraction of the holes produced by the absorption of the photons reaches the surface of the
electrode. Here, there are available, in alkaline solution, OH^-
ions which donate their electrons to the holes. The OH radicals,
produced on the surface, combine to give oxygen. The electron
produced by the advent of the photon in the valence band then
migrates round the circuit and arrives at the other electrode,
where, if conditions are appropriate, it will emit to a proton in
solution in contact with the cathode, giving rise to hydrogen
atoms on the electrode surface and then hydrogen as a result of
chemical combination.

The question of what the current density will be, and therefore the degree of conversion of the light to hydrogen, depends
upon the kinetics of the process. Electrochemical consideration
of the penetration of the electrons from the OH^- of the barrier

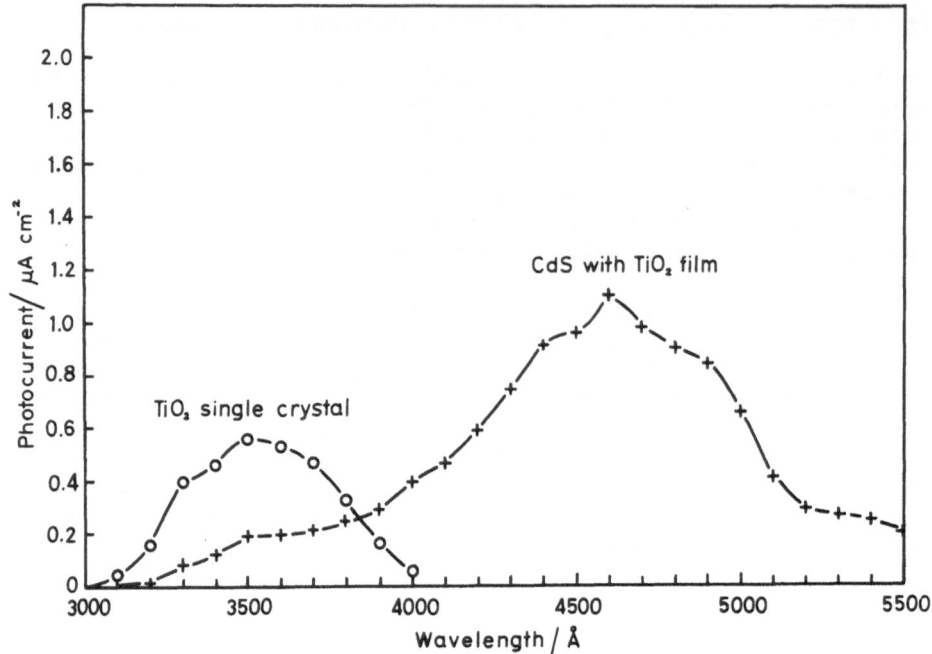

Fig. 4   The anodic photo-electrochemical current with irradiation
of TiO_2, with and without cladding (Bockris and Uosaki
(37)).

between the solution and the electrodes: and the corresponding
penetration at the cathode. These are quantum mechanical processes
and the only practical way to treat them has been to employ the
usual approximations for tunnelling processes (38).

Oxygen anodes must be n-type semiconductors. At first, this
surprises, because in semiconductor electrochemistry n-type substances usually act as cathodes but in an n-type substance there
are few holes in the valence band. The advent of light which
causes the holes gives rise to a negligible extra supply of
electrons but a substantial difference in the rate of supply of
holes. It is holes which make themselves available for the
donation of electrons from the solution, i.e. *anodic* electrochemical
reaction.

If one attempts to obtain a photo-anode from a p-type semiconductor, little change would be noticed on irradiation, because
p-type electrodes have plenty of holes in the dark. But they have
few electrons and hence are good photo-*cathodes*.

Optimization

The final equation for the current of a given electrode is quite complex and consists of 3 terms, the first for the fluctuation of the barrier to acceptors (considering p-type electrodes); the second for the penetration of the barrier to barrier at the double layer to form solvated electrons; and the third for the electrons which go over the barrier.

An application of this type of equation to both cathode and anode allows us to work out the hypothetical function of a cell (Fig. 5).

From these considerations it is possible to find the optimal conditions desirable, independent of any actual semiconductors which exist. Some optimised conditions are in Table 1.

The possibilities of corrosion in the solution and anodic breakdown have to be taken into account in applying these criteria.

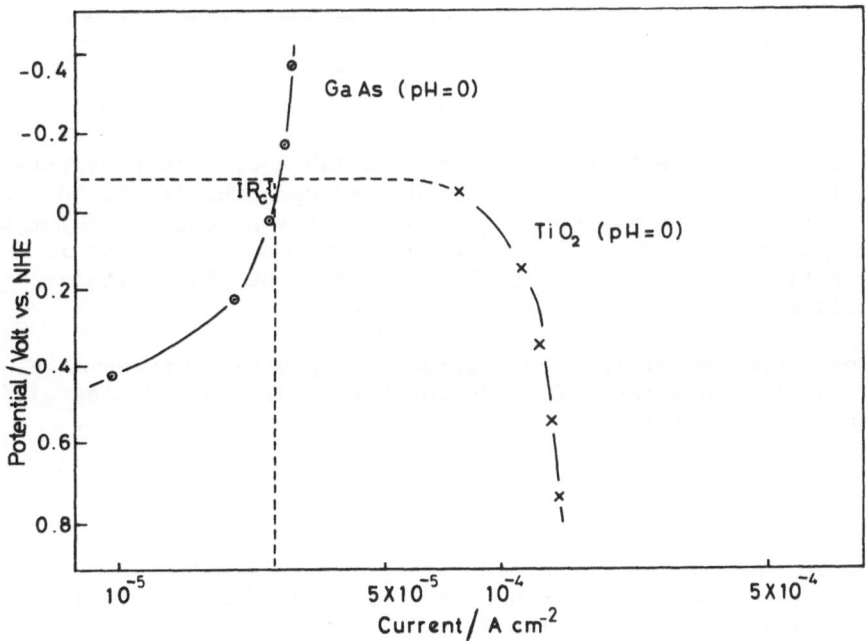

Fig. 5   Results of a calculation, cathodic and anodic currents in a photo-electrochemical generator of hydrogen and electricity. Parameters used for anode: those of TiO_2; for cathode: partly those of GaAs (Bockris and Uosaki (39)).

TABLE 1

Optimum Condition of Semiconductor Electrodes*

	p-Type	n-Type
Electron Affinity	The smaller, the better Minimum *ca.* 3.5eV	$\chi + E_g > 5.3eV$ $\chi < 5.3eV$
Life Time, Mobility	$\mu \times \tau > 10^{-8}(cm^2/V)$	
Energy Gap	The less, the better down to a minimum of about 1.2eV	
Flat Band Potential	$> 0.4V(NHE)$	$< 0.0V(NHE)$
Surface State	—	Very Important

* Table due to K. Uosaki

Some likely substances are shown in Table 2. The prospects for the photo-electrochemical method depend upon the efficiency at each electrode. The maximum efficiencies (for H_2 and O_2 evolution) reported so far are 2.4%. This would not be acceptable and if higher efficiencies (8%) cannot be obtained, the photovoltaic-electrolyser combination will be preferred.

For large commercial application, the photo-electrochemical method has the advantage of producing both hydrogen and electricity from the same reactors.

PHOTO-ASSISTANCE?

Could photo-assisted anodes reduce the cell potential in hydrogen reactors? For instance, suppose that the potential needed to drive hydrogen/oxygen cells is reduced by 0.6V, 30 to 50% of the electricity would be saved, i.e. the price of hydrogen will be reduced by that amount. However, the maximum current density would be about 10 mA cm^{-2} on the basis of a 10% efficiency and maximum isolation, so that the area of plants would be immensely increased. Photo-assistance would not be practical.

TABLE 2

Properties of Semiconductors*

Semiconductor	E_G (eV)
Ge	0.66
Si	1.09
GaAs	1.35
CdTe	1.5
CdSe	1.74
CuO	1.95
CdO	2.2
GaP	2.35
CdS	2.4
ZnSe	
ZnO	3.3
Cu_2O	2.02
NiO	4.0
V_2O_5	2.23
$CoTiO_3$	~2.2
*Stable anodes***	
TiO_2	3.0
SnO_2	3.7
$SrTiO_3$	3.2
$KTaO_3$	4.6
Fe_2TiO_5	~3.0
Fe_2O_3	2.0
ZrO_2	4.99
WO_3	2.7

* Table due to J.F. McCann
** This is a relative description, e.g. it has shown
that the photo-current decreases after about a
day (56).

A RADIOLYTIC APPROACH

Uranium salts in water give 11% of their energy in hydrogen
and oxygen evolved by radiolysis. The loss is due to recombination.
Methods might be evolved which would reduce this loss. However,
if only wastes were used as an origin of the radioactive energy,
the loss would be less important.

The hydrogen/oxygen mixture could be passed without separation

over appropriate electrodes in a fuel cell, the hydrogen being
picked up by the anode and the oxygen passing onto the cathode.

IMPROVEMENTS IN DIRECT WATER ELECTROLYSIS

Hybrid electrolysers may be preferable to direct water electrolysers. Were the potential at which hydrogen could be produced at
a reasonable current density (10-100 mA cm^{-2}) to be lowered to
1.6 V or less, there would be advantages in using straight hydrogen/
oxygen electrolysers: simplicity and the avoidance of extra capital
involved in treating chlorine, etc.

The 1.6 V could be reached but not without the development
work needed to scale up research devices. Thus:

Electrocatalysts. There are no catalysts used to reduce overpotential in classical electrolysers. Catalysts are used to some
extent in research ones (40).

Bubbles. At high current densities, bubbles exist between cathode
and anode and cause an IR-drop and waste energy. These bubbles
could be removed by sucking them into porous electrodes and allowing
the gases to be evolved in passages outside the electrodes, not
in the passage between them.

Flow. Flow of the electrolyte between smooth electrodes could be
made so that bubble evolution would not mix after the electrolyte
had passed out of the area between cathode and anode to a metal
separator held above them. Bubbles would be swept away before
they had grown, reduce IR drop caused by them and by the normal
membrane.

Porous electrodes. These have been used in some experimental
electrolysers (41). Porous electrodes in hydrogen evolution are not
obviously as valuable as in fuel cells. Nevertheless, enhanced
rates at a given overpotential seem to be found.

THE 1976 COST OF LARGE SCALE ELECTRICITY USED IN THE U.S.A.

Elucidation of the cost of electricity in the United States at
this time is less easy than in the recent past. This is because
the prices of energy at source - to be distinguished from that of
the medium, electricity - now vary very significantly. Thus, there
is "old" oil at cheaper prices; newly (licensed) oil at higher
prices; regulated natural gas at lower prices; nuclear heat, etc.
Surprisingly, hydro-electric sources also vary greatly in price.
The reason relates to the date at which the plant was built. Older
ones have amortized their plants and hence the cost of the hydro-

electric power is low, not burdened with capital repayment and interest charges. Newer plants must pay these. For example, present day hydro-electric plants cost about $1000 per kw to build. With 8736 hours per year, and 12% money, the cost per kwh for the finance alone is, then, 1.3 cents per kwh. To this must be added maintenance, insurance, manpower costs, etc.

In estimating present costs below, we shall eliminate taxes and dealer profit. Our price of electricity, and the corresponding cost of H_2, will then be comparable (per energy unit, e.g. 10^6 BTU) with the ex-refinery cost of gasoline ($2.75 average for the U.S.A. in mid-1976).

Let it be further said that the statement of *exact* numbers (including those for gasoline) would be unrealistic. The price of gasoline and of electricity wholesale (zero tax), varies from place to place by at least ± 25% for the same product. We shall discuss a range: and the prices we discuss here are not of course for the householder or other small user. They are for the large plant, where large means more than some 10 megawatts. It is further assumed that the plant is going to be run for long periods, i.e. several thousand hours per year.

Under these circumstances, the cost at Pittsburgh, Phila-delphia, for example, would be 0.99 cents per kwh plus 0.49 cents per kwh charge for increased fuel costs, i.e. 1.48 cents per kwh.

The cost in other areas for big supplies could reach 1.8 cents per kwh and could fall to as little as 0.6 cents per kwh for rather rare ("old") hydro-electric electricity.

The price obtainable for low use-period power (9pm - 8am plus Saturdays, Sundays and vacations) is about 1.2 cents per kwh from fossil fuel plants and 0.4 - 0.7 cents per kwh from the (rarely available) nuclear ones.

Thus, the U.S. range at this time can be taken as 0.6 - 1.8 cents per kwh. However, it would not be realistic to take a mean of these figures because the average cost would be nearer to 1.4 cents per kwh and this must be recalled when extremes of costs are stated below.* (Nuclear electricity in the 1980's, in 1976 dollars, will be 2-3 cents per kwh).

Costs in other relevant countries are both lower and higher than those in the United States. In Canada, for example, the load factor for increased fuel costs is lower than in the United States,

---

* The practicality of 1.4 cents per kwh may involve the acceptance of a condition that the power is available for only 80% of the time.

because of the greater content of nuclear and hydro-electric power
(the main energy source in the United States is coal). Thus, the
Canadian figure, corresponding to the 1.4 cents per kwh mentioned
here for the best estimate of relevant U.S. costs, would be 1.1
cents per kwh. In Japan, electricity comes largely from imported
Middle East oil, and consequently prices will be at least twice
those in the United States.

These prices are only partly rational. A rational price is
one which is determined by the manufacturer costs + 10-20% profit.
Middle East oil is the best example of a price based on a mono-
polistic position ("irrational"). However, at this time, a degree
of price fixing (not only too high, as for the Middle East oil,* but
also too low as for regulated natural gas), invades all energy
prices. Thus, the price of electricity from coal has roughly
doubled since 1974 (whereas the cost of ex-refinery gasoline has
increased only by about 25%). The expected rational price increase
would have been the inflational increase for labour, insurance,
maintenance, etc. However, coal backs oil and when oil moves up -
in the absence of a meaningful degree of cheaper nuclear power -
then coal and hence electricity can be moved up as well. The
price is always the maximum at which one can sell, i.e. just less
than the price of the nearest alternative.

Finally, the rate of inflation (and hence the cost of money)
dominates all financial estimates of new products. Were this
"zero" (or, in practice, 1-2%), the cost of money would be 3-4%.
The effect on the price of goods is obvious, just as it is obvious
that a sufficiently high inflation rate makes the price of all new
(financed) products unacceptably high (thereby causing the government printing of more and more unbacked notes called "one dollar",
etc.).

THE COST OF HYDROGEN

In the following, projections will be made of the cost of
hydrogen by certain methods with the economics which would pertain
were these methods in a production plant in 1976. In reviewing
these, and contrasting them with the ones predicted in earlier
publications, the sharply increased cost of electricity (about 50%
between mid-1974 and mid-1976) is to be noted. This increase comes
out of the "fuel loading factors" which the electricity companies

---

* The use of the term irrational for Middle East oil can be argued.
  It is the principal remaining Energy Source of the Western
  World. Raising its price (i.e. causing inflation and depression
  in the West) may be rational in terms of allowing more time for
  Research and Development on collection and conversion of the
  inexhaustible, clean sources.

in the U.S.A. have added to their prices to correspond to increases
in prices of uncontrolled sections of the fossil fuel market. In
contrast to the electricity price, the average price of ex-refinery
gasoline between September 23, 1974 and May 10, 1976 increased by
about 12%. Apparently there is more political difficulty in an
increased petroleum price, in comparison with the (in-effect) mono-
polistic utility price.

Thus, the price of 10^6 BTU of gasoline in mid-1976, U.S.A.,
ex-refinery (no taxes, transportation costs, dealer mark up) is:
$2.75 (*cf*. $2.50 in September 1974).

The equation for the price of 10^6 BTU of hydrogen, in terms
of the cell potential, E, and the cost of 1 kwh of electricity, c,
has been quoted as:

Cost (U.S. cents) = 229Ec + 40.   (5)

The 229 is a phenomenological factor. E is in volts and is the
total cell potential for production of gaseous H_2 at a given rate
(less than 100 A per sq. ft. is not practical). The cost, c, has
been discussed above. The factor 40 (cents) already contains future
optimism. It arises out of a calculation of 80 cents due to Gregory
(24) for the cost of H_2 in terms of insurance, maintenance, manpower,
etc. This was halved on the assumption that current density could
be double in a future plant and hence half the volume etc. used,
with decrease in maintenance, and so on. This seems an acceptable
proportion. However, Gregory's figure is a 1972 figure and must be
corrected for inflational changes. At 10% per year for 4 years,
this is:

$$40 (1 + 0.1)^4 = 58 \text{ U.S. cents}$$ (6)

Hence,

Cost (U.S. cents) of 10^6 BTU of gaseous H_2 fuel = 229Ec + 58
(7)

Values of E for simple electrolysis vary from 1.3 to 2.4 V. Values
of c (*v.s.*) vary from 0.6 through 1.1 to 1.8 cents. Substituting
these figures, the range is: $2.45 to $3.87 to $10.48 per 10^6 BTU.
Thus, for straight electrolysis at room temperatures, the cost
compared with gasoline would be between about -10 and +280%.

The fact that hydrogen could be produced more cheaply than
gasoline at this time is not to be taken as too exciting because
the amount of this 6 mil power is small. But it would *begin* an
H_2 Economy. H_2 is intended as a medium coupled to solar and
nuclear sources and we are quoting prices mainly from fossil-based
electricity. It is, however, conceivable that we may have a coal-

based H_2 Economy (instead of one based on synthetic CH_4 and liquids from coal) because of the preferential pollutional prospect.

We now turn to other methods of producing H_2.

Solid Electrolyte Production from Steam

A detailed calculation has been given (25). The precise numbers depend on the fraction of the energy which is Carnot-limited (i.e. electricity) and that fraction which comes *directly* from atomic heat, avoiding the Carnot-limited electricity. A difficulty is that knowledge of the cost of this will depend on the precise arrangement ($ per 10^6 BTU will be assumed). Here a temperature of 1500°C (about 50% heat) will be assumed. With E = 1.3 V, c = 0.6 cents, and assuming that 50% of energy is obtained directly as heat at $1 per 10^6 BTU, the minimum cost would be about $1.97. However, in this is included the 58 cents per 10^6 BTU, the general figure covering maintenance, finance charges, etc. It is better to raise this by 50% to project an increase cost due to the use of more expensive membranes, and a realistic minimum price becomes $2.26. For c = 1.4, and the other assumptions the same, the cost is $3.03. Using the maximum electricity cost of 1.8 cents, this becomes $4.05.

Anode Depolarized Electrolysis

This can be assumed, from the work of Juda and Moulton (23) to lead to a diminution of the E to about 1/3 its original value. Thus, taking a cell potential of 0.8V, c as 0.6 - 1.4 - 1.8 cents and 58 cents as other costs, we get: $1.67 to $2.96 to $3.87. These values neglect the effect of selling the product (e.g. H_2SO_4) and are hence maxima. A minimum cost in 1976 collars of about $1.40 seems thus a reasonable goal with this method.

The Photo-Electrochemical Method

This now has a little more to go on. The electrode materials could be TiO_2 and CdTe, i.e. cheap materials, in very thin (10^{-4} cm) layers. Using the formula for cost given in Bockris (42), with t = 0.01 (to allow for a base material), ρ = 3, c = $2 per 1b, p = 12%, ε = 3, one gets:

$$\text{Cost} \simeq \$1 \text{ per } 10^6 \text{ BTU.} \tag{8}$$

The research position has been reviewed above.

Aerogenerator Approach

The following Table is from 1975 (43):

TABLE 3

Cost Estimates for a Wind-Based Hydrogen Energy System

Item	$ per kw
Wind generators	100-300
Electricity generator	50-150
Hydrogen production	100
Undersea storage facility	50
Fuel cell	150-350
Totals	490-950

For 1976, let us increase the costs by 10% to $485 - $1040. We add 10% for maintenance, insurance and manpower ($533 - $1050) and take money at 12%. The yearly cost of owning 1 kw of productive capacity is about 0.5 - 1 cent, which means $1.50 - $3.00 per 10^6 BTU (Table 4).

COMPETITORS OF ELECTROLYTIC HYDROGEN

Methane from coal must be cheaper than hydrogen from coal-based electricity. This arises from the Carnot argument.

TABLE 4

Costs of Projected Method of Hydrogen Production in 1976 Terms

Method	Cost Range ($) (See Text)
Electrolysis of water	2.45 - 3.87 - 10.48
Electrolysis of steam	2.26 - 3.03 - 4.05
Anode depolarized	1.67 - 2.96 - 3.87
Photo-electrochemical	*ca.* 1
Wind	1.50 - 3.00
Unweighted average	3.37
Minimum reasonable estimate in limited quantities	1.50
Estimate of most probably attainable minimal cost	2.61
Cost in 1976 of U.S. gasoline ex-refinery	2.75

One could produce hydrogen from coal. A new energy distribution system would be needed (44,45). Hydrogen from coal would avoid pollution. The CO_2 produced at central plants could be wasted into the sea. Conversion could be gradual. As coal is exhausted, solar or atomic energy could produce hydrogen from water and feed into the H_2-carrying network.

THE PRESENT PROSPECTS OF A HYDROGEN ECONOMY

So much enthusiasm has met the suggestion of a Hydrogen Economy (46,47) that it is worthwhile balancing with the negative side.

There are two considerations which must be got over: what is the time frame in which the Hydrogen Economy will begin and what will the dominant source of energy be at the time?

Confusion has been caused by the implication of enthusiasts that the Hydrogen Economy should begin "now". Whilst methane from coal is available, it will be cheaper than hydrogen, because of the Carnot cycle loss of producing electricity from coal; and analogous difficulties in the energy for water splitting by the cyclical method.

However, this does not mean that we shall have to wait until the middle of the next century, when world coal resources will be exhausted, if we continue an expansive economy using coal. We may never drive coal to exhaustion: producing sufficient coal mines to meet the energy crunch of 2000 is improbable (48). Pollutive dangers of the use of coal would be very hard to avoid, and if we burned up coal resources by the middle of the 21st century, the present world temperature decline would be reversed and the "greenhouse effect" attained (49). Finally, a realization of the lack of wisdom in burning up our dwindling reserves of C-H bonds may prevail.

A Hydrogen Economy will not be produced before the supply of energy from coal becomes limited. Let us suppose that coal does not supply more than 25% of our energy (perhaps 10% will be atomic) after the year 2000: with only a quarter of a century to go before a synthetic fluid fuel to use with atomic and other reactors is required, the needed practical development work on a Hydrogen Economy to couple with wind, solar (and atomic) installations should be initiated at once. (Hitherto, the studies have been mainly systems analyses). Thus, consideration of the time of implementaion of a Hydrogen Economy (50) suggests that "immediate" work would give the possibility of widespread building not before the year 2010.

Comparisons of hydrogen costs should be made with alternative fuels in an atomic- or solar-based economy. These could be electricity or methanol.

Shall we have largely an atomic or a solar base? An atomic base coupled with hydrogen has often been suggested (51) but one could couple an atomic technology with an electrochemical industry (52).

With a solar-based economy, hydrogen will be a necessity, not only because of the desirability of having gaseous or liquid fuels for transportation and industry but because the transmission of hydrogen over great distances would be more important in a solar-based economy, because good solar sources tend to be within 3000 km of the equator, and are far greater in the Southern Hemisphere. Hydrogen is optimal in flow characteristics.

Methanol is attractive because it is less dangerous and has an excellent consumer image. However, methanol must come from hydrogen and carbon-dioxide: its cost would thus be greater than that of hydrogen.

The *disadvantages* of a Hydrogen Economy must be borne in mind.

(i) "Hydrogen is too expensive". The cost of oil will go up due to exhaustion in the time a Hydrogen Economy could be built. What is true is that hydrogen from coal would be about the same price (per unit of energy) as methane but that the latter fuel contains three times more energy per unit volume than hydrogen. There are four reasons for introducing hydrogen before the exhaustion of coal.
   (1) Coal will not be available in sufficient amounts.
   (2) CH_4 will give a CO_2 effect if used in sufficient amounts.
   (3) We should conserve coal for use in food and textile manufacture.
   (4) H_2 couples with atomic and solar (the understood future sources) so that amortisation prospects are better (53).

(ii) "A double Carnot difficulty". Would it not be better to drive cars by electricity, the production of which has involved one Carnot cycle, rather than to utilize hydrogen and then get energy from H_2 by putting it (in an i.c. engine) through another Carnot loss?

This objection can be examined by comparing the cost of hydrogen per 10^6 BTU with that of refined gasoline and of methanol in a time frame past the year 2000; and weighing the pollutive aspects of these fuels, along with the capacity of electro-chemical storage systems. At present, the prospects are clouded by the near absence of research funding for new battery

systems; and the lack of development of fuel cells for transportation. There is plenty of indication that suitable new
battery systems, (e.g. Fe-H$_2$) *could* be developed, i.e. by
using readily available materials with the required characteristics. But in the U.S. there is negligible research funding
in this field outside the automotive companies who themselves
do not wish the development. Hydrogen, on the other hand,
could be used as a fuel for present cars with small modifications.

The double Carnot difficulty could in any case be avoided by
using electrochemical reconversion to electricity by non-
Carnot limited fuel cells.

(iii) "Hydrogen is dangerous". Hydrogen is more dangerous than
      methane, the correct comparison material, because of its
      lower flash point, and greater power of diffusion. An
      appropriate technology for handling hydrogen has been worked
      out. The experience of the U.S. space programme with the
      massive use of H$_2$ over many years gives confidence in the
      safety precautions which could be taken in a hydrogen tech-
      nology.

The extent of a Hydrogen Economy will be a matter of economics
in comparison with electricity and methanol. We shall evolve an
economy which at first uses all three fuels. Depending on the
dominance of (nearby) fusion plants and (far off) solar sources, it
seems likely that the final fuel will be electricity or hydrogen,
respectively.

SUMMARY

H$_2$ will come only from water. Sea water will have to be
treated cautiously because of Cl$_2$.

Thermal splitting has no efficiency advantage over electrolysis and in all other directions there are disadvantages.

High-temperature electrolysis processes convert heat to
hydrogen avoiding the Carnot cycle for the thermal part of the
energy used. U$_3$O$_8$-Y$_2$O$_3$ offers an advantage because it has a high
electronic and ionic conductivity.

Photosynthesis of H$_2$ from H$_2$O is a complex and low efficiency
path. It is little researched. 0.3% is an observed yield of H$_2$
from H$_2$O.

The photo-electrochemical method would give electricity and
hydrogen. Hitherto its efficiency is a maximum at only 3%. Cells

using both a p- and an n-type electrode are essential.

Electrolysers in the Laboratory can work at 100 mA cm^{-2} and < 1.5 V. Modern electrochemical principles have not yet been applied to large scale electrolysers.

Electricity costs relevant to hydrogen production on a large scale have to be taken with the usual distribution costs eliminated; and using low use-time periods. The present cost of block contracts for this type of interruptable power in the U.S.A. would be about 1.1 cent per kwh. Present gasoline ex-refinery is \$2.75 per 10^6 BTU. The costs of various processes of producing 10^6 BTU of H_2 from water in 1976 dollars ranges from about \$1 for the laboratory stage photo-electrochemical method to about \$10 for classical electrolysis. Aerogenerators would seem able to produce limitless H_2 gas at about the same price as present U.S. gasoline.

REFERENCES

1.   J.O'M. Bockris, *Energy: The Solar-Hydrogen Alternative*, Australia and New Zealand Book Co., Sydney, 1975; (also *Cornell International Symposium on the Hydrogen Economy*, 1973, p. 143.)
2.   J.O'M. Bockris, *THEME Conference*, Miami, Florida, 1974, p. S9-1.
3.   J.O'M. Bockris, *First World Hydrogen Energy Conference*, Miami, Florida, March, 1976.
4.   A.Y.M. Ung and R.A. Back, Can. J. Chem., 42 (1964) 753.
5.   B.I. Eastland and W.C. Gough, *162nd Meeting of Amer. Chem. Soc.*, Boston, April, 1972.
6.   Described at First World Hydrogen Energy Conference, Miami, Florida, 1976.
7.   J.O'M. Bockris, Ref. (1), p. 196.
8.   THEME Conference, Miami, 1974, p. S5-1 to S5-41.
9.   J.O.'M. Bockris, *First World Energy Conference on Hydrogen Energy*, Miami, Florida, 1976.
10. J.O'M. Bockris, Ref. (1), pp. 59-206.
11. A. Fujishima and K. Honda, Nature, 238 (1972) 38.
12. A. Mitsui, *THEME Conference*, Miami, Florida, 1974, S5-41.
13. G. Neil, G. Nicholas, J.F. McCann and J.O'M. Bockris, *First World Energy Conference on Hydrogen*, Miami, Florida, 1976.
14. J.O'M. Bockris, Ref. (1), pp. 166-174.
15. C. Marchetti, *Progress Report No. 1*, EURATOM Joint Nuclear Research Centre, Ispra, Italy, 1972.
16. J.E. Funk, W.L. Conger and R.H. Carty, *THEME Conference*, Miami, Florida, 1974.
17. J.O'M. Bockris, Ref. (1), pp. 162-163.
18. J. Appleby, Nature, 253 (1975) 257.
19. R. Bidard, Int. J. Hydrogen Energy, 1976, in press.
20. J.O'M. Bockris, Ref. (1), p. 163.

21. A. Meinel, Physics Today, February, 1972, p. 44.
22. J.O'M. Bockris, Ref. (1), pp. 188-191.
23. W. Juda and D.McL. Moulton, Chem. Eng. Symp. Series, p. 59.
24. D. Gregory, assisted by P.J. Anderson, R.J. Dufour, R.H.
    Elkins, W.J.D. Escher, R.B. Foster, G.M. Long, J. Wurm and
    G.G. Yie, *The Hydrogen-Energy System*, prepared for American
    Gas Association by I.G.T., 1973.
25. J.O'M. Bockris, J. Adv. Energy Conv., 14 (1975) 81.
26. General Electric Co., Internal Memorandum, 1966.
27. J. Bevan, H. Singh and J.O'M. Bockris, Flinders University,
    1976.
28. *Solar Energy Research in Australia*, Australian Academy of
    Science, 1973.
29. J.O'M. Bockris, Ref. (1), p. 203.
30. H. Gaffron & G. Rubin, J. Gen. Physiol., 1942, p. 219.
31. E. Kessler in W.D.P. Stewart (Ed.), *Algal Physiology and Biochem-
    istry*, Botanic Monograph, Volume 10, London, 1974.
32. G.E. Fogg, W.D.P. Stewart and A.E. Walsby, *Blue-Green Algae*,
    Academic Press, London, 1973.
33. G. Neil, D.J.D. Nicholas, J.O'M. Bockris and J.F. McCann, Int.
    J. Hydrogen Energy, 1 (1976) 48.
34. K. Honda and H. Fujishima, J. Electrochem. Soc., 122 (1975)
    1482.
35. Y. Naketo, T. Ohuishi and H. Tsobmure, Chem. Letters, 1975,
    p. 883.
36. B. Tseung, F.T. Bacon, private communication,1976.
37. J.O'M. Bockris and K. Uosaki, Energy, 1 (1976) 95.
38. J.O'M. Bockris and R.K. Sen, Chem. Phys. Letters, 18 (1973)
    166.
39. J.O'M. Bockris and K. Uosaki, Int. J. Hydrogen Energy, in
    press.
40. J.O'M. Bockris and J. MacHardy, J. Electrochem. Soc., 120
    (1973) 54.
41. Oil and Gas Journal, May, 1976.
42. J.O'M. Bockris, Ref. (1), p. 193.
43. J.O'M. Bockris, Adv. Energy Conv., 14 (1975) 87.
44. J.O'M. Bockris, Ref. (1), pp. 140-157.
45. E. Justi and J.O'M. Bockris, Memo to Westinghouse Company,
    1962.
46. J.O'M. Bockris, Environment, 13 (1971) 51; Chemical Engineering
    News, Oct. 1972, p. 51.
47. J.O'M. Bockris, Ref. (1), pp. 51-55.
48. J.O'M. Bockris, Ref. (1), Fig. 44, p. 54.
49. J.O'M. Bockris, Ref. (1), pp. 333-341.
50. A. Weinberg and P. Hammond, Am. Sci., 58 (1970) 412.
51. P. Hammond in E. Kovac (Ed.), *Technology of Efficient Energy
    Utilization*, Report to NATO Scientific Conference, Les Arcs,
    October 1973.
52. J.O'M. Bockris, Ref. (1), pp. 307-332.
53. J.O'M. Bockris, Ref. (1), p. 333.

54. J.O'M. Bockris, Ref. (1), pp. 21-25.
55. J.O'M. Bockris and A.K. Reddy, *Modern Electrochemistry*, Plenum Press, New York, 1973, p. 1281.
56. L.A. Harris and R.H. Wilson, General Electric Technical Information Series, Report No. 75CD228, Oct. 1975.

POWER SOURCES AND SOME BASIC PROBLEMS IN ELECTROCRYSTALLISATION AND PASSIVITY

H. R. Thirsk
Electrochemistry Research Laboratories
School of Chemistry, The University
Newcastle upon Tyne, England

INTRODUCTION

The intention of this paper is to survey some of the progress made in the study of problems, basic to the electrochemical behaviour of batteries, which have been the particular interest of the author and his colleagues over a number of years. To this end it is convenient to classify the systems from which examples can be taken by the following scheme.

	System	Example
1.	Metal anodes; dissolution and reforming (electrodeposition) passivity.	Zn, Cd, In
2.	Metal/solution gas electrodes	H_2, O_2 electrodes $Pb/PbSO_4$
3.	Metal/metal salt systems	$Ag/AgCl$ Hg/Hg_2Cl_2
4.	Metal/metal salt/oxide systems	$Pb/PbSO_4/PbO_2$
5.	Metal/metal oxide systems	Hg/HgO Ag/Ag_2O, AgO
6.	Metal/metal oxide/hydroxide system	$Ni/NiO(OH)$

These relate to

- Aqueous electrolytes
- Non-aqueous electrolytes
- Fused salt electrolytes
- Solid electrolytes, including those

> operational at room temperature (e.g. silver rubidium iodide) and those
> operational at elevated temperature (e.g. β-alumina.)

A major problem with the systems classified above comes under the general description of (a) electrocrystallisation since they operate predominantly through the formation of a solid phase or the transformation of a solid phase to one with the cation in a higher or lower valency state, and (b) passivity since the free solubility of metal anodes during electrolysis is of course essential and the growth of non-conducting phases is also a passivating process.

Attention has also been given in recent years to a more quantitative assessment of the porosity of electrodes.

In the space available for this review, an attempt will be made to refer essentially to the very considerable literature shedding light on battery electrode processes which have followed from the application of potential step and a.c. perturbations to potential-controlled investigations. The first publication referring to the use of potential step methods indicated in the literature would seem to be about 1952 (1) and an interesting a.c. investigation on primary cells using complex plane plots appeared in 1957 (2).

To identify the problem, if we assume that a new phase appears by nucleation and subsequent growth, the growth of a new phase in terms of the electrical and physical parameters, concentration of species at the surface and the area of the growing phases, i must be of the form

$$i = f(C_s, \eta, S) \tag{1}$$

where C_s is the surface concentration, η the applied potential and S the active surface area (3) from which

$$\frac{di}{dt} = \left(\frac{\partial i}{\partial \eta}\right)_{C_s, S}\left(\frac{\partial \eta}{\partial t}\right) + \left(\frac{\partial i}{\partial C_s}\right)_{S, \eta}\left(\frac{\partial C}{\partial t}s\right) + \left(\frac{\partial i}{\partial S}\right)_{C_s, \eta}\left(\frac{\partial S}{\partial t}\right) \tag{2}$$

$(\partial\eta/\partial t)$ is the imposed function, $(\partial C_s/\partial t)$ is probably determinable as a solution of Fick's equation. Only with a constant potential, however, is the electrochemical rate constant and the rate of nucleation constant. Thus, although di/dt gives direct information about the growth of the phase $\partial S/\partial t$, the problem of setting up models can be difficult. A large number of solutions have been reviewed and the cited references are helpful in setting out a number of examples (5 - 8).

The basic mechanism for the electrode process is that the phase change is taking place as a solid-state reaction with the formation of nuclei which continue to grow as electrochemically controlled procedures and eventually overlapping as a physical hindrance to growth.

Thus

$$i = nFkS \tag{3}$$

where k is an electrochemical rate constant $(\text{mols}^{-2}\ \text{sec}^{-1})$ and S is an area at which growth is taking place. In the absence of overlap

$$nkS = \frac{\rho}{M}\ nF\left(\frac{dV}{dr}\right)\left(\frac{dr}{dt}\right) \tag{4}$$

where V is the volume of the nucleus, ρ the density and M the molecular weight. The growth parameter, r, is one dimensional in character.

We also have the nucleation rate being potential dependent, the number of nuclei being given by the expression

$$N = No\ 1 -(\exp(-At)) \tag{5}$$

where A is a potential-dependent term. This expression enables calculation of the current for free growth, which is given by

$$i = \int_{o}^{t} i(u)\left(\frac{dV}{dt}\right)_{t-u} du \tag{6}$$

where i(u) is developed from Eq. (3) and (4) for an assumed geometry. Overlap can be fairly readily accounted for (4, 5), if nucleation is randomly cited, by a theory due to Avrami.

If physical reality can be given to the models proposed for the system under examination, and sometimes an appropriate form of structural examination by microscopy or electron microscopy can do this, the electrochemical rate constants may be determined and suggestions for the electrochemical mechanism substantiated.

In the development of this work, in our laboratories, the first application of potential step techniques to phase changes in electrochemical reactions was to the $PbSO_4/PbO_2$ oxidation (9-11). Subsequently the Hg/Hg_2Cl_2 system with the formation of a solid on mercury was examined (12): we then undertook a number of studies of the formation of solid phases on mercury and mercury/amalgam substrates to substantiate the nucleation/growth mechanisms in two and three dimensions (13, 14).

A substantial summary of potentiostatic methods as applied to electrochemically-formed phases considered from the point of view of nucleation and growth have been given in refs. (4) and (5).

A.C. METHODS

In the further development of our work, other techniques were desirable in addition to potential step procedures.

Following Randles' classical paper in 1947 (15) there was a continuing interest in the employment of a.c. perturbations to the treatment of galvanic cells as having an electrical equivalent in a simple network of resistances and capacities, plus the effect of diffusion polarisation in a component termed the Warburg impedance (15). Thus, the experimental data obtained was that of an impedance with dependence on changes in concentration and frequency.

The interest was confined to systems for which there was no problem of solid phase formation. Further development occurred when Sluyters (16) treated the cell impedance as a vector in the complex plane instead of treating the system. Sluyters and collaborators developed the treatment and the applications in a substantial manner (see ref. (18)) but not for the formation of solid phases although they were aware of its application to a battery system (2).

Application of a.c. methods to systems in which adsorption and solid phases were formed began in our laboratories in 1965 - initially on the mercury and mercury amalgam substrates for which we had a good deal of experience. Work on solid surfaces followed later. The major developments in these laboratories have been with R. D. Armstrong and collaborators.

Associated with the experimental work there has also been considerable development in instrumentation with attention being focussed on sinusoidal perturbations. Square wave signals as employed by Leach for anodic systems have not been used (17).

As with other workers, (18) the Wien bridge was used for frequencies up to 10 kHz. Such bridges can be used to measure resistances with precision up to 500 kHz but for higher frequencies transformer ratio-arm bridges are employed (19) with a phase sensitive detection system. A summary of the general experimental procedure over a wide range of frequencies as used by ourselves is given by Armstrong *et al.* (24).

The examination of an electrochemical system over a wide range of frequencies and the subsequent analysis of the impedance data for complex plane plotting is time consuming. Steps have therefore been taken to automate using a Solartron 1170 series frequency response analyser, a plotter interface, X-Y recorder and teletypewriter. A frequency range of 0.1 m Hz to 10 kHz is covered by this system with point-to-point measurements.

Sinusoidal perturbations have been of particular value in the following: adsorption, dissolution and passivity of metals (where the work was developed from the early papers of Frumkin and Melik-Gaikazyan (20) and Lorenz (21, 22)), the behaviour of porous electrodes, (the work of de Levie, is of early importance (23)), and studies of batteries employing solid electrolytes with respect to study both of the electrolyte and the interface between the electrodes and the solid electrolyte.

As a guide to the development in our own laboratories of the experimental and theoretical treatment, a number of references are listed under the above groupings, together with a number of essentially theoretical papers (refs. (25) to (52)). They have been selected as being related to battery problems rather than electro-deposition and more general kinetics.

To revert to the actual problem of the formation of solid deposits; it must not be forgotten that a much older theory for phase growth when an electrode is being consumed with the eventual formation of a solid phase was proposed by Müller (53). This essentially involved dissolution, followed by precipitation when a critical concentration of dissolving metal as ions or complexes is reached and a sparingly soluble salt is formed with an anion in the electrolyte. A number of workers supported this mechanism combining the electrochemical investigation with an often sophisticated optical examination of the surface (55); unfortunately the theory has required revision and this will be referred to below.

There are many examples of electrode systems where fairly massive deposits are formed and where a more thorough experimental treatment is desirable in order to clarify the contribution to the overall process, of the solid-state type of growth with adherent growth centres and eventual overlap and that from dissolution and precipitation with the identification of the dissolving species. In the latter case, the solid material is loosely attached to the surface or held in the interstices of the more porous electrodes. We have therefore, in our own work, continued to examine these problems.   For further experimental control rotating disc and ring-disc electrodes have been frequently used.

Initially attention was devoted to a clarification of conditions under which the two kinds of layer formation could take place with attention to the more theoretical aspects (56), taking into account earlier criticism of the theory employed in ref. (55). (See, for example, (57)). In the dissolution-precipitation theory, a characteristic time, τ, is introduced as the time at which indication of a reaction between the dissolving metal anion or complex and an anion in the solution leading to precipitation is experimentally observable.   The correct result may be exemplified by the following typical mechanism for the behaviour of a bivalent metal forming an hydroxide.

$$M \longrightarrow MOH^+ \underset{k+}{\overset{k-}{\rightleftharpoons}} M^{2+} + OH^- \tag{7a}$$

$$MOH^+ + OH^- \longrightarrow M(OH)_2 \tag{7b}$$

$$i\tau^{\frac{1}{2}} = \frac{1}{2}(D\pi)^{\frac{1}{2}}nF(1+K)C_a - \frac{1}{2}K\pi^{\frac{1}{2}}i(k_++k_-)^{-\frac{1}{2}}erf((k_++k_-)\tau)^{\frac{1}{2}} \tag{8}$$

$$K = k_+/k_- \tag{9}$$

D is the diffusion coefficient, i the current density and C_a is the concentration of the dissolving species or reacting complex near the anode, and the negative sign before the second term is the correction to be noted.

For more recent studies on phase formation following the above arguments, reference can be made to work on the Pb/PbSO$_4$ system undertaken in our laboratories which has recently (60) been examined by a.c. and potential step methods using a rotating disc lead electrode.

The nickel/cadmium secondary battery still shares a pre-eminence with the lead acid battery as a secondary battery power source.   The chemistry and electrochemistry of the nickel electrode is still a demanding subject for investigation.   Potentiostatic studies of the electrochemistry of the oxidation reaction have contributed to the understanding while major studies have been made with galvanostatic techniques (61-64).

Potentiostatic studies in these laboratories centred first on the formation of NiOOH in buffered acetate solutions indicating the growth of a thin oxide layer by three-dimensional centres (65) and a further study was made of the electrochemical oxidation of α-Ni(OH)$_2$ to γ-NiOOH (66).   This latter study was of considerable interest since it confirmed that methods applicable to the PbSO$_4$/PbO$_2$ system (67) and the Ag$_2$O/AgO system (68) were also equally useful with this chemically much more difficult electrode. However, the nickel system is much more complex because of the existence of diffusion controlled processes of a greater magnitude than reported for the rather analogous MnO$_2$ system (69, 70).   The results of our study (66) agree with those obtained in a parallel study by MacArthur (71).

The possibility of proton mobility in such systems has been the subject of speculation for a long time and both this species and the foreign cation (e.g. K$^+$, Na$^+$, Li$^+$) have clearly an important role in the functioning of the electrode system which must be clarified.   A valuable summary of the position at this moment will be found in ref. (72).

Reference will again be made to the nickel electrode later in this paper when discussing newer structural investigations.

The dissolution of metals

Many power systems depend on the dissolution of a metal anode. Our own interests have been rather specifically restricted in the battery field to the behaviour of zinc, cadmium, indium and indium alloys in alkaline solutions.   The first two have of course been very widely studied and recent reviews, for example, have been given by Armstrong and co-workers (73, 74).

The elucidation of the zinc reactions, as is seen in the review, has followed from a broad application of techniques and extends to the study of porous electrodes.   Nevertheless two areas of uncertainty remain which if resolved would greatly increase the applicability of power sources employing these anodes.   The first is in the passive region of the electrodes.   Work from these laboratories points to a two-dimensional model of passivation (75), a mechanism which we have studied in detail in other systems and which is a very common crystal growth habit, and also elsewhere by adsorption models (76).   In addition, although dissolution-precipitation models are often proposed, particularly by workers employing constant-current techniques, these studies are not conclusive in themselves.

The second problem: in principle the zinc electrode could
be made the anode for a secondary battery system but here a
severe problem inhibits the use.   This is the dendritic metallic
growth on reforming the anode during charging.   A solution would
appear to require a much more profound knowledge of the process
of electrodeposition.

Problems with the cadmium electrode are mainly centred on the
formation of the hydroxide.   Our own work, both electron optical
and potentiostatic, favours the solid state mechanism.
Unfortunately, because the studies assuming dissolution -
precipitation models gave results apparently agreeing with an
incorrect theory, the applicability of the mechanism is very much
in doubt.   On the structural side, the predominating phase would
seem to be β-Cd(OH)$_2$ with substantially less evidence for the
presence of γ-Cd(OH)$_2$ and the oxide CdO.

Cells with solid electrolytes

Some years ago we considered that it would be of interest to
ourselves, as well as having possible value for systems development, to initiate fundamental studies on cells based on solid
electrolytes.   Two systems were selected; the high temperature
Na/β-Al$_2$O$_3$/S system (as an interesting comment if not too relevant,
the first reasonably complete phase diagram for the Na/S system
was obtained in Newcastle (77)) and alternatively systems working
at room temperature of the Ag/Ag$_4$RbI$_5$/I$_2$ type.

The application of a.c. methods, coupled in some cases with
potentiostatic perturbations, has been extremely useful in these
studies.   It is predominently the a.c. investigations that are
now briefly summarised.

The breakdown of the β-alumina ceramic with the development
of electronic conductivity was followed by observations of the
impedance spectra over a range of frequencies of 0.01 Hz to
10 kHz during d.c. polarisation (78).   A theory for the breakdown of the ceramic due to the dendritic growth of sodium metal
was advanced but in the absence of mechanical data for the ceramic
it was not possible to test the theory quantitatively.   The
problem of the impedance of powdered and sintered solid ionic
conductors has also been examined (79).

Departures from the expected impedance plot for surface roughness at the blocking electrode/solid electrolyte were postulated.
Experimental results were obtained for different preparations both
of β-alumina and silver rubidium iodide and compared with theory.
A particular warning is made in that an apparent 'Warburg'

impedance can arise due to a rough surface and care must be
exercised in interpretation;   there is only one mobile species.

The work was extended to an impedance study of a single
crystal of β-alumina (32) and indicates a precise value from the
impedance plot for the specific conductivity.

The impedance spectra has been determined over a wide
frequency range for metal/solid electrolyte/metal cells and the
behaviour at high frequencies would seem to be of particular
interest (82) with a negative dielectric constant and a frequency
dependent conductivity.   It can be explained by a theoretical
model in which ions can be thermally excited into free ion-like
states.   Furthermore, the occurrence of an apparent negative
dielectric constant is predicted and demonstrated.   The interfacial
impedances for a number of electrode systems are listed in this
paper.

There is little reported work on metal deposition from solid
electrolytes (81) but some further contributions to this problem
have been conducted on the Ag/Ag_4RbI_5 interface (82) and on the
Cu/Cu^{2+} electrode (83) using electrolytes arising from the incorporation of tetra alkyl ammonium cations into cuprous bromide (84).

As is generally understood, the study of the β-alumina system
at the moment is the most beneficial from a power sources point of
view.   Systems based on electrolytes operating at the room
temperature are very specialised and the most rewarding studies
would seem to be of the high frequency behaviour of the electrolyte
and are not really relevant to power sources.

The final section of this article refers to structural problems
and is the area where the author has served his longest apprentice-
ship and from which, in point of fact, derived the stimulus to turn
to electrochemical studies.   The reference is to work in the late
thirties on the examination by electron diffraction of metal
deposition on single crystals of metals and the epitaxy and
orientation of metals and metal oxides on inorganic single crystal
substrates.

Two important points emerge from the ideas and work following
on from this primary study.   Firstly, the effective use of single
crystal substrates is necessarily limited to specialised experiments
requiring a great deal of ancillary equipment for surface studies;
they at best are associated with initiating surface reactions of a
catalytic or epitaxial type.   Of more importance is the fact that
oxides created at a gas/solid interface are nearly always very
different materials from those formed at an aqueous electrolyte/
solid interface.   Anodic oxidations forming oxides and hydroxides

yield complex products often with foreign cations present which
have an electrochemical significance.

Because of this, structural examination by diffraction
methods offers particular difficulties.

There are at the moment many techniques available for study;
we ourselves employ ellipsometry, optical microscopy, low and high
energy electron diffraction, electron microscopy, scanning
electron microscopy and the related chemical analysis from X-ray
excitation by the electron beam, Auger spectroscopy and X-ray
photoelectron spectroscopy or ESCA (electron spectroscopy for
chemical analysis).   An elementary survey of these techniques
relevant to electrochemistry has been published (85).   With the
exception of light reflectance techniques none of these
are *in situ* methods so clearly each investigation should be
carefully scrutinised.

The benefits that have followed from the application of
X-ray methods and the less common use of electron diffraction
have been very considerable.   The admirable study by Burbank
and collaborators using a number of techniques on the lead acid
battery (86) is a good example of an integrated study of a
classical system.

At the present time it would appear that a greater exploitation
of spectroscopic techniques giving direct chemical information
would be most helpful in the examination of problems of power
sources.   One new technique in particular is especially worth
mentioning – the application of ESCA to electrode problems.   We,
for example, are using ESCA in the continuation of work on the
nickel electrode reinforcing our extensive experience from X-ray
and electron diffraction studies by Briggs and collaborators.
The method has also been applied to a number of other electrode
systems and solid electrolytes with some publication of results –
a study of the gold and platinum electrodes with oxide formation (87);
silver and copper compounds forming solid electrolytes and cation
sites in β-alumina (88);   and the reduction of a redox system containing Cr(VI) on gold (89).   Somewhat parallel experimental work
has begun on the passivation of certain metals, by Auger spectroscopy.

In all, we await with considerable interest both from our own
and other laboratories further work of this type on the structure
of poorly characterised materials in electrode systems.

CONCLUSION

This contribution has of necessity been somewhat superficial but it is hoped that it has also been sufficient, with the use of the selection of references, to substantiate the idea that developments in both experimental and theoretical methods have greatly strengthened the possibilities in the understanding and development of both old and new forms of electrochemical batteries. In particular, the survey has been predominantly by reference to work in the Electrochemical Research Laboratories at Newcastle since it was felt that the invitation to present this review also conferred some licence to do so.

Finally, reference must be made to the author's indebtedness to many collaborators both past and present. It would be difficult to list these by name and as fully as is deserved but hopefully this identification is readily deduced from the references quoted.

REFERENCES

1. H. R. Thirsk and M. Fleischmann, 5th Meeting of CITCE, Stockholm, published by Butterworths, 1953.
2. J. Euler and K. Dehmelt, Z. Elektrokhim., 61 (1957) 1200.
3. W. Davison, J.A. Harrison and J. Thompson, Faraday Society Discussions of the Chemical Society, No. 56 (1973) 171.
4. M. Fleischmann and H. R. Thirsk in P. Delahay (Ed.), *Advances in Electrochemistry and Electrochemical Engineering,* Vol. 3, Interscience, New York, 1963, p. 123.
5. J. A. Harrison and H. R. Thirsk in A.J. Bard (Ed.), *Electroanalytical Chemistry,* Vol. 5, Dekker, New York, 1971, p.67.
6. M. F. Bell and J. A. Harrison, J. Electroanal. Chem., 41 (1973) 15.
7. W. Davison and J. A. Harrison, J. Electroanal. Chem., 44 (1973) 431.
8. J. A. Harrison, J. Electroanal. Chem., 36 (1972) 71.
9. M. Fleischmann and H. R. Thirsk, Trans. Faraday Soc., 51 (1955) 71.
10. M. Fleischmann and H. R. Thirsk, *1st International Symposium on Batteries,* Bournemouth, 1958.
11. (a) M. Fleischmann and H. R. Thirsk, Electrochim. Acta, 1 (1959) 146.
    (b) M. Fleischmann and H. R. Thirsk, Electrochim. Acta, 2 (1960) 22.
12. (a) A. Bewick, M. Fleischmann and H. R. Thirsk, Trans. Faraday Soc., 58 (1962) 1975.
    (b) R. D. Armstrong, M. Fleischmann and H. R. Thirsk, Trans. Faraday Soc., 61 (1965) 574.

13.   M. Fleischmann, I. M. Tordesillas and H. R. Thirsk, Trans.
      Faraday Soc., 58 (1962) 1865.
14.   M. Fleischmann, K. S. Rajagopalan and H. R. Thirsk, Trans.
      Faraday Soc., 59 (1963) 741.
15.   (a) J.E.B. Randles, Disc. Faraday Soc., 1 (1947) 11.
      (b) H. Erschler, Disc. Faraday Soc., 1 (1947) 269.
      (c) H. Gerischer, Z. Physik. Chem., 198 (1951) 286.
16.   J. H. Sluyters, Req. Trav. chim. Pays Bas, Belg., 79 (1960)
      1092.
17.   J.S.Ll. Leach, J. Inst. Metals, 88 (1959) 24.
18.   (a) D. C. Grahame, J. Amer. Chem. Soc., 71 (1949) 2975.
      (b) H. Gerischer, Z. Phys. Chem., 202 (1953) 302.
      (c) R. Parsons, Trans. Faraday Soc., 56 (1960) 1340.
      (d) M. Sluyters-Rehbach and J. H. Sluyters, Req. Trav. chim.
          Pays Bas, Belg., 82 (1963) 636.
19.   (a) G. H. Nancollas and C. A. Vincent, J. Sci. Instr., 40
          (1966) 306.
      (b) J. O'M. Bockris, E. Gileadi and K. Müller, Chem. Phys.,
          44 (1966) 1445.
20.   A. N. Frumkin and V. I. Melik-Gaikazyan, Dokl. Akad. Nauk.
      SSSR, 77 (1951) 855.
21.   W. Lorenz, Z. Elektrochem., 62 (1956) 507.
22.   W. Lorenz and F. Mockel, Z. Elektrochem., 60 (1956) 507.
23.   R. de Levie in P. Delahay (Ed.), *Advances in Electrochemistry
      and Electrochemical Engineering*, Vol. 6, Interscience, New
      York, 1967, p.329.
24.   R. D. Armstrong, W. P. Race and H. R. Thirsk, Electrochim.
      Acta, 13 (1968) 215.

Adsorption and adsorption kinetics

25.   R. D. Armstrong, W. P. Race and H. R. Thirsk, J. Electroanal.
      Chem., 16 (1968) 517.
26.   R. D. Armstrong, J. Electroanal. Chem., 22 (1969) 49.
27.   R. D. Armstrong and M. Henderson, J. Electroanal. Chem., 39
      (1972) 81.
28.   R. D. Armstrong and M. F. Bell, J. Electroanal. Chem., 58
      (1975) 419.
29.   R. D. Armstrong and K. Edmondson, Electrochim. Acta, 18 (1973)
      937.
30.   R. D. Armstrong, M. F. Bell and R. E. Firman, J. Electroanal.
      Chem., 48 (1973) 15.
31.   R. D. Armstrong and R. E. Firman, J. Electroanal. Chem., 45
      (1973) 3.
32.   R. D. Armstrong, R. E. Firman and H. R. Thirsk, Faraday Disc.
      of Chem. Soc., No. 56 (1973) 244.

33. R. D. Armstrong and R. E. Firman, J. Electroanal. Chem., 45 (1973) 257.
34. W. Davison, J. A. Harrison and J. Thompson, Faraday Disc. of Chem. Soc., 56 (1973) 171.

Dissolution and passivation

35. R. D. Armstrong, L. J. Pearce and H. R. Thirsk, Electrochim. Acta, 14 (1969) 949.
36. R. D. Armstrong, W. P. Race and H. R. Thirsk, J. Electroanal. Chem., 23 (1969) 351.
37. R. D. Armstrong and K. Edmondson, J. Electroanal. Chem., 53 (1974) 371.
38. R. D. Giles, J. A. Harrison and H. R. Thirsk, J. Electroanal. Chem., 22 (1969) 375.
39. R. D. Giles and J. A. Harrison, J. Electroanal. Chem., 24 (1970) 399.
40. R. D. Giles, J. Electroanal. Chem., 27 (1970) 11.
41. R. D. Armstrong and M. Henderson, J. Electroanal. Chem., 40 (1972) 121.
42. R. D. Armstrong, M. Henderson and H. R. Thirsk, J. Electroanal. Chem., 35 (1972) 119.
43. R. D. Armstrong, R. E. Firman and H. R. Thirsk, Corrosion Science, 13 (1971) 409.
44. R. D. Armstrong and A. C. Coates, J. Electroanal. Chem., 50 (1974) 303.

Porous electrodes

45. R. D. Armstrong, D. Eyre, W. P. Race and A. Ince, J. Applied Electrochem., 1 (1971) 179.
46. R. D. Armstrong and K. Edmondson, J. Electroanal. Chem., 63 (1975) 287.
47. M. Sluyters-Rehbach and J. H. Sluyters in A. J. Bard (Ed.), *Electroanalytical Chemistry*, Vol. 4, Dekker, New York, 1970, p.1.
48. R. Parsons in P. Delahay (Ed.), *Advances in Electrochemistry and Electrochemical Engineering*, Vol. 7, Wiley, New York, 1970, p.177.

Solid electrolytes

49. R. D. Armstrong and R. Mason, J. Electroanal. Chem., 41 (1973) 231.

50. R. D. Armstrong, T. Dickinson and P. M. Willis, J. Electro-
     anal. Chem., 53 (1974) 389.
51. R. D. Armstrong and K. Taylor, J. Electroanal. Chem., 63
     (1975) 9.
52. R. D. Armstrong, T. Dickinson and P. M. Willis, J. Electro-
     anal. Chem., 67 (1976) 121.
53. W. J. Müller, Z. Elektrochem., 33 (1927) 401.
54. W. J. Müller, Trans. Faraday Soc., 27 (1931) 737.
55. (a) A.K.N. Reddy, M. A. Devanathan and J. O'M. Bockris,
         J. Electroanal. Chem., 6 (1963) 61.
     (b) A.K.N. Reddy, M. A. Devanathan and J. O'M. Bockris,
         Proc. Roy. Soc., A279 (1964) 327.
     (c) A.K.N. Reddy and B. Rao, Canad. J. Chem., 47 (1969) 2687
         and 2693.
     (d) M.A.V. Devanathan and S. Lakshmanan, Electrochim. Acta,
         13 (1968) 667.
56. (a) R. D. Armstrong, Corrosion Science, 11 (1971) 693.
     (b) R. D. Armstrong and J. A. Harrison, J. Electroanal. Chem.,
         36 (1972) 79.
57. B. Behr and J. Taraszewska, J. Electroanal. Chem., 19 (1968)
     373.
58. G. Archdale and J. A. Harrison, J. Electroanal. Chem., 34
     (1972) 21.
59. (a) G. Archdale and J. A. Harrison, J. Electroanal. Chem.,
         39 (1972) 357.
     (b) G. Archdale and J. A. Harrison, J. Electroanal. Chem.,
         43 (1973) 321.
60. A. N. Fleming and J. A. Harrison, Electrochim. Acta, in the
     press.
61. B. Conway and P. L. Bourgault, Canad. J. Chem., 37 (1959) 292.
62. P. L. Bourgault and B. Conway, Canad. J. Chem., 38 (1960) 1557.
63. B. Conway and P. L. Bourgault, Trans. Faraday Soc., 58 (1961)
     593.
64. B. Conway and E. Gileadi, Canad. J. Chem., 40 (1962) 1933.
65. G.W.D. Briggs and M. Fleischmann, Trans. Faraday Soc., 62
     (1966) 3217.
66. G.W.D. Briggs and M. Fleischmann, Trans. Faraday Soc., 67
     (1971) 2397.
67. M. Fleischmann and H. R. Thirsk, Trans. Faraday Soc., 51
     (1955) 71.
68. M. Fleischmann, D. Lax and H. R. Thirsk, Trans. Faraday Soc.,
     64 (1968) 3137.
69. A. B. Scott, J. Electrochem. Soc., 107 (1960) 941.
70. J. P. Gabano, J. Seguret and J. R. Laurent, J. Electrochem.
     Soc., 117 (1970) 147.
71. D. M. MacArthur, J. Electrochem. Soc., 117 (1970) 729.
72. G.W.D. Briggs, The Chemical Society S.P.R., *Electrochemistry*,
     Vol. 4, 1974, p.33.
73. R. D. Armstrong and M. F. Bell, *ibid.*, p.1.

74. R. D. Armstrong, K. Edmondson and G. D. West, *ibid.*, p.18.
75. R. D. Armstrong and G. M. Bulman, J. Electroanal. Chem., 25 (1970) 121.
76. (a) M. N. Hill, J. Ellinson and J. E. Toni, J. Electrochem. Soc., 117 (1970) 192.
     (b) M. N. Hill and J. E. Toni, Trans. Faraday Soc., 67 (1971) 1128.
77. T. G. Pearson and P. L. Robinson, J. Chem. Soc., 14 (1930) 73.
78. R. D. Armstrong, T. Dickinson and J. Turner, Electrochim. Acta, 19 (1974) 187.
79. R. D. Armstrong, T. Dickinson and P. M. Willis, J. Electroanal. Chem., 57 (1974) 589.
80. R. D. Armstrong and K. Taylor, J. Electroanal. Chem., 63 (1975) 9.
81. R. D. Armstrong, T. Dickinson, H. R. Thirsk and R. Whitfield, J. Electroanal. Chem., 29 (1971) 301.
82. R. D. Armstrong, T. Dickinson and P. M. Willis, J. Electroanal. Chem., 59 (1975) 281.
83. R. D. Armstrong, T. Dickinson and K. Taylor, J. Electroanal. Chem., 57 (1974) 157.
84. T. Takahashi, O. Yamamoto and S. Ikeda, J. Electrochem. Soc., 120 (1973) 1431.
85. H. R. Thirsk and J. A. Harrison *A Guide to the Study of Electrode Kinetics*, Academic Press, 1972, Chap. 4.
86. J. Burbank, A. C. Simon and E. Willihnganz in C. W. Tobias (Ed.), *Advances in Electrochemistry and Electrochemical Engineering*, Vol. 8, Interscience, New York, 1971, p.157.
87. T. Dickinson, A. F. Povey and P.M.A. Sherwood, The Chem. Soc., Faraday Trans. I, 71 (1974) 298.
88. T. Dickinson, A. F. Povey and P.M.A. Sherwood, J. Solid State Chem., 13 (1975) 237.
89. T. Dickinson, A. F. Povey and P.M.A. Sherwood, in the press.

SODIUM-SULFUR BATTERY RESEARCH

Manfred W. Breiter

General Electric Corporate Research and Development

P.O. Box 8, Schenectady, New York 12301

I. INTRODUCTION

High energy density batteries are required for peaking and load leveling systems by the electric utilities and for electrical vehicle propulsion. The battery design depends upon the intended use since the energy density has to be optimized for the first application and the power density for the second application. The sodium-sulfur cell (1) is one of several electrochemical systems under development for the said applications.

Research on sodium-sulfur batteries for peaking and load leveling systems is carried out in the General Electric Company Research and Development Center. The approach concentrated on the construction of a prototype 17 Ah cell. This cell serves as a means of testing the performance of cell components and fabrication methods. By applying suitable analytical techniques, experiments with this cell permit the evaluation of predicitons regarding the electrochemical behaviour of the anode and cathode as well as of the effects of design changes on performance and life characteristics.

Both the reactants and reaction products are liquid in the sodium-sulfur cell (*cf.* the schematic diagram of a laboratory cell in Figure 1). Sodium and sulfur/polysulfides are separated by a suitable piece of sodium ion-conducting ceramic, beta-alumina. On discharge from sodium and sulfur, elemental sodium is oxidized at the sodium/beta-alumina interface, and sodium ions migrate through the solid electrolyte into the sulfur compartment which is filled with electronically conducting carbon felt. Initially Na_2S_5 is formed. Since Na_2S_5 is not soluble in sulfur, two liquid

127

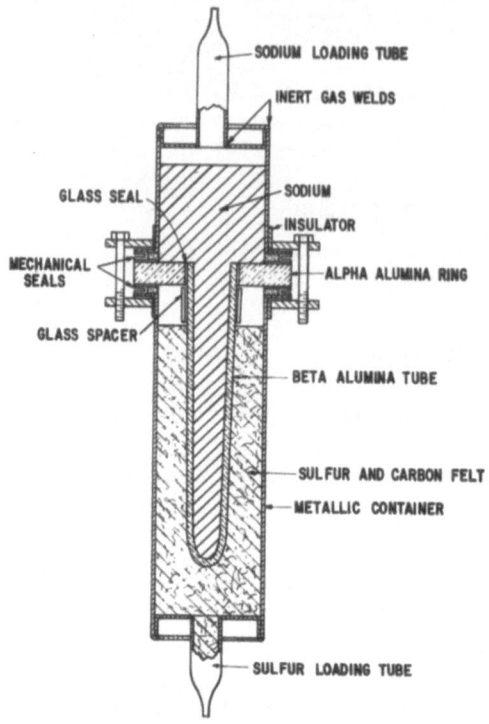

Fig. 1:   Schematic diagram of laboratory sodium-sulfur cell

phases exist throughout *ca* 60% of discharge.   After all the sulfur
is converted to Na_2S_5, the pentasulfide is transformed to trisulfide.
Since Na_2S_5 and Na_2S_3 are miscible, the latter part of discharge
involves a one phase liquid.   In order to keep all the phases
liquid throughout the compositional range, the sodium-sulfur cell
is operated above 275°C, typically between 300 - 375°C.   The final
reaction product Na_2S_3 corresponds to 100% discharge.   During
charge, the reactions are reversed.

II. PROTOTYPE CELL

Cell Design

     The cell which is successfully used (2,3) in laboratory testing
is schematically shown in Figure 1.   It has a glass seal between
the alpha alumina ring and the beta alumina tube.   The alpha
alumina ring serves as an insulating separator between the sodium
and the sulfur containers.   The containers are made from suitable

metals.   A mechanical seal holds the cell together.   The sulfur
is contained in carbon felt.   The loading tubes are crimped after
filling of the cell with reactants.

The basic features of the sodium-sulfur test cell may be summarized as follows:
   a) tubular design for ease of manufacturing;
   b) good ratio of surface of beta-alumina tube to volume of
      reactants;
   c) seals not in contact with polysulfide;
   d) high purity condition for vacuum bake-out and filling with
      reactants, if desired.
More technical details concerning cell construction are given in
references 2 and 3.   Additional important considerations which
determine the feasibility of the sodium-sulfur system from an
economic point of view are:
   a) use of low-cost materials;
   b) application of industrially feasible assembly technology.

Production of Beta-Alumina Tubes

The development of beta alumina tubes for sodium-sulfur cells
is guided by three criteria:
   a) satisfactory ionic conductivity;
   b) sufficient mechanical strength;
   c) negligible attack of the ceramic under cycling conditions.
Experimental evidence suggests that the three conditions are not
independent of each other.   However, quantitative relationships
have not yet been established.

The main features of the beta alumina fabrication procedure
are (3):
   a) use of Bayer process beta alumina;
   b) forming by electrophoretic deposition;
   c) sintering by the stoker or pass-through method.
Beta alumina derived from the Bayer process is relatively inexpen-
sive and available at the 99.99% purity level.   The forming or
shaping step which is carried out by electrophoretic deposition was
already described (4) in detail.   In this process beta alumina
particles are put into suspension in n-amyl alcohol.   The particles
become negatively charged during the milling of the powder when the
particle size is reduced to about 1 micron.   Using fields of about
700 V/cm, the beta alumina is deposited on a mandrel.   The green
deposits are removed from the mandrel and sintered in an oxidizing
atmosphere near 1700°C.   With the stoker method, the furnace is
preheated to the sintering temperature and a train of sintering
boats containing the greenware is continuously pushed through the
furnace at a constant rate.   The sintering strongly affects the
properties of the beta alumina tubes.

An analysis of the resistivity of beta alumina tubes, containing 0.5% yttria as a sintering aid, was carried (5) out on the basis of the analogue circuit in Figure 2.   In this model, R_c designates the resistance associated with sodium ion motion in the interior of randomly oriented grains.   Since the geometric capacity in parallel to R_c is negligible in the studied frequency range (10 Hz to $5 \cdot 10^5$ Hz), it is not included in Figure 2. R_b and C_b are the resistance and capacity due to sodium ion motion across grain boundaries.   The components of the analogue circuit were determined from the experimental frequency dependence at a given temperature. R_c and R_b were roughly equal at 300°C for beta alumina with yttria addition.   For ceramics with other additives, or without additives, the resolution of the measured resistance into single components meets with great difficulty, even when the frequency range is extended (6) by pulse techniques up to about 10^8 Hz.   Another analysis (3) using "resistivity distribution plots" yields the separate values of R_c and R_b in this case. Since R_b is comparable or larger than R_c, a large portion of the voltage drop across the wall of the beta alumina tube is dissipated across grain boundaries.   Assuming the grain boundaries to be about 10 Å thick, the electric field at grain boundaries is estimated to be about 10^5 V/cm.   Such fields give rise to concern about breakdown processes that might occur in the grain boundaries.

The sintering conditions determine (2,3) the grain size distribution of the beta alumina ware.   From the point of mechanical strength and cycling life of beta alumina tubes, the grain size should be uniform and small.   Although tubes with larger grains possess a lower resistivity, this favorable feature is outweighed

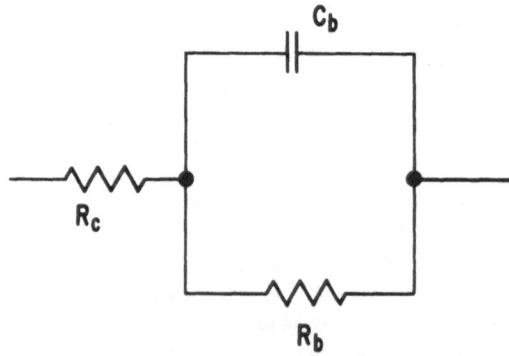

Fig. 2:   Analogue circuit for AC measurements of the resistance of beta alumina

by the negative features of decreased mechanical strength and
shorter life.   The tendency for a few grains to grow to very large
size during the sintering has to be overcome since it leads to the
existence of areas with smaller resistivity where the local current
density becomes large.   Breakdown is likely to initiate in these
areas.

Seal Development

    At least two types of seals are incorporated into a sodium-
sulfur cell with metal containers for the reactants:
    a) seal between beta alumina tube and insulating separator;
    b) seals between insulating separator and metal cans.
The type of seal depends upon the materials used in the cell construction.   Since the insulating separator consists of an alpha
ring in the present test cell (See Figure 1), a glass seal between
beta alumina tube and separator proved satisfactory.   This seal
which is made in a platinum furnace has to withstand the attack by
liquid sodium at about 300°C for a sufficiently long time.

    Originally a glass seal was also used (2) between the alpha
alumina ring and the flanges of the metal cans (Rodar alloy in this
case).   However, this glass seal showed a tendency to crack on the
sodium side.   It is likely that sodium in contact with the metal
oxides in the sealing area gradually reduced the oxides to metal.
Since the bond between glass and metal oxide is responsible for the
strength of the seal, the oxide reduction leads to cracking and
seal failure.   A mechanical seal with Inconel O rings of C-shaped
cross section (*cf*. Figure 1) was subsequently developed.   This
seal can be made helium-leak tight and has stood up so far under
cycling.   The mechanical seal has the great advantage over a glass
seal in that other metals besides Rodar alloy can be used for the
metal compartments.   Glass seals between metal and ceramic require
matching expansion coefficients for the two materials.   Since the
expansion coefficient of the separator material in the test cell
does not match those of other metals like stainless steel or
aluminum, a glass seal between separator and metal cans severely
restricts the choice of materials for the cell construction.

    An actual cell is shown in Figure 3.   Stainless steel collars,
bolted together by screws, exert pressure of about 200 pounds on
the Inconel O rings.   Both metal containers consist of aluminum.
Since aluminum becomes covered with an insulating layer of aluminum
sulfide in the presence of sulfur, the inner walls of the sulfur
container were treated with an electronically conductive coating.
Electric leads and voltage probes are separately attached to each
of the metal containers.   To prevent short-circuiting of the cell,
insulating tape covers the cans in the area of the collars, and thin

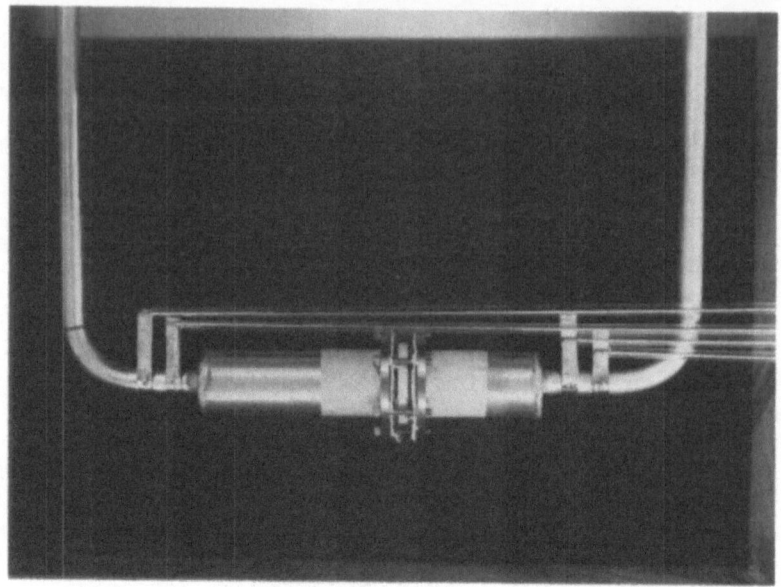

Fig. 3:   Photograph of actual sodium-sulfur cell

fiberglas rings insulate the collars from the flanges of the cans.
Liquid reactants can be introduced via two tubes at either end of
the cell.

Filling Methods

Two different techniques have been employed for the loading of
the cell with reactants:
   a) loading with liquid reactants from special filling apparatus;
   b) filling of a half cell, consisting of sodium container, alpha
      alumina ring and beta alumina tube, with electrolytic sodium
      by electrolysis in molten sodium nitrate and subsequent cell
      assembly with a precast plug of sulfur-carbon.
While filling technique (a) represents the more general approach,
applicable to any cell construction, loading by method (b) is
restricted to cells with mechanical seals.   Technique (b) is more
rapid for laboratory testing and is therefore used at present.
However, a cost analysis demonstrates that electrolytic filling
from molten sodium nitrate becomes too expensive for practical
sodium-sulfur batteries.   The testing of the prototype cells do
not indicate an influence of the loading techniques on cell performance and life.

III. CELL TESTING

To approximate actual conditions for load leveling with batteries, the cell is discharged at constant current (at present 50
mA/cm^2, referred to the geometric surface area of the beta alumina
tube) to 1 or 1.2 V, depending upon the cell resistance.   Then it
is automatically switched to charging at 50 mA/cm^2.   Charging is
stopped at 3V, and the next discharge is started.   Cycling is
carried out at 300°C in an air furnace.

The cell voltage is recorded as a function of time during
cycling.   A typical trace is shown in Figure 4.   The cell voltage
decreases gradually with time during discharge.   The opposite
behaviour is observed for charging.   The voltage change is due to
the mentioned transitions between one-phase liquid and two-phase
liquid.   In fact, the cell voltage-time trace consists of two
waves for cells which can be recharged into the region of the two-
phase liquid.   Poor cells which can only be cycled in the region
of the one-phase liquid display only the first wave in their voltage-
time plot.   The transitions are clearly demonstrated by the open
cell voltage which was simultaneously recorded at intervals of 10
minutes.   Relays periodically interrupt the current for a short time.

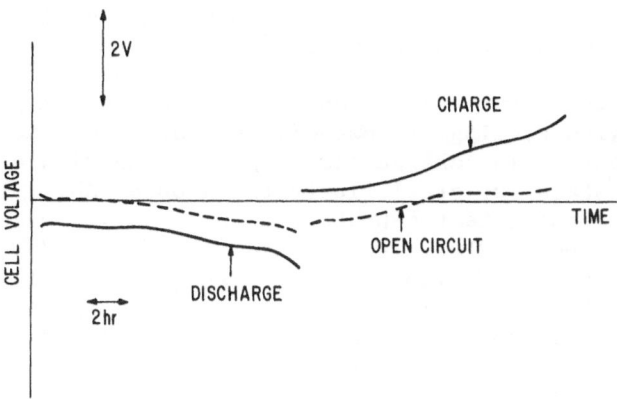

Fig. 4:   Cell voltage-time trace during cycling of sodium-sulfur
          cell at constant current between voltage cut-off limits

Rapid measurements of current voltage curves at a given melt composition yielded linear relations between current and cell voltage. The losses during cycling are largely ohmic. The latter result was confirmed by AC measurements of the cell impedance with a small superimposed signal of 3 mV_{eff} at frequencies $(\omega/2\pi)$ between 10 and 10,000 Hz during cycling. In general, the ohmic component R_s of the impedance in an analogue series circuit is much larger than the capacitive component $(1/\omega Cs)$. The capacitive component becomes only noticeable at the end of discharge or charge respectively for cells with acceptable performance. Data acquisition was done by recorder in the past, but has been changed to recording by magnetic tape with subsequent computerized data reduction and printout of results. The large number of data necessary for a statistical evaluation of cell performance requires a computerized approach.

IV. ANALYSIS OF CELL PERFORMANCE

The cell performance during cycling is well characterized by the type of plots shown in Figure 5, Figure 6, and Figure 7. The upper plots give the number of amp hours for each of the successive cycles. In addition, the points of transition from the two-phase liquid to the one-phase liquid determined from the open cell voltage, are marked by solid symbols in the upper plots. The first discharge is counted as the first cycle, the subsequent charge as the second cycle and so on. It is found that the first discharge corresponds to the nominal cell capacity, computed on the basis of the initial amounts of sodium and sulfur. The lower plots in Figure 5 to Figure 7 give the average cell resistance as a function of cycles. This resistance is determined from the difference of cell voltage with and without current.

Both cell 79 (Figure 5) and cell 62 (Figure 6) displayed a capacity loss with cycling. However, the nature of the capacity loss is different. The charge limit for cell 79 with a sulfur compartment of stainless steel fluctuates around 5Ah. This cell could be charged into the two-phase region to a limited extent only. Cell 79 operated with an efficiency of about 70%. The reduced efficiency is attributed to the isolation of polysulfide from the carbon felt. Cell 62 with a sulfur compartment of Rodar alloy showed a rapid change of the discharge limit and a slow change of the charge limit. The cycling behavior of cell 62 reflects the influence of two factors:
   a) combination of sulfur as corrosion product with the
      container;
   b) isolation of polysulfide from the current collector.
The first effect reduces the amount of sulfur available for the formation of polysulfides and moves the discharge upwards. It was confirmed by measurements of the thickness of the corrosion layer

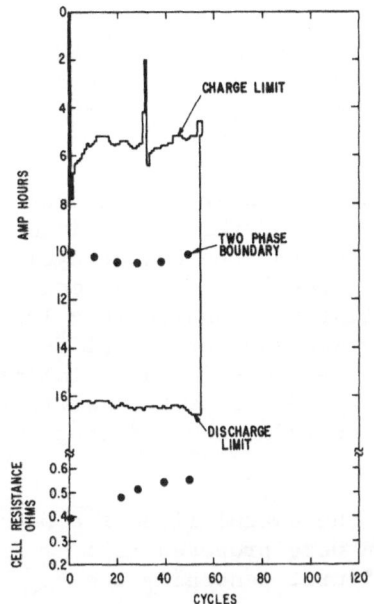

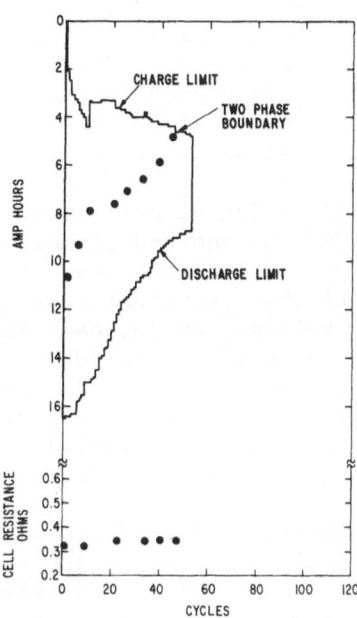

Fig. 5: Cycling record for cell   Fig. 6: Cycling record for cell
         79 with sulfur compart-   62 with sulfur compart-
         ment of stainless steel.   ment of Rodar alloy.

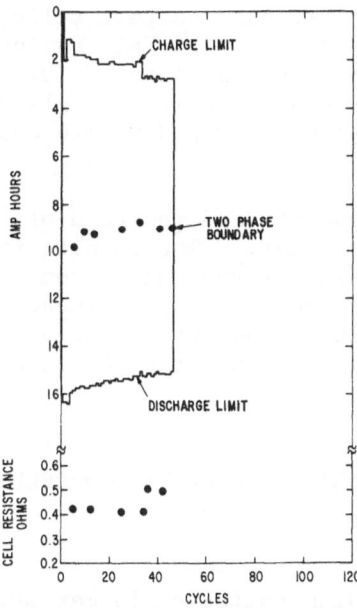

Fig. 7: Cycling record for cell 60 with sulfur compartment of nickel

on the walls of failed cells.   The second factor is recognizable
in Figure 6 by a similar displacement of the two-phase boundary and
the discharge limit.   Shortly before abrupt failure, cell 62 cycled
with small efficiency in the two-phase region only.   Because of the
poor performance of cells with a sulfur compartment of Rodar alloy,
this material has to be ruled out for cell construction.

The cycling of cell 60 (Figure 7) was exemplary.   The cell
discharged and charged steadily at about 84% of the theoretical
capacity.   The slight loss of capacity towards the end of cycling
is due to the concurrent increase in cell resistance.   At constant
cut-off voltage, an increase in cell resistance results in a loss
of cell capacity.   The sulfur container was made of nickel for
cell 60.   In general, cells with nickel cans displayed a similar
behavior with regard to cell capacity.   However, nickel has to be
ruled out as a cell material because it corrodes badly during
cycling and is too expensive.

In general, three ways of defining the useful life of a sodium-
sulfur cell in load leveling application were proposed (3):
   a) time to abrupt or catastrophic failure, generally due to
      internal short circuits;
   b) time to reach an uneconomic capacity;
   c) time to reach an uneconomic resistivity.
Abrupt failure may result from cracking of the beta alumina tube,
of the alpha alumina ring, of the glass seal between beta alumina
tube and alpha alumina ring and from failure of the external seal.
From inspection of the first hundred test cells, cracking of beta
alumina is the predominant cause for cell failure at present.   The
best four cells had (3) an average life of 104 Ah/cm^2.   The life
is measured in Ah/cm^2 here because it strongly depends upon the
current density.   It decreases rapidly with current density above
100 mA/cm^2.

Decrease of cell capacity and increase of cell resistance with
time occur frequently together.   The results from cells with
different metal containers for the sulfur compartment demonstrate
that corrosion products have a strong influence on both effects.
However, the total cycling behaviour is not understood at present.
It is sure that other factors than the corrosion products are also
involved.

V. EXAMINATION OF FAILED CELLS

Techniques

Malfunctions in sodium-sulfur cells may arise for many reasons.
As discussed before, the causes for the gradual decrease of the

performance can sometimes be deduced from the electrical characteristics of the charge-discharge cycle and attributed to insufficient
conversion of the reactants due to isolation or formation of corrosion products.   Other identifiable causes of malfunctions are:
improper cell assembly, leaks in seals, cracks in the alpha alumina
separator, cracks or deep pits in the beta alumina tube.   Three
techniques have been used for the examination of failed cells:
    a) cell disassembly;
    b) cell sectioning;
    c) X-ray radiography.
Since each of the three techniques is applied after the failed cell
has been cooled down to room temperature, the possibility that the
cooling process leads to additional effects has to be considered
carefully.   For instance the beta alumina tube sometimes cracks
radially at or near the seal to the alpha alumina ring during
cooling.   The freshness of the appearance of the cracks allows a
decision in such a case to be made on which of the cracks was
responsible for cell failure.

     Since the arrangement of carbon felt and the polysulfide/sulfur
mixture in the sulfur compartment are destroyed by the washing with
methanol and water after the cap of the sulfur can is cut off, the
principal information obtained by cell disassembly concerns the
condition of the beta alumina electrolyte, the alpha alumina separator, and the glass seal between them.   Cell sectioning for
autopsy is difficult and expensive.   Since sodium and sodium
polysulfides are highly reactive, sectioning must be carried out in
an inert atmosphere of argon or nitrogen.   In addition, important
structural and chemical features may be lost, altered or obscured
when the reactants and components are separated by cutting with a
diamond wheel for observation.   In contrast, x-ray radiography has
been found to be a useful non-destructive method.   Information
about improper assembly is easily obtainable before cell testing.
Changes produced by cycling can be found from a comparison of
pictures taken before and after cycling.   The reader is referred
to reference 3 in which numerous examples are given for the application of x-ray radiography.   At present a distinction between sulfur
and polysulfide for a determination of their spatial distribution
has not been achieved.   It is envisaged (3) that future work with
x-ray analysis may provide information on the behaviour of the
sulfur electrode during cycling.

Failure of Beta Alumina

     The role of the solid electrolyte tube which keeps sodium and
sulfur apart is crucial to the operation of the sodium-sulfur cell.
It allows for a long shelf life of the cell because self discharge
is absent.   The cracking of the tube is the most common cause of

abrupt failure.   Two types of cracks have been identified in our
work:
   a) radial cracks at or near the seal to the alpha alumina ring;
   b) longitudinal cracks running along part of the length of the
      tube
As pointed out in the preceding section, radial cracks may result
as an artifact during cooling.   However, discoloration of the
whole area of the radial crack was observed in certain cases.
The discoloration suggests that the radial crack opened while the
cell reactants were still molten.   Thus, radial cracks must be
regarded as a possible mode of tube failure and probably result
from stresses induced during the sealing of the beta alumina tube
to the alpha alumina ring.

     Examination of the fractured area revealed discoloration in
most cases of longitudinal cracks.   The cracks were the cause for
abrupt failure.   In agreement with other work (7), the removal of
sintering aids reduced the frequency with which this failure mode
occurs.   Since the test cells are electrolytically filled with
sodium, the detrimental effect on life of the incorporation of
potassium ions into the beta alumina lattice is minimized.

     Pitting of the outer surface of the beta alumina tube is also
observed, often in combination with one of the modes of cracking.
Sometimes the pitting penetrates the entire tube thickness, and
frequently a crack passes through the pitted area.   It is considered likely that pitting is caused by localized areas of high current
density, due to inhomogeneities either in the solid electrolyte or
in the contact between the solid electrolyte and carbon felt.
Experimental evidence exists for both causes of an inhomogeneous
current distribution.   Large grains which are observed visually
represent areas of higher conductivity.   Nonuniformity of the
contact between carbon felt and beta alumina was detected in some
cells by x-ray radiography and cell sectioning.

     Four different mechanisms for the degradation of beta" alumina
(β"-Al_2O_3) were recently distinguished (8).   Similar mechanisms
should apply to beta alumina.   Since the question of the degradation mechanism has not been resolved for β or β" alumina, a discussion of these processes is not included here.   However, it
should be pointed out that the experimental evidence for the existence of strained areas in the beta alumina tube, for instance close
to the glass seal, is mounting.   Stress corrosion cracking of the
beta alumina may be involved as a failure mode.

Failure of Alpha Alumina Separator and Glass Seal

     Radial cracking of the beta alumina tube below the seal to the
alpha alumina separator was found to occur sometimes during the
sealing process when very dense and nearly 100% alpha alumina ceramic rings were used.   The mismatch of expansion coefficients is

responsible for the radial cracking.   In order to reduce the failure
rate, alpha alumina rings with additives (designation AT 1075) were
used subsequently.   These rings have practically the same expansion
coefficient as beta alumina.

While this material did not deteriorate under exposure to
sodium, liquid sodium penetrated into the pores of the AT 1075
ceramic and led to cracking of the separator in long-lived cells.
Experimental investigations are required to determine the amount of
additives that can be tolerated in the ceramic.

Only one failure of the glass seal between beta alumina tube
and alpha alumina ring has been found during the testing of about
100 cells.   Test results for longer periods than about three
months are still lacking.

VI.   THE SULFUR ELECTRODE

As pointed out in Section IV above, the sulfur electrode is
responsible for the rechargeability of the cell.   Both the sulfur
electrode and the resistance of the beta alumina tube determine
the internal cell resistance.   Since the complex operation of the
sulfur electrode is not yet understood, subsequent discussion is
of a qualitative nature.

Melt Composition

It appears reasonable to describe the two-phase liquid as consisting of sulfur and sodium pentasulfide.   However, since the
threshold temperature for the onset of chain formation in sulfur
depends upon the addition of inert diluents, the viscosity of the
sulfur-rich phase may significantly differ from that of sulfur.

The one-phase liquid can contain Na_2S_5, Na_2S_4 and Na_2S_3.
The latter is the final reaction product during discharge.   Numerical values for the density, conductance, viscosity and surface
tension were obtained (9) for different compositions of the polysulfide by interpolation of results in references 10 and 11.
The conductivity increases from about 10^{-8} Ω^{-1} cm^{-1} for sulfur to
$0.18\Omega^{-1}$ cm^{-1} for $Na_2S_{5.1}$ and to 0.41 Ω^{-1} cm^{-1} for Na_2S_3 at $300°C$.
However, since sulfur and Na_2S_5 are immiscible and the pentasulfide
appears to be produced (8) in small droplets during discharge, the
conductivity of the two-phase liquid will remain smaller than the
electronic conductivity of the carbon felt.   In contrast, the
conductivity of the one-phase liquid and that of the carbon felt
are comparable (12).   The density increases from 1.66 g/cm^3 for S
to 1.86 g/cm^3 for Na_2S_5 at $350°C$.   The respective increase of the
surface tension by about a factor of 3 is considerable.

Transference experiments (11) demonstrated that sodium polysulfide melts in the composition range of the one-phase liquid do not contain elemental sulfur, e.g. the melt is dissociated into Na^+ ions and polysulfide ions.   Since S_3^{2-} appears (8) to be unstable above $106^{\circ}C$, the undiluted sodium polysulfide melts in the $300^{\circ}C$ - $400^{\circ}C$ range should contain mainly S_3^{2-}, S_4^{2-}, and S_5^{2-}.   The relative amount of the anion species depends upon the melt composition. Other work (13,14) confirms this assignment.   Raman spectroscopy of glossy polysulfides may yield (8) further information.

Electrochemical Reactions

The subsequent results were obtained under idealized conditions and not in actual cells.   It is not clear at present what the observed influence of additives on the performance of the sulfur electrode means with regard to the electrode process.   A distinction has also to be made between the electrochemical reactions occurring in the composition range of the one-phase and two-phase liquid respectively.

A special cell which allowed observation of the sulfur compartment by microscope was used (8) to study the electrochemical reactions at slow rates of discharge in a cell containing sulfur. After a few hours of discharge a finely divided polysulfide phase appeared near the β" alumina interface.   The polysulfide globules were estimated to be a few microns in diameter.   More droplets of Na_2S_5 were present after longer times of discharge.   The droplets which are not formed at carbon fibers away from the ceramic are slowly removed from the surface of the β" alumina by convection. Based on these results the following initial discharge reaction was postulated:

$$2\ Na^+ + 5S + 2e^- = Na_2S_5 \tag{1}$$

Electrons are transported to the reaction sites along the carbon fibers.   The reaction sites are those places where the fiber touches the surface of the β" alumina.

Studies of the mechanism of the sulfur electrode on vitreous carbon (8,9,16,17) graphite (8,9,15) and platinum (15) have been carried out using voltammetric (8,9,15,16) chronopotentiometric (16) and rotating disc (9,17) techniques and also by taking measurements of steady-state current/potential curves (17) in polysulfide melts within the composition range of the one-phase liquid.   Both anodic and cathodic potential scans, starting at the rest potential of the test electrode in Na_2S_5, Na_2S_4 or Na_2S_3, display (16) a similar shape at $250^{\circ}C$.   After a first wave the current does not change

much with potential over a wide potential range (0.5 to 1 V).   It
looks as if a limiting current is reached.   Then the current rises
again.   It has been known for sometime (16) that the formation of
different films inhibits the respective electrode reactions and
leads to a current decrease after the first wave.   Nevertheless,
conclusions on the mechanism have been drawn (8,9,15,16) from the
dependence of the peak current of the first wave upon the square
root of the sweep rate.   The use of the simple Randles – Sevcik
analysis is not justified under these circumstances (17) since it
is well known that the interpretation of voltammetric data is more
complicated when film formation is involved.

A schematic diagram of a steady-state current/potential curve
and of the processes occurring in certain potential ranges on
vitreous carbon is shown in Figure 8.   This interpretation is
reported in reference 17 and represents the most plausible one at
present.   More experimental work is required to fill in details.

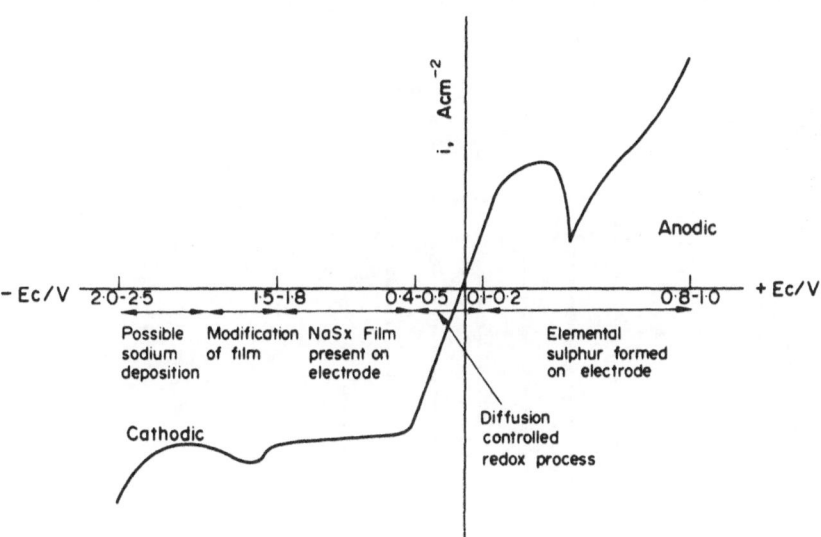

Fig. 8:   Schematic diagram of steady-state current/potential curve
           on vitreous carbon in polysulfide melts at 350°C and of
           corresponding electrode processes

Studies of the Sulfur Electrode In Sodium-Sulfur Cells

To obtain information on the mechanism of the sulfur electrode,
measurements of the impedance of the laboratory cell, illustrated
in Figure 1 and Figure 3, were carried out with a small AC signal
of 3 mV_{eff} at frequencies between 10 Hz and 10,000 Hz during
cycling.   It was observed that the capacitive component $1/\omega C_s$ in
a series analogue circuit remained much smaller than the ohmic
component R_s during most of the discharge or charge cycle.   The
capacitive component became larger, but still remained smaller
than R_s, towards the end of discharge or charge.   The ohmic component was practically independent of frequency.

In Figure 9, the charge and the ohmic component at 1000 Hz are
plotted during each of the successive cycles for a cell with good
rechargeability.   The resistance of the beta alumina tube was
independently determined during the electrolytic loading with
sodium.   It is plotted as a straight line in Figure 9 under the
 assumption that an ohmic component of resistive layers which may
form on the outer surface of the beta alumina tube during cycling
is considered to be contribution of the sulfur electrode.

An analysis of the behaviour of R_s during discharge reveals
that major changes only occur at the beginning and end of a cycle.

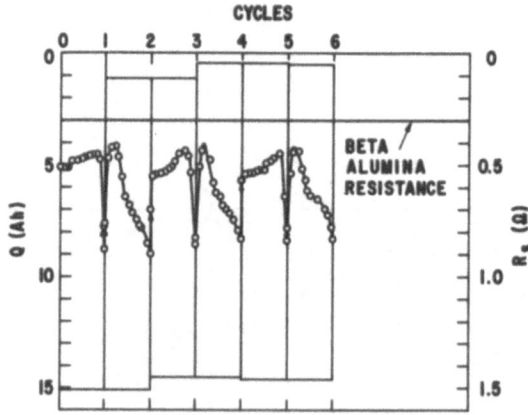

Fig. 9:   Cycling record and ohmic component of the impedance of
          cell 129

The transition from the two-phase liquid to the one-phase liquid is reflected by a slight decrease of R_S. The dependence of R_S upon charge is more pronounced during the charging cycle. It has a parabolic shape in the region of the two-phase liquid. In the one-phase region the ohmic component R_S increases steadily with charge. When the current is interrupted at the end of a cycle, the resistance R_S decreases rapidly. The latter decrease is relatively large at the end of the charging cycle.

The dependence of the cell impedance upon frequency and charge suggests the formation and removal of two different resistive layers during cycling. The first layer is produced towards the end of the discharge cycle. Two simultaneous effects are considered responsible:

  a) formation of an inhibiting layer of Na_2S_2 on the carbon fibers;
  b) presence of a resistive layer at can wall, probably because of loss of contact between can wall and carbon fibers when polysulfides are produced there.

Evidence for the first process is presented by the kinetic studies of the sulfur electrode (*cf*. Figure 8). A small effect of the second type could be demonstrated for Rodar alloy cans by incorporating a separate probe and monitoring the voltage between can wall and probe. The first resistive layer is quickly removed during the subsequent charge cycle.

The second resistive layer begins to form when the charging has progressed beyond about 30% during the charge cycle, e.g. close to the transition from the two-phase liquid to the one-phase liquid. It is attributed to preferential formation of a resistive layer of sulfur in the vicinity of the beta alumina/carbon fiber interface. Upon switching from charge to discharge, the second layer is rapidly removed by conversion of sulfur to polysulfide, probably Na_2S_5.

ACKNOWLEDGEMENT

This work is co-sponsored by the Electric Power Research Institute under Contract RP128-3.

Contributions to the sodium-sulfur project at the General Electric Research and Development Center by R.R. Dubin, R.H. King, S.P. Mitoff, O. Muller, R.W. Powers, W.L. Roth and E. Szymalak are acknowledged.

REFERENCES

1.   J.T. Kummer and N. Weber, U.S. Patent 3404035, Oct. 1968
2.   Sodium–Sulfur Battery Development for Bulk Energy Storage, EPRI 128-0-0, Final Report, July 1974.
3.   Sodium–Sulfur Battery Development for Bulk Energy Storage, EPRI 128-2, Interim Report, September 1975.
4.   R.W. Powers, J. Electrochem. Soc. 122 (1975) 490.
5.   R.W. Powers and S.P. Mitoff, J. Electrochem. Soc. 122 (1975) 226.
6.   G.C. Farrington, Report No. 75CRD146, General Electric Company, Schenectady, New York, July 1975, submitted for print to J. Electrochem. Soc.
7.   Y. Lazennec, C. Lasne, P. Margotin and J. Fally, J. Electrochem. Soc. 122 (1975) 734.
8.   Research on Electrodes and Electrolyte for the Ford Sodium–Sulfur Battery, Contract No. NSF–C805(AER–73–07199), July 1975.
9.   Research on Electrodes and Electrolyte for the Ford Sodium–Sulfur Battery, Contract No. NSF–C805, July 1974.
10. B. Cleaver, A.J. Davies and M.D. Hames, Electrochim. Acta 18 (1973) 719.
11. B. Cleaver and A.J. Davies, Electrochim. Acta 18 (1973) 727.
12. J.G. Gibson, J. Appl. Electrochem. 4 (1974) 125.
13. E. Rosen and R. Tegman, Chemica Scripta 2 (1972) 63.
14. B. Cleaver and A.J. Davies, Electrochim. Acta 18 (1973) 741.
15. S.M. Selis, Electrochim. Acta 15 (1970) 1285.
16. K.D. South, J.L. Sudworth and J.G. Gibson, J. Electrochem. Soc. 119 (1972) 554.
17. R.D. Armstrong, T. Dickinson and M. Reid, Electrochim. Acta 20 (1975) 709.

ELECTROCHEMICAL INJECTION OF IONS INTO NON-STOICHIOMETRIC ELECTRODES

B.C.H. Steele

Dept. of Metallurgy & Materials Science

Imperial College, London, S.W.7. England

INTRODUCTION

Selected non-stoichiometric compounds can function as a host lattice for the incorporation of many elements such as hydrogen, lithium, sodium, oxygen, which can function as the electro-active species in a variety of electrochemical devices. Providing the kinetics of dissolution into the solid compounds are rapid and providing that relatively large quantities of the solute can be incorporated, then these non-stoichiometric solids can be used as solid solution electrodes   (S.S.E.) in appropriate electrochemical cells. The characteristics of a solid solution electrode (S.S.E.) are somewhat different from those normally associated with traditional electrode materials and these differences can be illustrated with reference to the following two electrochemical cells:

$$\text{Na (1)} \quad / \quad \text{Na} - \beta - Al_2O_3 \quad / \quad \text{S (1)}$$
$$(\Delta \bar{G}'_{Na}) \qquad\qquad\qquad (\Delta \bar{G}''_{Na})$$

$$\text{Na (s)} \quad / \quad \begin{matrix}\text{organic}\\ \text{electrolyte}\\ (Na^+)\end{matrix} \quad / \quad TiS_2 \text{ (s)}$$

The emf developed by all these cells is given by the relationship,

$$E = \frac{\Delta \bar{G}'_{Na} - \Delta \bar{G}''_{Na}}{F} = \frac{\Delta \bar{G}''_{Na}}{F}$$

as the activity of sodium at the left hand electrode is unity. The factors governing $\Delta \bar{G}''_{Na}$ can be shown on the relevant free energy composition diagrams for these systems (Figs. 1 and 2). As the electro-active species (Na^+) is transported through the cell the

145

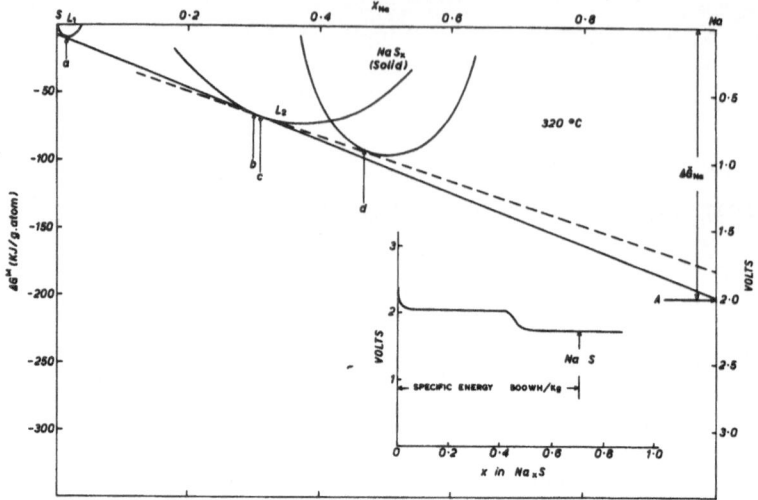

FIG. 1   Schematic free energy/composition diagram for Na-S system

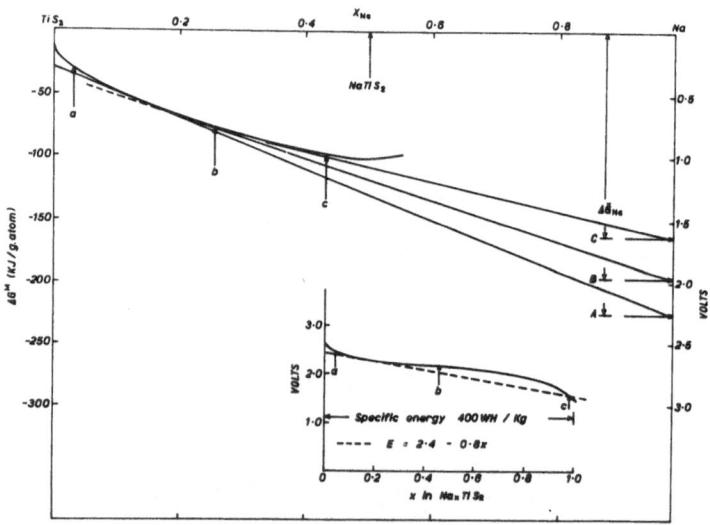

FIG. 2   Schematic free energy/composition diagram for psendo binary
           Na - TiS$_2$ system.

ratio Na/electrode component will gradually increase and the relevant value for $\Delta \bar{G}''_{Na}$ is given by the intercept of the tangent to
the free energy composition curve at the appropriate composition.
For the first cell the transport of sodium to the right hand electrode results in the formation of two immiscible liquids (Fig. 1)
which establishes a constant value for $\Delta \bar{G}''_{Na}$ between the compositions
Na$_{0.01}$S (point a) and Na$_{0.39}$S (point b). The introduction of

more sodium will result in a steady decrease of cell voltage until
the composition (c) is attained which is inquilibrium with the
solid phase Na_2S_2 at (d). The tangent c – d now establishes the
constant value for the open circuit voltage of the cell. As
liquid phases are necessary to permit large current densities the
operation of the cell is thus usually restricted to the composition
limit of $Na_{0.67}S$ (Na_2S_3), which corresponds to a maximum theoretical
specific energy density of approximately 500 WH/Kg.

An alternative approach is provided by the application of
solid solution electrodes and the thermodynamic data for the system
$Na_x TiS_2$ shown in Fig.2 provides a suitable example. In this
system the integral ΔG_M curve can be represented approximately
by a shallow curve and so the slope of the tangents and thus the
associated $\Delta \bar{G}''_{Na}$ values change with composition. For this system
the incorporation of 1 gm atom of sodium ($TiS_2 \longrightarrow Na TiS_2$) is
accompanied by a voltage change of 2.4 – 1.4V. Assuming that $\Delta \bar{G}''_{Na}$
can be represented by a linear function of x, i.e. $\Delta \bar{G}_{Na}$ =
–240,000 + 80,000 x (J), then the theoretical energy is given by

$$\int_{x = o}^{x = 1} \Delta \bar{G}_{Na} \; dx,$$

and thus the associated theoretical specific energy density is
400 WH/Kg. This value is comparable to that calculated for the
Na/S battery. This type of thermodynamic analysis does suggest,
therefore, that S.S.E.'s could be incorporated into viable secondary
batteries provided that the rate of mass transport within the non-
stoichiometric solid is high enough for the required current densities.

SPECIFICATIONS FOR SOLID SOLUTION ELECTRODES

The desirable properties of solid solution electrodes have
been discussed by Steele (1) and are summarised below.
(i)   Large range of homogeneity.
(ii) Partial molar free energy ($\Delta \bar{G}$) of the electro-active species
      should be relatively constant over a wide range of composition.
(iii) Good electronic conductor.
(iv) Rapid chemical diffusion of the electro-active species within
      the non-stoichiometric electrode.
(v)   No significant interaction with the electrolyte phase
(vi) Low weight materials, economical to fabricate.

MEASUREMENTS OF THERMODYNAMIC AND TRANSPORT PROPERTIES

AS SOLID SOLUTION ELECTRODES

Solid solution electrodes have often been used in high temperature electrochemical cells incorporating solid electrolytes, and
reviews (2, 3, 4, 5) are available describing the measurement of
the relevant thermodynamic and kinetic properties. Coulometric
titrations enable the partial molar free energy ($\Delta \bar{G}$) of the

electro-active species in the S.S.E. to be measured as a function
of composition. Potentiostatic, galvanostatic and other electrochemical relaxation techniques (6, 7) can then be employed to investigate how the chemical diffusion coefficient $(\tilde{D})$ varies as a
function of composition. If appropriate, the chemical diffusion
data can be supplemented by measurements of ionic conductivity
using blocking electrodes, and by determination of self diffusion
coefficients (D^*) using radio-active tracers, and other techniques
such as N.M.R. (8). The various diffusion coefficients should be
consistent with relationships of the type,

$$\tilde{D}_i = D_i^* \frac{d\ln a_i}{d\ln x_i}$$

and so a check on the measurements is often possible. The electronic
properties of the S.S.E. can also be determined by well established
four-point techniques, and it is often possible to combine these
measurements with investigations into the e.m.f./composition relationship using the Van der Pauw method (9). Finally, the performance of the S.S.E. has to be assessed in the appropriate electrochemical cell.

PROPERTIES OF SELECTED SOLID SOLUTION ELECTRODES

a. Incorporation of Oxygen

The non-stoichiometric fluorite phases, UO_{2+x}, CeO_{2-x} were
among the first S.S.E. electrodes to be investigated. The e.m.f./
composition data was first reported (10) in 1962 and the associated
chemical diffusion data measured in 1970 (11) using solid state
cells of the following type at elevated temperatures (800 - 1000°C),

$$Cu, Cu_2O / Zr_{0.85}Ca_{0.15}O_{1.85} / UO_{2+x}$$

the relevant data are collected together in Fig. 3 and attention is
drawn to the large values of $\tilde{D}_o$ and their variation with composition
At these high temperatures there was sufficient plastic flow to ensure good interfacial contacts and the rate limiting step was usually the ionic flux through the solid electrolyte. Whilst these
data are of importance for the fabrication of ceramic nuclear fuel
elements, they are not applicable to the problem of electrochemical
energy storage.

More relevant are the investigations of Kudo et al. (12). Examination of their paper indicates that the perovskite solid solutions $Nd_{1-x}Sr_xCoO_{3-\delta}$ can behave as solid solution electrodes in alkaline solutions at 25°C. A typical voltage/composition curve obtained under dynamic conditions is depicted in Fig. 4., and the
effective diffusion coefficients were in the range $10^{-11} - 10^{-14}$
cm²/s. The low oxygen capacity of the system and the relatively
low diffusion coefficients reported make this non-stoichiometric
oxide unsatisfactory for an electrochemical storage electrode.

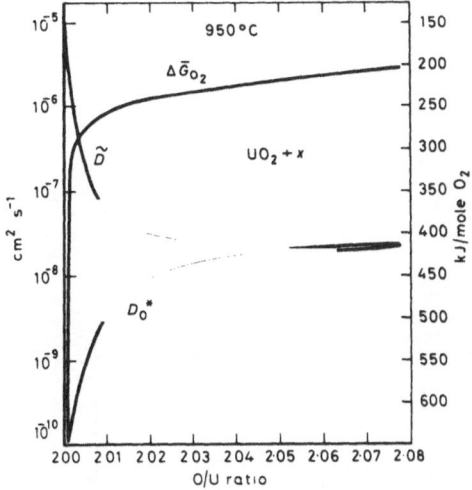

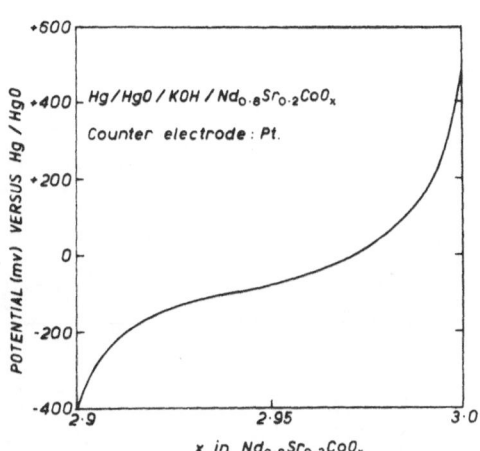

FIG. 3 Thermodynamic and transport
properties of oxygen in UO_{2+x} as
a function of composition.

FIG. 4   Potential/composition
curve for $Nd_{0.8}Sr_{0.2}CoO_x$

This type of behaviour however, has obvious implications for the
proposed application (13) of this material as an oxygen electrode.

b.   Incorporation of Hydrogen

It is well known that large quantities of hydrogen can dissolve
in certain metals and alloys and often the hydrogen diffusion coefficient is of the appropriate magnitude (10^{-7} - 10^{-6} cm^2/s) for
the system to be considered as a viable solid solution electrode.
Many of the simple binary metal hydrides are too stable thermodynamically to be used in electrochemical systems incorporating
aqueous electrolytes. However, using nickel as an alloying element
it is possible to produce ternary hydride systems in which $\Delta \bar{G}_{H_2}$
has values appropriate for use in aqueous electrolytes.
$La Ni_5 H_x$, for example, dissolves 6 atoms of hydrogen per mole $LaNi_5$
at 25°C and the corresponding hydrogen pressure is only 2.5 atm.
Moreover, NMR investigations (8) suggest a high proton mobility.
This material and other rare earth metal alloys are being investigated as hydrogen storage electrodes (14) but detailed results remain to be published. Information, however, is available for another system $Ti_2NH_{2.5}$ which has been successfully employed as an
electrode (15). In the presence of small quantities of the second
phase TiNiH, approximately two atoms of hydrogen per mole Ti_2Ni can
be rapidly and reversibly incorporated into the electrode. The pot-

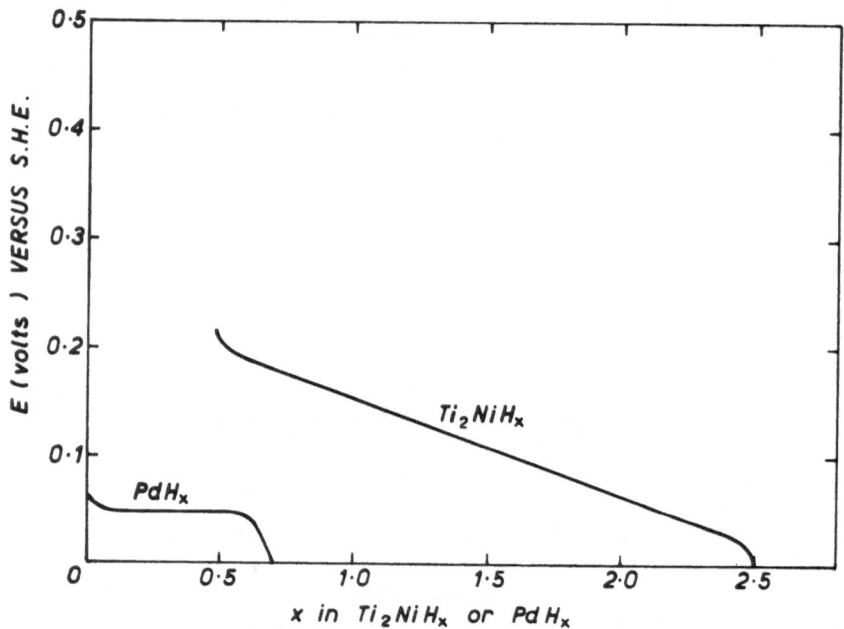

FIG. 5. Potential/composition curve for Ti_2Ni H_x and Pd_x.

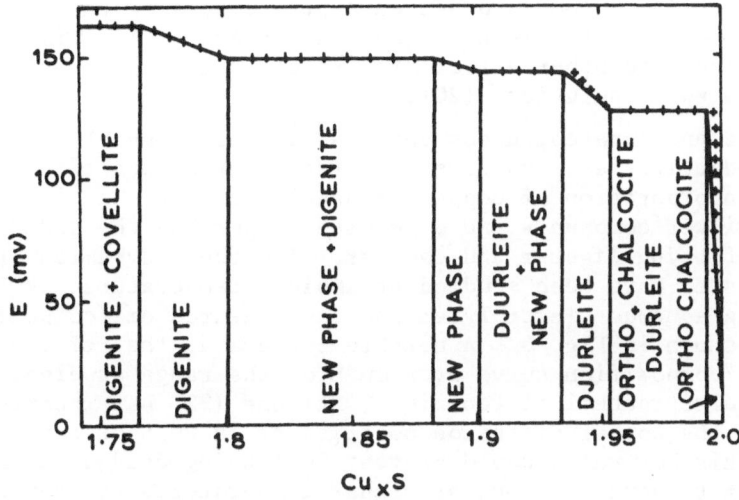

FIG. 6. Potential/composition curve for Cu - S system at 25°C.

ential / composition curve for this electrode is shown in Fig. 5 which also includes for comparison the emf/composition curve for palladium hydride which also exhibits a high proton diffusivity $(5 \times 10^{-7} \text{ cm}^2/\text{sec})$ at room temperature (6).

Protons can also dissolve in non-stoichiometric compounds such as oxide but very little quantitative data are available (16). It is known for example that protons can be incorporated into MnO_{2-x} (17), and Gabano (18) has measured a value of D_H in MnO_{2-x} of 10^{-19} cm^2/s at 25°C. Higher values $(10^{-10} - 10^{-11}$ cm^2/s) have been reported (19) for proton diffusion coefficients in nickel hydroxide.

c.   Incorporation of Copper and Silver

The use of Ag_2S as a solid solution electrode has been exploited for many years particularly by Wagner and Rickert (4) to provide many elegant experiments which illustrate the various measurement techniques possible in solid state electrochemistry. The high temperature form of Ag_2S can be deformed relatively easy by small stresses and so excellent solid/solid interfaces (e.g. Ag_2I/Ag_2S or $RbAg_4I_5/Ag_2S$) can be prepared. This facility when combined with the remarkably high D_{Ag} $(10^{-2} - 10^{-1}$ cm^2/s) ensures that very high currents can be passed across this solid/solid interface. However, although the transport properties of Ag_2S are excellent it has a poor storage capacity because of the limited range of stoichiometry in Ag_2S ($Ag_{2.000}S$ - $Ag_{2.002}$). Similar comments apply to other silver electrodes, e.g. Ag_2Se, and Ag - S-Se - P - O solid solutions (20).

The copper chalcogenides can also exhibit high electronic and ionic conductivities. Moreover, the ranges of composition available for the incorporation of copper is usually higher than in the corresponding silver compounds and of course copper has the added advantage of being lighter and cheaper. The thermodynamics of the Cu - S system have been studied at ambient temperatures by Rickert (21) using aqueous electrolytes and the measured emf/composition curve is shown in Fig. 6. A notable feature is the relatively flat potential/composition curve over much of the range studied. Also using an electrochemical technique, Etienne (22) has reported a value for the copper diffusion coefficient in Cu_2S of approx. 10^{-10} cm^2/s. This low value would prevent Cu_2S being used as an S.S.E. at ambient temperatures but at higher temperatures the copper mobility is certainly sufficient.

d. Incorporation of Alkali Ions

i) <u>Transition metal bronzes with one dimensional channels</u>. The alkali metal tungsten and vanadium bronzes (e.g. Na_xWO_3 and $Na_xV_2O_5$) have structures which contain channels of various diameters in which the alkali ions reside. The channels in the tetragonal and hexagonal structures should be large enough to permit translational motion of the alkali ions and so these materials have been investigated as possible S.S.E. as they also exhibit high electronic conductivities. Thermodynamic and kinetic measurements made with cells of the type :

$$\text{Na / Propylene carbonate (NaI) / Na}_x\text{WO}_3$$
$$\text{or} \quad \text{Na / beta alumina} \quad / \quad \text{Na}_x\text{WO}_3$$

quickly revealed (23) that although the $\Delta \bar{G}_{Na}$ values were very promising (see Fig. 8), the chemical diffusion coefficients were very low ($\sim 10^{-15}$ cm^2/s) which prevented their application as S.S.E. It appears that transport along the tunnels can be impeded by the presence of impurity ions, or more likely crystallographic imperfections exist which effectively block the tunnels. High resolution electron micrographs (24) of other tetragonal bronze structures reveal complex distorted arrangements of the WO_6 oxtahedra in contrast to the simple idealised structure originally proposed for these materials. It is relevant to note that disappointing results have also been obtained for ionic transport in other structures having relatively larger tunnels such as hollandite ($K_{1.6}Mg_{0.8}Ti_{7.2}O_{16}$).

ii) <u>Graphite and Fluorographite</u>. It is well known that graphite can intercalate alkali metals and emf data (25) for the potassium graphite system are included in Fig.8. The maximum potassium content is represented by the formula C_8K, and attention is drawn to

the high activity of potassium in this phase. In contrast the in-
tercalation compound C_8CrO_3, prepared by Armand (26), behaves as a
very deep electron sink and consequently lithium atoms incorporated
into this structure have a very low activity ($\sim 3.9V$). The range
of stoichiometry in this compound, $Li_x C_8CrO_3$ appears to be very
large and the lithium diffusion coefficient approaches 10^{-6} cm^2/s.
The principal problem with this type of compound is the additional
incorporation of solvated species into the compound when the material is used as an electrode in electrochemical cells incorporating
organic electrolytes such as propylene carbonate. It should also
be noted that the fluorographite $(CF)n$ electrode probably behaves
as a solid solution electrode during the initial stage of its reaction with lithium. This suggestion (27) could explain why the
voltage developed by $Li_x \cdot CF_x$ cells is considerably less than that
calculated for the Li - LiF couple; and also why the discharge
characteristics are markedly influenced by the organic electrolyte
used in the cells (28).

iii) <u>Dichalcogenides</u>.   The metal dichalcogenides (MX_2) of
Groups IV, V, VI adopt a layer structure (CdI_2) in which slabs of
XMX layers are held together by Van der Waals bonding. Recent
work (29, 30, 31) has confirmed that alkali metals can be rapidly
and reversibly intercalated within the Van der Waals layer at ambient temperatures and emf/composition results (31) for the incorporation of sodium into single crystals of Ti_xS_2 are depicted
in Fig. 7. The emf results shown were obtained after 3-4 cycles
and were slightly dependant upon the Ti/S ratio in the original
crystals. The Ti/S ratio, however, had a profound influence upon

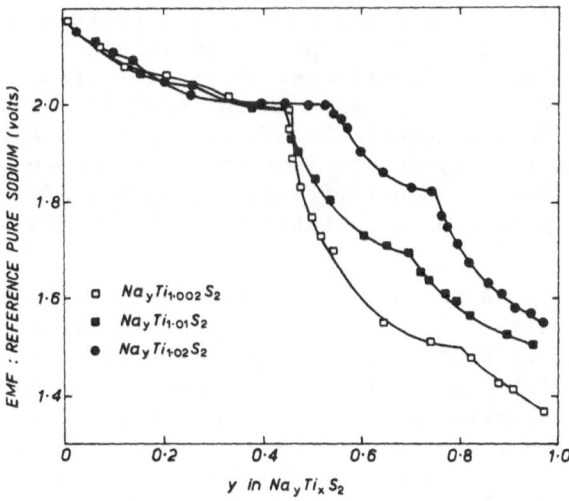

FIG. 7. Potential/composition curves for $Na - Ti_xS_2$ system.

the alkali metal chemical diffusion coefficients as measured by
electrochemical techniques. The excess titanium residing in the
normally vacant Van der Waals layer must develop strong bonds between the adjacent sulphur layers as well as blocking sites for
the transport of the alkali metal.

Lithium can also be rapidly and reversibly electrointercal-
ated into TiS_2 (31). It is interesting to note that the voltage
differential for the incorporation of 1 g-atom of lithium is only
0.5 V (2.4 - 1.9V) and thus the theoretical specific energy density
(500 WH/Kg) is even more favourable than that calculated for the
Na - TiS_2 system (400 WH/Kg). It is interesting to note that TiS_2
has recently been successfully employed (32) as a reversible solid
solution cathode in the high temperature cell LiAl/KCl -LiCl/TiS_2.
Niobium diselenide has also been used (29) as a solid solution
electrode for lithium in similar cells using organic electrolytes.

The transition metal trichalcogenides MX_3 also intercalate
lithium (29, 33) to form compounds of the type Li_3MX_3. At present
detailed emf and transport data for these compounds are not available but certain of these components appear to be able to function
as cathodes in secondary cells (29) using propylene carbonate as
the electrolyte.

iv) Beta ferrites. Potassium beta ferrite ($K_{Hx}Fe_{11}O_{17}$) is one of
the isomorphous iron analogues of beta alumina and exhibits electronic as well as ionic conductivity. Additional potassium ions
can be incorporated into the structure and charge is compensated by
the formation of Fe^{2+}. Emf/composition relationships have been obtained using either propylene carbonate (25 - 60°C) or beta alumina
(50 - 300°C) as the electrolyte phase (9). At this temperature the
current density ($\sim$ 100 uA/cm^2) used in the coulometric titrations
was limited by transport of the potassium ions across the ceramic
electrolyte/electrolyte interfaces. These interfaces are difficult
to prepare because the components do not exhibit the plastic de-
formation which is characteristic of, for example, AgI/Ag_2S interfaces. Although this system is interesting insomuch as one can
investigate the ionic conductivity as a function of charge carrier
density it is not of course suitable for a S.S.E. in a high energy
density battery.

vi. Other oxides. A veriety of non-stoichiometric oxides and
sulphides have been employed as cathodic depolarizers in organic
electrolyte primary batteries usually as additions to graphite and
other positive plate materials. There have been reports in which
certain of these materials e.g. MoO_3, MnO_2 V_2O_5 (e.g. 34) have been
the only constituent present in the electrode and typical results
are shown in Fig. 8. The conventional electrochemist's interpretation of these results is to assume a cathode reaction of the type:

$$MoO_3 + 2Li \longrightarrow MoO_2 + Li_2O,$$

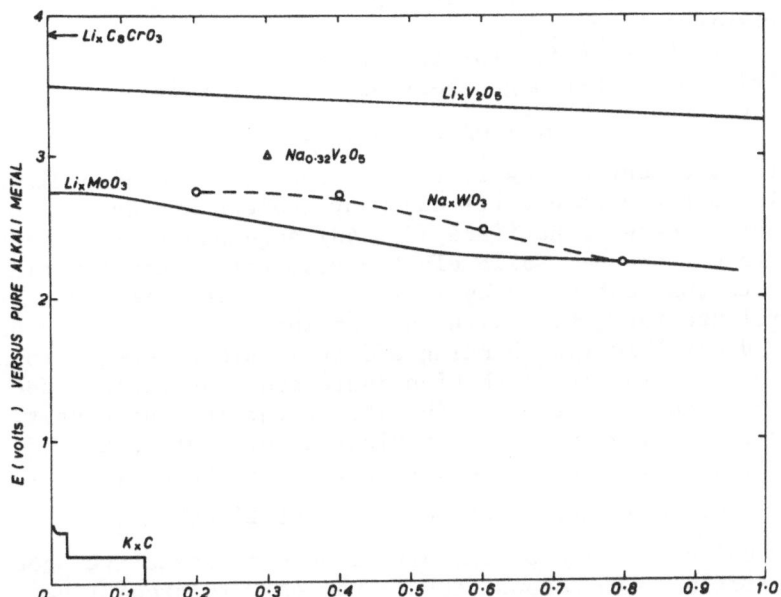

FIG. 8 Potential/composition curves for alkali metal incorporated
in selected non-stoichiometric electrode materials.

However, thermodynamic data indicates that the maximum voltage to
be expected for reactions of this type is about 2V. A more plausible
explanation is to assume that lithium is dissolving in the MoO_3
host lattice and the cell voltage reflects the activity of lithium
in this non-stoichiometric phase. The transport of lithium in these
compounds appears to be sufficiently high for their application as
S.S.E. although there is little information about the ability of
these materials to tolerate repeated cycling. It is also worth
mentioning that V_2O_5 and MoO_3 may be regarded as layer structures
which obviously favour the rapid transport of lithium.

DISCUSSION

The concept of using S.S.E. in secondary batteries is relatively
new and very little data are available to assess their future development. The favourable properties of hydrogen in selected alloys
and of lithium in titanium disulphide are already being exploited
in the development of new battery systems, and it appears likely
that other technologically interesting materials will be prepared.
It is appropriate therefore to conclude with a few comments about
the theoretical models relevant to the specifications suggested in
section 2 for S.S.E.

a. Electronic Conductivity

It is impossible to predict a priori the band structures of

complex oxides and sulphides. The approach of Goodenough (35) based on chemical bonding and crystallography however, does provide a framework which is useful for the preliminary selection of materials likely to exhibit high electronic conductivities.

b. Ionic Transport

The factors controlling fast ion transport have been discussed in the Belgirate conference (1) and many reviews are available (36, 37, 38). Recent experience (39, 40) suggests that it is very difficult to prepare new solid electrolytes with conductivities comparable to that exhibited by $\beta-Al_2O_3$. The restrictions are somewhat relaxed for S.S.E. with the possibility of introducing covalent and metallic type bonding and it should be easier to prepare compounds exhibiting both high ionic and electronic conductivity. It should be noted that the possibility of varying the density of ionic charge carriers in S.S.E. provides an elegant method of testing many theoretical models for fast ion transport.

c. Free Energy/Composition Relationships.

The magnitude of $\Delta \bar{G}$ of the dissolved electro-active species and its variation with composition is not easy to predict because of lack of knowledge of both the relevant electronic levels and the ion-ion interactions. A variety of statistical thermodynamic models have been proposed and recent reviews are available (41)

d. Interfacial Problems

The detailed mechanism of the injection of ions from a liquid or solid electrolyte into a solid solution electrode, and vice-versa, remains obscure and it certainly requires systematic investigation in which the environment of the electro-active species is varied both in the electrolyte and electrode phase. Recent relevant reviews on the topic include (42, 43, 44).

Finally, it should be noted that the present survey has been principally concerned with the application of S.S.E. in secondary batteries. Other applications include the use of S.S.E. in electrochromic passive light display devices. Cells of the type Ag/Rb Ag_4I_5/MoO_3, SnO_2, can be used to reversibly inject ions (H^+, Ag^+, Li^+) into thin films of MoO_3 or WO_3 electrodes. The colour centres produced can be used as a display unit (45) requiring only a small energy input (cf. liquid crystals).

REFERENCES

1. B.C.H. Steele in W. Van Gool (Ed.), *Fast Ion Transport in Solids*, North Holland, Amsterdam, 1973, p. 103.
2. B.C.H. Steele in G.R. Belton and W.L. Worrell (Eds.), *Heterogeneous Kinetics at Elevated Temperatures*, Plenum, New York, 1970, p. 135.
3. J. Hladik (Ed.), *Physics of Electrolytes Volumes 1 and 2*, Academic, New York, 1972.
4. H. Rickert, *Einfuhrung in die Elektrochemie fester Stoffe*, Springer, Berlin, 1973.
5. F.A. Kroger, *The Chemistry of Imperfect Crystals Volume 3*, North Holland, Amsterdam, 1974.
6. H. Zuchner and N. Boes, Ber.d.Bun. Gesellschaft, 76 (1972) 783.
7. W.F. Chu, H. Rickert and W. Weppner in W. Van Gool (Ed.), *Fast Ion Transport in Solids*, North Holland, Amsterdam, 1973, p. 181.
8. T.K. Holstead, J. Solid State Chem., 11 (1974) 114.
9. G.J. Dudley, B.C.H. Steele and A.T. Howe, J. Solid State Chem., to be published.
10. T.L. Martin and R.J. Bones, U.K.A.E.A. Report A.E.R.E. R 4042 1962.
11. B.C.H. Steele and C.C. Riccardi in O. Kubaschewski (Ed.), *Proceedings of the International Symposium on Metallographical Chemistry*, H.M.S.O., London, 1972, p. 123
12. T. Kudo, H. Obayashi and T. Gejo, J. Electrochem. Soc., 122 (1975) 159.
13. V.S. Patent 3,804,674 (16/4/74).
14. P.A. Boter, *Symposium on Novel Electrode Materials*, September 25-26th, Brighton, U.K., Abstract No. 8.
15. M.A. Gutjahr, H. Buchner, K.D. Beccu and H. Säufferer in D.H. Collins (Ed.), *Power Sources 4*, Oriel Press, Newcastle, 1973, p. 79
16. L. Glasser, Chem. Rev., 75 (1975) 21.
17. J. McBreen, Electrochim. Acta, 20 (1975) 221.
18. J.P. Gabano, J. Seguret and J.F. Laurent, J. Electrochem. Soc., 117 (1970) 147.
19. D.M. MacArthur, J. Electrochem. Soc., 117 (1970) 729.
20. T. Takahashi, E. Nomura and O. Yamamoto, J. Appl. Electrochem., 3 (1973) 23.
21. H. J. Mathiess and H. Rickert, Z. Physik. Chemie, N.F. 79 (1972) 315.
22. A. Etienne, J. Electrochem. Soc., 117 (1970) 870.
23. B.C.H. Steele in A.R. Cooper and A.H. Heuer (Eds.), *Mass Transport Phenomena in Ceramics*, Plenum, New York, 1975, p. 269.
24. S. Iijima and J.G. Allpress, Acta Cryst. A 30 (1974) 22.
25. S. Aronson, F.J. Salzano and D. Bellafiore, J. Chem. Phys., 49 (1968) 434.

26. M.B. Armand in W. Van Gool (Ed.), *Fast Ion Transport in Solids*, North Holland, Amsterdam, 1973, p. 665.
27. M.S. Whittingham, J. Electrochem. Soc., 122 (1975) 526.
28. R.G. Gunther in D.H. Collins (Ed.), *Power Sources 5*, Academic, London, 1975, p. 729.
29. J. Broadhead and F.A. Trumbore in D.H. Collins (Ed.), *Power Sources 5*, Academic, London, 1975, p. 661.
30. M.S. Whittingham and F.R. Gamble, Mat. Res. Bull, 10 (1975) 363.
31. D.A. Winn and B.C.H. Steele, Mat. Res. Bull., 11 (1976) to be published.
32. D. Inman and Y.E.M. Mariker, *Symposium on Novel Electrode Materials*, September 25-26th, Brighton, U.K., Abstract No. 2.
33. R.R. Chianelli and M.B. Dines, Inorg. Chem., 14 (1975) 2417.
34. F.W. Dampier, J. Electrochem. Soc., 121 (1974) 656.
35. J.B. Goodenough in H. Reiss (Ed.), *Progress in Solid State Chemistry Volume 5*, Pergamon, Oxford, 1971, p. 145.
36. W. Van Gool, Am. Rev. of Materials Science, 4 (1974) 311.
37. B.C.H. Steele and G.J. Dudley, *M.T.P. International Review of Science, Series Two, Inorganic Chemistry Volume 10*, Butterworths, London, 1975, p. 181.
38. M.S. Whittingham, Electrochim. Acta, 20 (1975) 575.
39. J. Singer, H. Kautz, W. Fielder and J. Fordyce, NASA Technical Memo, TMX - 71753, May, 1975.
40. W.L. Roth and O. Muller, NASA Technical Memo, CR - 134610, April, 1974.
41. B.E.F. Fender, *M.T.P. International Review of Science, Series One, Inorganic Chemistry Volume 10*, Butterworths, London, 1972, p. 243.
42. D.O. Raleigh in A.J. Bard (Ed.), *Electroanalytical Chemistry, Volume 6*, Dekker, New York, 1973, p. 88.
43. A.D. Franklin, J. Am. Ceram. Soc., 58 (1975) 465.
44. M. Kleitz (Ed.), *Electrode Processes in Solid State Ionics*, Reidel, Dordrecht, to be published.
45. I.F. Chang, B.L. Gilbert and T.I. Sun, J. Electrochem. Soc., 122 (1975) 955.

THE ELECTRICAL DOUBLE LAYER ON SILVER IODIDE

J. Lyklema

Laboratory for Physical and Colloid Chemistry
Agricultural University
De Dreijen 6, Wageningen, Netherlands

INTRODUCTION

Silver iodide and mercury are very different substances, but they have one feature in common. Both are *model substances* for interfacial electrochemists. Mercury is a favourite object for the study of double layers and electrode kinetics. Its popularity derives, *inter alia* from its high polarizability, its high conductivity and its easily renewable surface. Silver iodide is the classical model substance of the colloid chemist, mainly because durable sols can be made from it. However, it is also a suitable substance for studies on double layers. Various techniques (e.g. potentiometry, direct measurement of double layer capacitance, electrophoresis) are in use to estimate the composition of electrical double layers around dispersed AgI particles. The outcome of such studies is obviously important for the colloid scientist, since AgI sols are *electrocratic*, i.e. they exist only by virtue of electrostatic repulsion between the particles. It also has bearing on the mercury work since comparison of data about double layers from the two systems helps us to assess how general the "mercury results" are.

This paper reviews some basic features and results of the AgI system and, where expedient, compares them with corresponding results of the Hg system.

SOME BASIC ASPECTS

Thermodynamically, the major difference between the two systems is that the Hg solution interface is *polarizable*

whereas the AgI solution interface is *reversible*. In the first
case an outer electric potential ψ_o can be applied across the interface. This is an independent variable and hence occurs in the Gibbs
equation

$$d\gamma = - S^\sigma dT - \sigma_o d\psi_o - \sum_i \Gamma_i d\mu_i \qquad (1)$$

where S^σ is the interfacial entropy, σ_o the surface charge and the
other symbols have their usual meaning. The summation extends in
principle over all components. Pressure is kept constant.

For the AgI solution interface, by contrast, ψ_o is *not* an
independent variable. Its value is the (dependent) consequence of
the amounts adsorbed of the *potential – determining (p.d.) ions*
Ag^+ and I^-. Hence, in this case

$$d\gamma = - S^\sigma dT - \sum_i \Gamma_i d\mu_i \qquad (2)$$

where the summation again extends over all components, the p.d.
electrolytes included. For example, in an aqueous solution of
$AgNO_3$, KI, KNO_3 and an organic adsorptive A

$$d\gamma = - S^\sigma dT - \Gamma_w d\mu_w - \Gamma_{AgI} d\mu_{AgI} - \Gamma_{AgNO_3} d\mu_{AgNO_3} - \Gamma_{KI} d\mu_{KI} -$$
$$- \Gamma_{KNO_3} d\mu_{KNO_3} - \Gamma_A d\mu_A \qquad (3)$$

The chemical potentials are not all independent. Firstly, the
Gibbs–Duhem relation in the solution allows us to eliminate, for
instance, μ_w, replacing any Γ_i by

$$\Gamma_i^{(w)} = \Gamma_i - \frac{x_i}{x_w} \Gamma_w \qquad (4)$$

where x_i is the mole fraction of i. Secondly, by virtue of the
chemical equilibrium $\quad KI + AgNO_3 \rightleftarrows KNO_3 + AgI$ we have

$$\mu_{KI} + \mu_{AgNO_3} = \mu_{KNO_3} + \mu_{AgI} \qquad (5)$$

allowing the elimination of the chemical potential of one of the
p.d. electrolytes. For example, if μ_{KI} is eliminated

$$d\gamma = - S^\sigma dT - (\Gamma_{AgI}^{(w)} + \Gamma_{KI}^{(w)}) d\mu_{AgI} - (\Gamma_{AgNO_3}^{(w)} - \Gamma_{KI}^{(w)}) d\mu_{AgNO_3}$$
$$- (\Gamma_{KNO_3}^{(w)} + \Gamma_{KI}^{(w)}) d\mu_{KNO_3} - \Gamma_A^{(w)} d\mu_A \qquad (6)$$

The third term on the right can be rewritten in terms of charge
and potential. The charge on the AgI surface can be defined as

$$\sigma_o \equiv F (\Gamma_{AgNO_3}^{(w)} - \Gamma_{KI}^{(w)}) \qquad (7)$$

With excess electrolyte, the change in the chemical potential of $AgNO_3$ can be written as

$$d\mu_{AgNO_3} = d\mu_{Ag+} + d\mu_{NO_3^-} \sim d\mu_{Ag+} = d\psi_o/F$$

according to Nernst's law. Hence

$$d\gamma = -S^\sigma dT - \sigma_o \, d\psi_o - (\Gamma_{AgI}^{(w)} + \Gamma_{KI}^{(w)}) \, d\mu_{AgI}$$

$$- (\Gamma_{KNO_3}^{(w)} + \Gamma_{KI}^{(w)}) \, d\mu_{KNO_3} - \Gamma_A^{(w)} d\mu_A \qquad (8)$$

of which usually the variant at constant T is needed ($d\mu_{AgI} = 0$)

$$d\gamma = -\sigma_o d\psi_o - \Gamma_{K+}^{(w)} d\mu_s - \Gamma_A^{(w)} d\mu_A \qquad (9)$$

realizing that $\Gamma_{K+}^{(w)} = (\Gamma_{KNO_3}^{(w)} + \Gamma_{KI}^{(w)})$. In eq (9) subscript s stands

for "salt" and $\Gamma_{K+}^{(w)}$ is the total amount of K^+ adsorbed; this can have either a negative or a positive value.

The most striking feature of eqs. (8) and (9) is, of course, the occurrence of the $-\sigma_o d\psi_o$ term, re-establishing the formal analogy with mercury (eq. (1)). Many analogies between the two systems can ultimately be attributed to the similarity between eqs. (9) and (1) (after elaboration of the sum in the usual way).

EXPERIMENTAL EVALUATIONS

Two problems are involved: experimental estimation of important parameters of double layers, such as the various surface excesses Γ, and identification of the parameters obtained with the corresponding ones in the thermodynamic equations. The latter problem is, of course, not typical of the AgI system.

With AgI suspensions, one has usually enough interfacial area available to measure surface excesses analytically. Because of eq. (7) the most important of them is obviously $(\Gamma_{AgNO_3} - \Gamma_{KI})$. This

quantity can be determined by potentiometric titration of the suspension with KI or $AgNO_3$ in a cell with an AgI and a reference electrode[1-4]. Recently, this technique has been automated in our laboratory by Mr A. de Keizer. The titration produces the excess $(\Gamma_{AgNO_3} - \Gamma_{KI})$ in mol kg^{-1} (relative to an arbitrary reference

point) as a function of pAg ($= -\log a_{Ag+}$) or pI, i.e. it produces a kind of adsorption isotherm. Three steps are needed to transform these data into a σ_o/ψ_o plot.

(1) The excess must be made absolute, requiring determination of the *point of zero charge* (pzc). For AgI, this is a problem of its own since electrocapillary data are not available. Van Laar was the first to study this systematically [2,5]. Nowadays the pzc is found by estimating the pAg at which σ_O does not depend on the activity a_s of the indifferent electrolyte. If there is no specific adsorption, this pAg value may be taken as "the" pzc, pAg^o. Actually, the procedure boils down to finding the point where the *Esin-Markov coefficient* β is zero [6]. Using eq. (9)

$$\beta = \left(\frac{\delta\sigma_o}{\delta\mu_s} \right)_{pAg} = \left(\frac{\delta\sigma_o}{\delta\mu_s} \right)_{\psi_o} = \left(\frac{\delta\Gamma_{K^+}^{(w)}}{\delta\psi_o} \right)_{\mu s} = 0 \ (pzc) \qquad (10)$$

Once pAg^o is found in the absence of specific adsorption, it can be easily determined in its presence by plotting the isotherm relative to the one without specific adsorption. In this way, we have found how the pzc shifts with adsorption of simple ions [4], uncharged organic additives [7,8], charged organic substances [9], and polymers [10]. Also its temperature dependence has been established [11, 12]. These shifts are as a rule qualitatively like the corresponding ones for mercury, a point to which we shall return below.

(2) Adsorbed amounts in mol kg^{-1} can be converted into values of Γ in $eq.m^{-2}$ if the *specific surface area* A_S is known. For mercury, this is no problem, but for dispersed particulate matter, it is. The particles are usually irregular and tend to 'age' upon standing. Direct measurement of average particle size underestimates A_S; adsorption from solution leaves the open question what molecular cross-section to assign to the adsorbate; and gas adsorption has the disadvantage that the particles need to be dried, with the inherent risk of altering the surface. We have found that negative adsorption is a suitable alternative, having none of these disadvantages. This will be discussed below.

Another electrochemical method is to equalize the differential capacitances C (in $\mu F\ cm^{-2}$) of AgI and Hg at the pzc in 10^{-3} mol dm^{-3} KNO_3. The double layer is then mainly diffuse and hence should be independent of the nature of the charge carrier. This approach, first put forward by Mackor [3], gives the same outcome as the negative adsorption method.

(3) Absolute potential differences between dissimilar phases are inaccessible quantities. It is customary to define the surface potential ψ_o through

$$\psi_o \equiv -RT(pAg - pAg^o)/F = RT(pI - pI^o)/F \qquad (11)$$

which implies that ψ_o is the Galvani potential difference with reference to the same at the pzc.

Points 1) - 3) summarize the main steps towards σ_o/ψ_o curves. However, there are a number of precautions that must always be taken. Most of them concern the fact that the solid-solution inter-face is not renewable and that a complete titration may take several days. For instance, the chemicals and water must be very pure and the suspension must be 'aged' before titration. For details, see Ref. 4) and 7). If the appropriate precautions are taken, repro-ducibilities of 0.1 μC cm^{-2} can be reached. In view of the diffi-culties inherent in working with particulate systems, this is quite satisfactory, though the precision is about a factor of 10 less than for mercury.

Double-layer capacitances on AgI can also be obtained directly, using an AC-bridge. The obvious problem is of course, how to deal with the Faraday current. This is more or less equivalent to the problem how to extrapolate as a function of frequency. After initial attempts by Oomen and Barchatky, reliable data have been obtained by Engel[13], Pieper and de Vooys [14]. The σ_o/ψ_o curves, obtained by this technique, agree satisfactorily with those from titration, confirming the correctness of both, including the sur-face area determination.

At present, the titration technique has a more routine charac-ter, especially after its automation. The results discussed below have all been obtained by titration.

DOUBLE LAYER IN THE PRESENCE OF SIMPLE ELECTROLYTES

Figure 1 gives a typical example. The range of surface poten-tial that can be covered is about 0.4 V, i.e. it is much smaller than on mercury. The general impression is like that for mercury, even if the double layer is compact. Compare the 10^{-1} M KF curve with the 10^{-1} M NaF one on mercury, taken form Grahame's capacitance data[15] after integration. The main difference is that on the extreme negative side silver iodide surfaces can be less well covered by I^- than mercury surfaces with electrons.

If the curves are differentiated, to give C/ψ_o curves, the splitting-up into two capacitances in series works satisfactorily.

A small shift on the pzc to the right is observed with in-creasing concentration of KNO_3. It suggests some specific adsorption of NO_3^-. The fluoride ion does not adsorb specifically. For anions, the order is

$$F^- < ClO_4^- < NO_3^- < SO_4^{2-} < HPO_4^{2-}$$

Towards the positive side, anion specificity increases markedly and capacitances progressively exceed the corresponding ones on

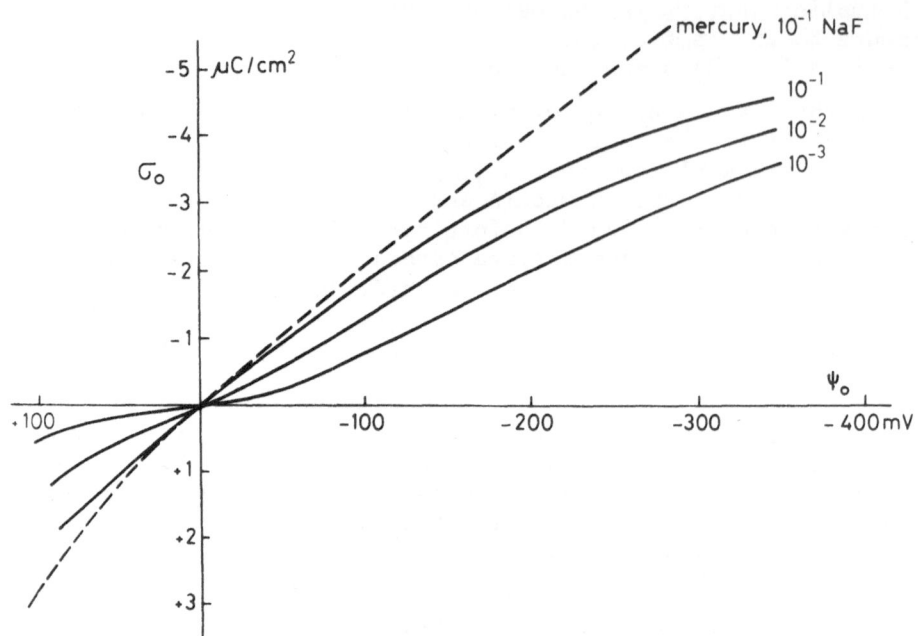

Fig.1   Electrical Double Layer on Silver Iodide in KF . For
comparison, the 10^{-1} mol dm^{-3} NaF curve for mercury is
also given. T = 25°C ψ_{o} is expressed with reference
to the pzc in 10^{-3} mol dm^{-3} KF or NaF.

mercury, probably because of association of the anion and the
adsorbed p.d. silver ions.
     Another typical feature of the silver iodide system is the
pronounced *alkali-ion specificity* at negative potentials. Table 1
illustrates this. On mercury, the mutual differences are of the
order of a few per cent. Such subtle trends would be experimentally
not detectable with silver iodide. It is only because the specificity is of the order of tens of per cent on silver iodide that it
can be measured. The absolute values of the capacitances are so
low that interpretation in terms of a model with the cation in
immediate contact with the surface is not appropiate. If the
counterions remain at some distance from the surface proper, say
with one solvent molecule between them, one is led to an interpretation in entropic rather than enthalpic terms. There are several
circumstantial indications for some structure formation in the

TABLE 1

Comparison of differential capacitances (in μF cm^{-2}) on mercury [39] in 10^{-1} mol dm^{-3} chlorides with those on silver iodide [4] in 10^{-1} mol dm^{-3} nitrates. T = 25°C. ψ_o = -220 mV (at this potential, the nature of the anion is irrelevant).

	Li^+	K^+	Rb^+
mercury	19.45	19.80	20.00
silver iodide	11.0	11.9	13.2

water adjacent to AgI surfaces, probably because of hydrophobic interaction. Cloud seeding experiments may be mentioned in this respect. The extent to which this structure is affected by counterion penetration then determines the entropic contribution to the free energy of sorption.

For further analysis, one rewarding approach is to exploit the fact that the stability of AgI sols has been extensively studied. This is one of the few advantages of the AgI system over the mercury system. The essential point is that sol stability is determined by the overlap of the diffuse parts of the interacting double layers. From stability data, the diffuse double-layer charge σ_d or the outer Helmholtz plane potential ψ_d can be evaluated. For silver iodide sols, a marked *lyotropic sequence* in the stability is observed [16]. The order is such that with the stronger adsorbing counterion (Rb^+ for the alkaline ions) the stability is lowest because the least charge is then available in the diffuse layer. For further analysis, two approaches are open: measurement of the *rate of flocculation* or of the *critical coagulation concentration* c_c. The first method is generally more informative, but the second one can be more easily formulated. According to DLVO theory [17,18], c_c is related to ψ_d, the Hamaker constant $A_{1(2)}$ of particle attraction and the relative dielectric constant ε of the medium by

$$c_c = \text{const.} \; \frac{\varepsilon^3 [\tanh(ze\psi_d/kT)]^4}{A^2_{1(2)} z^6} \sim \text{const.} \; \frac{\varepsilon^3 \psi_d^4}{A^2_{1(2)} z^2} \qquad (12)$$

where the constant follows from the theory. Eq.(12) underlines that c_c is sensitive to ψ_d, or conversely, the evaluation of ψ_d from c_c is a relatively reliable procedure. The various assumptions underlying this method have been discussed by Lyklema & de Wit [19]. We calculated σ_d from ψ_d, finding the Stern charge σ_m, from charge balance, $\sigma_m = - (\sigma_o + \sigma_d)$. We concluded that the Stern equation [20],

relating σ_m to c_s worked reasonably well with a specific free energy of adsorption of about 4 kT for K^+, which is comparable with
$T\Delta S^o_{tr}$ if ΔS^o_{tr} is the standard entropy of transfer of the ionic
species from water to other media[21].

The method of estimating the counter charge distribution over
σ_m and σ_d by combining titration and stability data has proved
extremely rewarding.

IONIC COMPONENTS OF CHARGE AND NEGATIVE ADSORPTION

As with mercury, it is possible to obtain the ionic components
of charge

$$\sigma_\pm = \pm z_\pm F\Gamma_\pm \tag{13}$$

from the Esin-Markov coefficients. In formula[22] for $z_\pm = 1$

$$\sigma_\pm = \pm F \int (\frac{\delta\sigma_o}{\delta\mu_s})_{\psi_o} d\psi_o - \frac{\sigma_o}{2} \tag{14}$$

In applying this equation to experimental data, a serious drawback
of the AgI system emerges. The range of potential and the accuracy
of the data do not allow the establishment of the platform value of
the negatively adsorbed species, say $\Gamma_{NO_3^-}$ on negative silver
iodide. This platform value is needed to find the integration constant in eq. (14), or, for that matter, the specifically adsorbed
amount at the pzc. Consequently, from eq. (14) σ_+ can be derived
only with reference to the pzc. Examples can be found in Ref.[22].

There is, however, also a point in favour, namely that the
negatively adsorbed amount Γ_- can be enough to become experimentally detectable, at least if certain conditions are met. The significance is that in the platform region Γ_- follows from theory in
eq. m^{-2} and from experiment in eq.g^{-1} , hence *negative adsorption*
can be used to estimate specific surface areas. The importance of
this possibility is not easily overestimated. Not only is it difficult to measure the 'electrochemical' area otherwise (see above)
but negative adsorption is also important in a number of different
fields, such as salt-sieving, membrane permeability and light scattering [23]. Moreover, experimental verification of Gouy theory is
possible. Work on negative adsorption with clays by Schofield *et*
al. [24-26] deserves note here. Negative adsorption on clays is
relatively easy because of their high areas (50 - 100 times that of
AgI suspensions) and high and constant surface charges.

The two main experimental problems inherent in the silver iodide system are
(1) the expelling areas are small, of the order of 1 $m^2 g^{-1}$, hence negative adsorption gives rise to only minute concentration increments,
(2) since the surface must be charged by adsorption of p.d. electrolyte, one always has to work with salt *mixtures*, i.e. a mixture of the p.d. electrolyte and the electrolyte whose negative adsorption is to be measured.
These problems call for careful experimental design. Work by van den Hul has shown that the method works well in $(1 - 1)$ + $(1 - 2)$ electrolyte mixtures and negative AgI surfaces, e.g. in a mixture of KI + K_2SO_4. The KI is used for charging and the increment Δc of the sulphate concentration due to the change is measured [27]. The following expression is found for the total expelling area S_t

$$S_t = \frac{V_t \frac{\Delta c_1}{c_1} \left(\frac{2e^2 N_{av} c_1}{1000 \varepsilon \varepsilon_o kT} \right)^{\frac{1}{2}}}{(3 + \frac{c_2}{c_1})^{\frac{1}{2}} - \frac{c_2}{2c_1} \ln\left\{ 2 + \frac{c_2}{2c_1} + (3 + \frac{c_2}{2c_1})^{\frac{1}{2}} \right\} + \frac{c_2}{2c_1} \ln\left\{ 1 + \frac{c_2}{2c_1} \right\}}$$

(15)

where c_2 is the equilibrium concentration of KI and V_t the total liquid volume. Under favourable conditions, the KI concentration can be chosen in such a way that on one hand enough I^- adsorbs to attain platform values of Γ_-, whereas on the other hand, so little KI remains in solution that $c_2 \ll c_1$. This would reduce the denominator to simply $\sqrt{3}$. Generally, also for other electrolyte mixtures, one can write

$$S_t = \frac{A\Delta V}{c_1^{\frac{1}{2}}}$$

(16)

where ΔV is the "expelled volume" = $V_t \Delta c_1$ and A is a constant, depending, for instance, on the nature of the mixture. A can be found from theory. Eq.(16) illustrates that negative adsorption can be envisaged as the complete expulsion from a volume ΔV around each particle, having an area S_t and a thickness $c_1^{\frac{1}{2}} A^{-1}$. This thickness is several times the Debye length κ^{-1}, implying that κ^{-1} is the 'yardstick' by which the surface area is measured. Experimental examples are given in Fig. 2 and 3. Fig. 2 shows that at sufficiently negative potentials (if this condition be fulfilled, the precise value of ψ_o is immaterial), eq. (16) applies very well. The expelling area is obtained as S_t = tgα where α is the angle with the abscissa axis. Fig. 3 gives the dependence of S_t obtained in this way on potential. S_t was independent of ψ_o if ψ_o was more

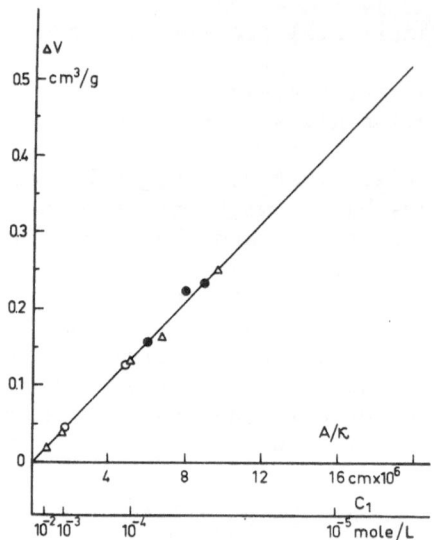

Fig. 2   Dependence of negative adsorption on substance concen-
          tration. $\psi_o \sim -300$ mV.   ($\bullet$, o) Negative adsorption of
          sulphate, (Δ) the same for phosphate.

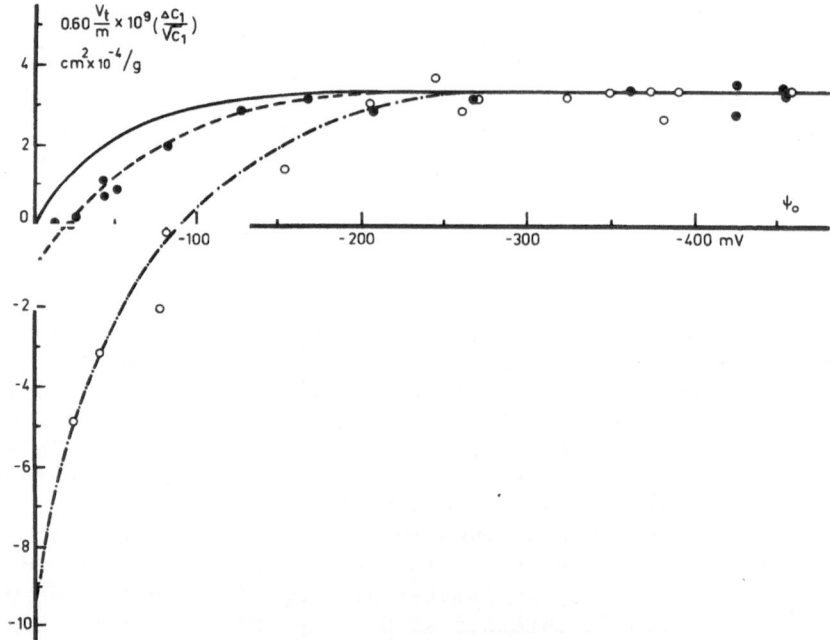

Fig. 3   Dependence of negative adsorption on surface potential.
          (o) sulphate, (Δ) phosphate. Drawn curve, computed from Gouy
          theory assuming no specific adsorption.

negative than - 200 mV for SO_4^{2-} and more negative than - 300 mV for HPO_4^{2-}. This constant value is the wanted area. If ψ_0 is not sufficiently negative, the maximum of Γ_- is no longer attained. Moreover, positive specific adsorption starts to occur, for phosphate stronger than for sulphate.

For silver iodide, the specific area obtained by negative adsorption agrees within 10% with the area obtained from capacitance but is higher than the BET area by a factor of 3 [28], perhaps because of area loss during drying. Incidentally, for a porous substance, like precipitated silica, negative adsorption from the pores is incomplete because of double layer overlap. In that case S(negative adsorption) < S (BET), the difference between the two being a measure of porosity [29].

EFFECT OF TEMPERATURE

As a counterpart to Grahame's work on the temperature dependence C(T) with mercury [30], Lyklema & Vincent [11, 12] studied the temperature dependence σ_o (T) on silver iodide. With increasing temperature, the experiments become progressively less reproducible, for instance because of the increasing solubility of the silver iodide. However, some firm conclusions could be drawn. One of them is that, as expected, cation specificity decreases with increasing temperature, both in surface charge and in stability. At 65 °C or more, all specificity was lost. The double layer was then apparently entirely diffuse, rendering AgI sols at high temperature an interesting model colloid.

From the temperature coefficient $(\delta\sigma_o/\delta T)_{pAg}$, the surface excess entropy could be calculated [12, 31]. Eq. (3) is the starting point. The elaboration is somewhat laborious because all chemical potentials are temperature-dependent, whereas the measurements are performed at constant concentrations, rather than at constant chemical potentials. The following pair of equations is obtained

$$(\frac{\delta\eta^\sigma}{\delta\psi_o})_{T,a_s} = - s^\alpha (\frac{C_{I^-}}{F}) + (\frac{\delta\sigma_o}{\delta T})_{pAg,a_s} +$$

$$- (\frac{\delta\sigma_o}{\delta\mu_s})_{pAg,T} (-s^o_s + R\ln a_s) - (\frac{C}{F})(-s^o_{Ag} + R\ln a_{Ag^+}) \qquad (17)$$

$$(\frac{\delta\eta^\sigma}{\delta\psi_o})_{T,a_s} = - s^\alpha(\frac{C_{Ag^+}}{F}) + (\frac{\delta\sigma_o}{\delta T})_{pI,a_s} +$$

$$+(\frac{\delta\sigma_o}{\delta\mu_s})_{pI,T} (-s^o_s + R\ln a_s) + (\frac{C}{F})(-s^o_{I^-} + R\ln a_{I^-}) \qquad (18)$$

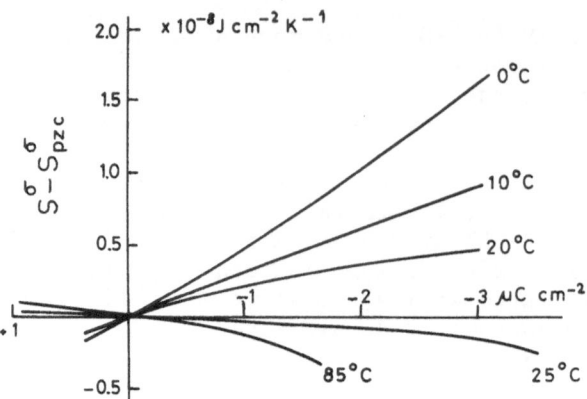

Fig. 4   Surface excess entropies for the double layer on silver
          iodide, relative to the pzc.

where

$$\eta^\sigma = S^\sigma - s_{AgI}\Gamma_{AgI} - s_w\Gamma_w,$$

the small s standing for molar entropy. C_{I^-} and C_{Ag^+} are the contributions of I^- and Ag^+ to the total capacitance, respectively. These
equations are equivalent but the precision of the two is different
in different ranges of potential. The obtained entropies are with
reference to the same at the pzc (Fig. 4). As for mercury [32],
they provide information on the extent of ordering of the water
near the surface . At and above room temperature, entropy is higher
on positive surfaces than on negative ones. This fits into the
general picture of much stronger specific adsorption of anions than
that of cations, with a concomitant lower ordering at the positive
side. At low temperatures, this trend is reversed. A possible
explanation is that some adjacent water pre-freezes and NO_3^- is
expelled. This is corroborated by a lowering of the double layer
capacitance [12].

EFFECT OF ADSORBED ORGANIC MOLECULES

The modification of the double layer by n- butanol has been
extensively studied [7, 31, 33]. Fig. 5 presents some results. Such
curves are typical, since analogous curves are obtained with other
butanols,[7] propanol, [7] ethylene glycol [8] and even with polymeric
alcohols [10]. The three main features are:

1) Adsorption of the organic substance reduces the slopes of the curves, i.e. it leads to a reduction of the differential capacitance
2) The pzc shifts to the left, that is, to the positive side
3) All curves pass through a (usually well-defined) common intersection

These three features are also observed with mercury [34, 35]. The differences between the two systems are quantitative rather than qualitative, pointing to similar molecular interpretations.

The reduction in capacitance could be due to a reduction of the integral capacitance $\varepsilon_i \varepsilon_o /d$ of the inner layer and/or displacement of specifically adsorbed counterions by the organic molecule.

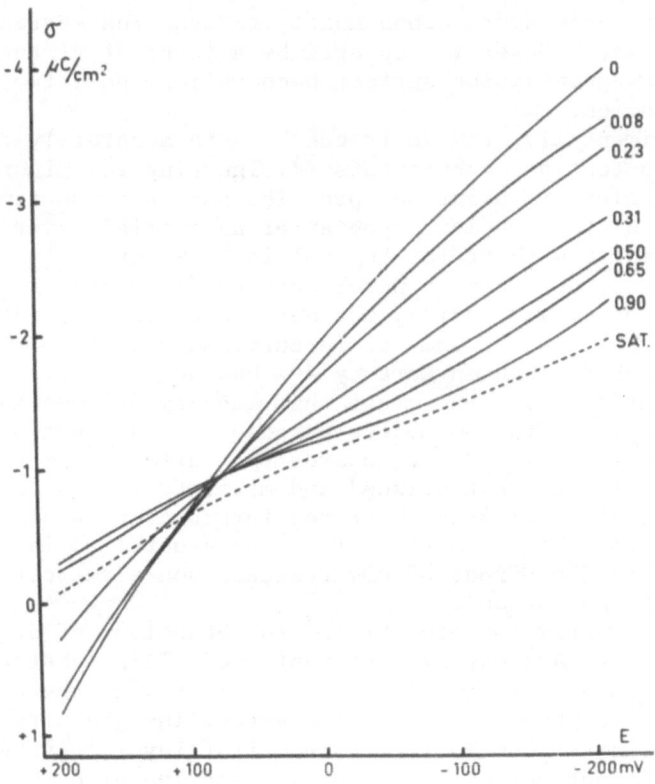

Fig. 5   Influence of n- butanol on the double-layer charge of silver-iodide suspensions. Electrolyte 10^{-1} mol dm^{-3} KNO_3.

Potentials E measured with reference to a 0.1 N KCl calomel electrode.

Here again one has an example where stability provides useful
additional data: at constant high pAg, i.e. at almost constant E, σ_o
continually decreases with increasing butanol concentration c_A but
the critical coagulation concentration c_c passes through a
maximum [33]. Increasing stability with decreasing σ_o can only mean
that the countercharge distribution becomes more diffuse, i.e. it
points to desorption of Stern ions. With ethylene glycol (EG), no
such maximum was observed [19]. The EG molecule is less surface-active
than butanol and its presence at the interface is not incompatible
with specifically adsorbed counterions. For Ba^{2+}, the specific free
energy of adsorption even increases with c_{EG}.

The positive shift of pzc can be explained in the same way as
for mercury: in the absence of the organic molecule, the water
dipoles have a net moment with the negative side towards the surface.
These dipoles are displaced by butanol molecules, adsorbing essentially with their hydrocarbon moiety towards the surface [36, 37].
Hence, the water layer is replaced by a layer of virtually zero
moment, consequently the surface becomes more positive with respect
to the solution.

Experimentally, the shift can be more accurately followed by
streaming potential measurements [7], ignoring the minor difference
between isoelectric point and pzc. The pzc as a function of c_A
curves are of the Langmuir type after an initial lower-energy
part. The maximum shift for butanol is + 190 mV. This is of the
same order as, but less than for mercury (+240 mV) [34]. Presumably,
at the pzc on silver iodide, the net moment of the adjacent water
layer is less than the same on mercury. Also with EG [8] a less
pronounced shift than on mercury [38] has been observed, but in this
case the difference is so great that one may presume that, in contrast to mercury, the EG molecules adsorb with a net negative component of their dipole moment towards the surface. All these trends
are now pretty well established and merit further attention. They
can also be of significance for the interpretation of the data on
mercury, especially with respect to the general validity of conclusions, and to the effect of any residual non-electrostatic interaction with the surface.

In the context of pzc shifts, the behaviour of dimethyl-
dodecylamine (DDAO) may also be mentioned. This substance adsorbs
onto silver iodide in two layers, the first with the polar head
towards the surface, the second inverted. The pzc first moves to the
left, then, after completion of the first layer, reverses. Interpretation is not yet complete but may involve also the orientation
of water dipoles with respect to the hydrocarbon moiety of the
adsorbate [9]. This matter has obvious biological significance.
Moreover, the sorption behaviour of substances like DDAO is
generally important in flotation.

The common intersection point (cip) is where the adsorption
of the organic substance is maximum as a function of σ_o or ψ_o.

This follows immediately from

$$(\Gamma_A)_{E_2} - (\Gamma_A)_{E_{cip}} = \int_{E_{cip}}^{E_2} \left(\frac{\delta \sigma_o}{\delta \mu_A} \right)_{E, \mu_s} dE \qquad (19)$$

where E_{cip} stands for the cell potential at the cip. The integral in Eq. (19) follows directly from eq. (8) by cross differentiation and subsequent integration. Because the pzc shifts upon addition of the organic component it is appropiate here to use E as the variable rather than ψ_o. The adsorption maximum is again a common feature for silver iodide and mercury. Again, the differences are quantative rather than qualitative. As a trend with silver iodide, the maxima are found at less negative surface charges than for mercury. (Table 2), perhaps because of the lesser orientation of the water dipoles at the pzc on silver iodide than on mercury. The theory of Damaskin [8] and Frumkin [40] relates the adsorption maximum to the capacitance and the pzc shift. This approach is in a sense phenomenological in that a relation is established with other experimental variables. On a molecular scale, interpretation is by no means complete and the silver iodide work may contribute to a solution. For example, in ethylene glycol $(\sigma_o)_{cip}$ depends markedly on the nature of the counterion: in 10^{-3} mol dm^{-3} Ba(NO$_3$)$_2$, it is -3.8 μC cm^{-2}. This may or may not be related to the higher specific free energy of adsorption of Ba^{2+} [19] as compared with K$^+$.

TABLE 2

Surface charge, $\sigma_{o,cip}$, at maximum adsorption for some alcohols on mercury and silver iodide. Values in μC cm^{-2} in 10^{-1} mol dm^{-3} KNO$_3$ or KCl

	Butanol	Ethylene glycol
Silver iodide	− 1.1 [7]	− 3.2 [8]
Mercury	−2.1 [34]	−3.7 [38]

EFFECTS OF ADSORBED POLYMERS

A systematic study of the modification of the double layer on silver iodide by adsorbing polyvinyl alcohol, PVA, has been made by Koopal [10]. Such double layer measurements proved to be a tool to establish the segment density distribution $\rho(x)$ of the polymer in the adsorbed state, provided additional measurements on electrophoresis were also available. The basic idea is that in 10^{-1} mol dm^{-3} KNO_3 the double layer is compact, and only train segments are 'seen', (trains are sequences of segments actually in touch with the surface). However, in 10^{-3} mol dm^{-3} KNO_3 the double-layer thickness is of the same order of magnitude as the effective polymer layer thickness.

The immobilization of the water by this layer leads to a reduction of the electrokinetic potential ζ and to a reduction of $(\delta\zeta/\delta\psi_o)_{pzc}$ which can be measured. After some interpretation, double layer data in 10^{-1} mol dm^{-3} KNO_3 lead to the amount Γ_{tr} adsorbed in trains and electrokinetic data in 10^{-3} mol dm^{-3} KNO_3 lead to an effective layer thickness. In combination with direct measurement of the total amount Γ_p adsorbed, this allows the construction of a picture for $\rho(x)$ [10, 43]. Such studies are obviously useful for the quantitative description of colloidal stability in the presence of adsorbed polymers [41 - 45] and to check polymer adsorption theories.

On mercury too, adsorption of PVA causes reduction in C [46]. As no electrophoretic data were provided, $\rho(x)$ relationships cannot be established. This is one of the cases where the additionally available information from the dispersed system makes the silver iodide system advantageous over mercury.

CONCLUSIONS

The double layer properties of silver iodide have been reviewed. This system is less amenable to straightforward sophisticated experiments than the mercury system, but this drawback is to a large extent compensated by the possibility of gaining extra information from dispersed systems. Generally double-layer features are qualitatively like the corresponding ones on mercury. The quantitative differences point to differences in the structural properties of the Stern-layer. The AgI-work has led to the development of some techniques of wider applicability. Examples are the measurement of specific surface areas by negative adsorption and the establishment of conformational properties of adsorbed polymers from a combination of double layer and electrophoresis measurements.

REFERENCES

1. H. de Bruyn, Recl. Trav. Chim., 61 (1942) 12.
2. J.A.W. van Laar, Thesis, State Univ. of Utrecht (1952); Dutch patent 79.472, November 15th (1955).
3. E.L. Mackor, Recl. Trav. Chim., 70 (1951) 763.
4. J.Lyklema and J.Th.G. Overbeek, J. Colloid Sci., 16 (1961) 595.
5. J.Th.G. Overbeek in H.R. Kruyt (Ed.), *Colloid Science Vol. I*, Elsevier, Amsterdam, 1952, p. 160.
6. J. Lyklema, J. Electroanal. Chem., 37 (1972) 53.
7. B.H. Bijsterbosch and J. Lyklema, J. Colloid Sci., 20 (1965) 665.
8. J.N. de Wit and J. Lyklema, J. Electroanal. Chem., 41 (1973) 259.
9. A. de Keizer and J. Lyklema, J. Ann. Real. Soc. Quim. Españ., Fis. i Quim., in press.
10. L.K. Koopal and J. Lyklema, Faraday Discuss. Chem. Soc., 59 (1975) in press.
11. J. Lyklema, Discuss. Faraday Soc., 42 (1966) 81.
12. B. Vincent and J. Lyklema, Special Discuss. Faraday Soc., 1 (1970) 148.
13. D.J.C. Engel, Thesis, State Univ. of Utrecht (1968).
14. J.H.A. Pieper, D.A. de Vooys and J.Th.G. Overbeek, J. Electroanal. Chem., 65 (1975) 429.
15. D.C. Grahame, Chem. Revs., 41 (1947) 441.
16. H.R. Kruyt and M.A.M. Klompé, Kolloid-Beheifte, 54 (1942) 484.
17. B.V. Deryagin and L. Landau, Acta Physicochim. URSS, 14 (1941) 633.
18. E.J.W. Verwey and J.Th.G. Overbeek, *Theory of the Stability of Lyophobic Colloids*, Elsevier, Amsterdam, 1948.
19. J. Lyklema and J.N. de Wit, J. Electroanal. Chem., 65 (1975) 443.
20. O. Stern, Z. Elektrochem., 30 (1924) 508.
21. M.H. Abraham, J. Chem. Soc. Faraday Trans. I, 69 (1973) 1375.
22. J. Lyklema, Trans. Faraday Soc., 59 (1963) 418.
23. J. Lyklema and H.J. van den Hul, *Proceedings of the IUPAC Symposium on Surface Area Determination*, Bristol, 1969. Butterworth, London,1970, 341.
24. R.K. Schofield, Nature, 160 (1947) 408.
25. R.K. Schofield and O. Talibuddin, Discuss. Faraday Soc., 3 (1948) 51.
26. R.K. Schofield and H.R. Samson, Discuss. Faraday Soc., 18 (1954) 135.
27. H.J. van den Hul and J. Lyklema, J. Colloid Sci., 23 (1967) 500.
28. H.J. van den Hul and J. Lyklema, J. Am. Chem. Soc., 90 (1968) 3010.
29. Th.F. Tadros and J. Lyklema, J. Electroanal. Chem., 17 (1968) 267.
30. D.C. Grahame, J.Am. Chem. Soc., 79 (1957) 2093.

31. B.H. Bijsterbosch and J. Lyklema, J. Colloid Interfac. Sci.,
    28 (1968) 506.
32. G.J. Hills and R. Payne, Trans. Faraday Soc., 61 (1965) 326.
33. B. Vincent, B.H. Bijsterbosch and J. Lyklema, J. Colloid
    Interfac. Sci., 37 (1971) 171.
34. E. Blomgren, J.O'M. Bockris and C. Jesch, J. Phys. Chem., 65
    (1961) 2000.
35. B.B. Damaskin, A.A. Survila and L.E. Rybalka, Elektrokhimiya,
    3 (1967) 146.
36. J.O'M. Bockris, M.A.V. Devanathan and K. Müller, Proc. Roy.
    Soc., A274 (1963) 55.
37. J.O'M. Bockris, E. Gileadi and K. Müller, Electrochim. Acta,
    12 (1967) 1301.
38. S. Trasatti, J. Electroanal Chem., 28 (1970) 257.
39. D.C. Grahame, J. Electrochem. Soc., 98 (1951) 343.
40. B.B. Damaskin and A.N. Frumkin, J. Electroanal. Chem., 34
    (1972) 191.
41. G.J. Fleer and J. Lyklema, *Proceedings of Fifth International
    Congress on Surface Active Substances*, Barcelona, 1968,
    Ediciones Unidas Vol. 2, (1969) 247.
42. J. Lyklema, *Proceedings of the IUPAC Symposium on Macromole-
    cules*, Jerusalem, 1975, in press.
43. G.J. Fleer, L.K. Koopal and J. Lyklema, Kolloid-Z.Z. Polymer,
    250 (1972) 689.
44. G.J. Fleer, L.K. Koopal and J. Lyklema, Kémiai Közlemények,
    38 (1972) 1.
45. G.J. Fleer and J. Lyklema, J. Colloid Interfac. Sci., 46
    (1974) 1.
46. W. Wojciak and E. Dutkiewicz, Roczniki Chem., 38 (1964) 271.

PREPARATION AND PROPERTIES OF MONODISPERSED METAL HYDROUS OXIDE

LATICES

Egon Matijević

Institute of Colloid and Surface Science and Department
of Chemistry, Clarkson College of Technology
Potsdam, New York 13676, USA

I. INTRODUCTION

Over the past few years considerable theoretical advances have
been made in developing an understanding of the problems of colloid
stability.   In order to test the various models, it is essential
to have colloidal dispersions well defined in terms of particle
size, shape, charge, and other characteristics.   Until recently,
essentially the only such systems have been the monodispersed
organic latexes consisting of spherical particles in micron and
sub-micron sizes.

We have now produced an entirely new family of inorganic
colloidal dispersions with particles of exceedingly narrow size
distribution and of varying shapes, which include spheres, cubes,
cube-octahedra, needles, ellipsoids, etc.   Chemically, all these
systems are metal (hydrous) oxides.

The novel sols are excellent models for the study of colloid
stability, since their surfaces are rather well-defined and because
their charge can be conveniently changed in magnitude and in sign
just by varying the pH.   The inorganic latexes to be described are
also of interest in various applications, such as pigments, catalysts
and catalyst carriers, coatings, fillers, etc.

In view of the importance of metal hydroxides it is no surprise
that countless studies have been reported on their precipitation
and characterization.   Yet, there is no good theory on the mechanisms
of formation and of the growth of such systems.   This is easily
understood if one considers that generation of metal hydroxides
depends on a large number of factors, the most important being pH,

temperature, aging, and the nature of the anions, in addition to
the parameters which affect any solid phase formation (e.g., the
concentrations of the reacting components, method of mixing, etc.).

The rather dramatic role exhibited by the pH and by the anions
in solution indicates that specific solute metal complexes precede
or even initiate nucleation of metal hydrous oxides.   The availability of monodispersed inorganic latices, which can be reproducibly
prepared, has now made it possible to elucidate the reactions which
lead to the solid phase formation and to study the mechanism of
particle growth.   The understanding of these processes is of
interest in many areas, but probably it is of greatest importance
in the explanation of corrosion of metals in general, and of iron
in particular.

In this paper we will describe the methods of preparation of
a number of monodispersed metal hydrous oxide sols including those
of chromium, aluminum, iron, titanium, thorium, and copper.   Furthermore, the effects of anions and of other additives on particle size,
shape, and composition will be illustrated.   Finally, at least in
the case of chromium, the mechanism of formation of its hydroxide
will be discussed.

II. METHODS OF PREPARATION OF MONODISPERSED METAL HYDROUS
OXIDE SOLS

The essential aspect of the procedures for the preparation of
metal hydroxide colloidal dispersions consists of forced hydrolysis
of metal ions in acidic (sometimes very highly acidic) aqueous
solutions of metal salts by aging the latter at elevated temperatures
$(75 - 180^{\circ}C)$ for varying periods of time (20 minutes to several
weeks).   Simultaneously, the nature and the concentration of the
anions have to be carefully controlled, as these exercise a
dominant effect on the composition, size, and shape of the final
sol particles.   It is noteworthy that sulfate and phosphate ions
seem to play the most important role in the formation of monodispersed
metal hydrous oxide sols.   For example, only in the presence of
these anions have spherical particles of chromium, aluminum, thorium,
and zirconium hydroxides been generated.   In their absence either
sols do not appear (chromium salt solutions) or particles of entirely
different shapes (rods and needles) precipitate on aging (aluminum
salt solutions).

The optimal temperatures, times of aging, and the concentration
of metal ions and anions vary from case to case.   For example,
~ 24 hours at $75^{\circ}C$ are needed to obtain a monodispersed chromium
hydroxide sol, whereas only 2 hours at $98^{\circ}C$ will give an excellent
ferric basic sulfate dispersion.   The rate of heating also plays a
role, but mostly with regard to the degree of uniformity.

In some instances, the hydrolysis may be assisted by homogeneous
generation of hydroxyl ions, such as happens when the aging is
carried out with urea added to the metal salt solutions.

III. EXAMPLES OF MONODISPERSED METAL HYDROUS OXIDE PARTICLES

By using the described procedure and by properly adjusting the
experimental parameters a number of metal hydrous oxides of narrow
size distribution have been produced. In particular, depending on
the anions present in the aged solutions, some metals will give
either spherical (amorphous) or crystalline particles. Hydrosols
consisting of spherical particles are especially useful as they
can be readily analyzed in situ for size distribution and number
concentration by light scattering. This information is pertinent
in the study of particle nucleation and growth mechanisms.

(a) Chromium

Figure 1 shows an electron micrograph of a chromium hydroxide
sol obtained by aging a chrome alum solution. All particles are
spherical and have a narrow size distribution; their modal radius
decreases as the concentration of the solution becomes lower (1).
Similar results are obtained if chromium nitrate, chloride, or
perchlorate are heated in the presence of a soluble sulfate salt,
but not in the absence of sulfate ions; in the latter case these
chromium salts solutions produced no sols on aging under similar
conditions. The particle size also depends on the chromium to
sulfate ratio; with increasing sulfate concentration the same
chromium salt solutions yield larger particles (1,2). The addition
of phosphate ions has a similar effect as sulfate ions on the
formation of chromium hydroxide sols (3).

It is important to recognize that the spherical particles are
amorphous and that they contain no sulfate (4), although the latter
species is essential to the solid phase formation. The role of
sulfate ions will be discussed below.

(b) Aluminum

Figure 2 shows an electron micrograph of a carbon replica of
an aluminium hydrous oxide sol obtained by aging an $KAl(SO_4)_2$
solution (5). Again, spherical particles form, but only in the
presence of sulfate or phosphate ions. Aluminum chloride, nitrate,
or perchlorate solutions yield on aging anisometric particles
(needles, rods, etc.). Unlike chromium hydroxide, the particles
of aluminum hydrous oxides prepared from aluminum sulfate solutions

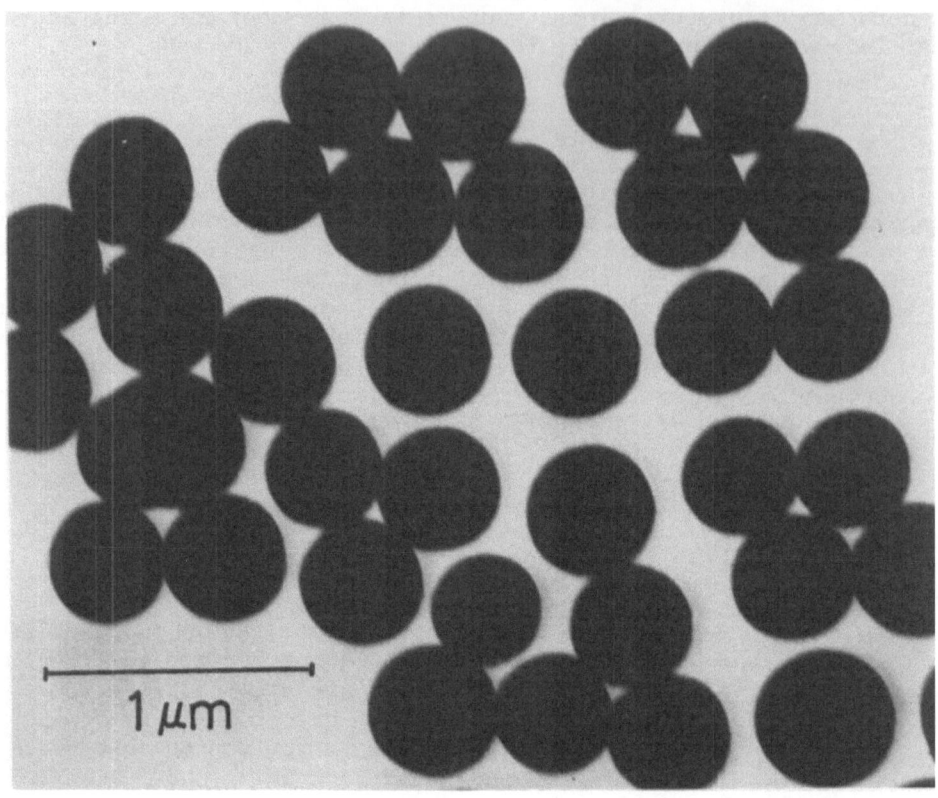

Figure 1.   Electron micrograph of chromium hydroxide sol particles
obtained by heating a 8 x 10^{-4}M chrome alum solution at 75°C for
26 hours.

contain appreciable amounts of sulfate ions, but these seem to be
weakly bound and are readily exchangeable.   On heating the
deionized sol with dilute base (at pH 9.7) the entire content of
sulfate in the particles is substituted by hydroxyl ions.   As a
consequence, the particles shrink a little but retain their
spherical shape (5).   Once the sulfate ions are removed, further
aging of the sols brings about crystallization of the particles
and consequently a change in shape.   Figure 3 shows such a
transition.   Originally all particles were perfectly spherical,
and if the secondary aging had been continued they would have all
appeared as crystalline platelets.   The change in the structure
can be confirmed by electron diffraction which shows clearly that
the spherical particles are amorphous whereas the diffraction

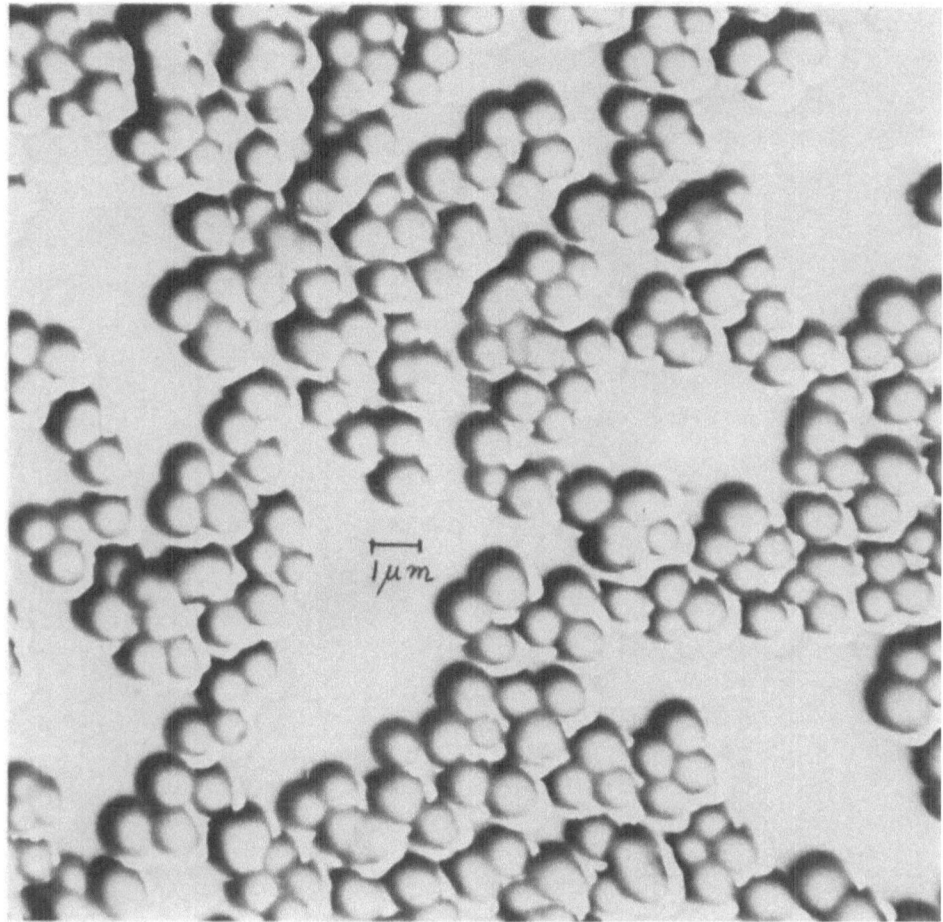

Figure 2. Electron micrograph of a carbon replica of aluminium hydrous oxide particles obtained by aging 1 x 10^{-3}M $KAl(SO_4)_2$ at 98°C for 24 hours.

pattern of crystallized material corresponds to boehmite, $AlO(OH)$.

It is now possible to prepare aluminum hydrous oxide sols in large concentrations.   On spray-drying these sols give powders consisting of spherical particles which can be most readily redispersed on addition of water.

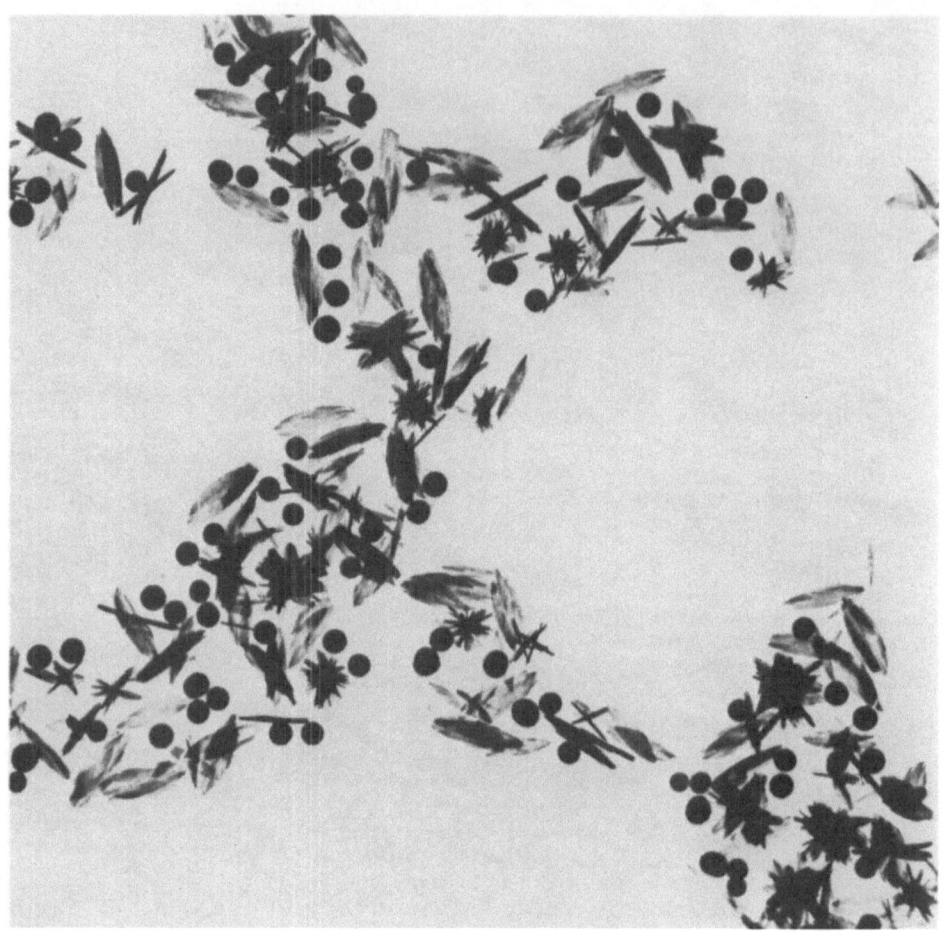

Figure 3.   Electron micrograph of a sol prepared by aging 1 x 10^{-3}M
$Al_2(SO_4)_3$ for 24 hours at 98°C followed by adjustment of pH to 9.7
by NaOH, deionization, repeptization at pH ∿ 8, and additional
aging at 98°C for 24 hours.

(c) Thorium

Most recently, sols consisting of spherical particles of
thorium hydrous oxide were prepared by aging of thorium nitrate
solutions in the presence of Na_2SO_4.   Significantly, no particles
appear in the aged solutions in the absence of sulfate ions.

(d) Iron

Of different metals, ferric ions are among the most reactive. Thus, the feasibility of the preparation of colloidal monodispersed ferric hydrous oxides seemed rather remote. However, we succeeded in establishing conditions which gave such particles of excellent uniformity in a variety of shapes. Again the composition of the particles and their morphology was strictly dependent on the nature and the concentration of the anions present in the solutions that were aged.

Figure 4 shows an electron micrograph of a colloidal ferric basic sulfate obtained by aging a solution containing $Fe(NO_3)_3$ and Na_2SO_4. Although the particles appear spherical when observed by transmission electron microscopy, they are actually of hexagonal symmetry as clearly seen from the replication of the same sample; the scanning electron micrograph (insert, Fig. 4) nicely confirms the morphology. When the ratio of ferric to sulfate ions is altered, particle shape also changes. Two kinds of particles were obtained: hexagonal and monoclinic and their structures were confirmed by X-ray analysis. The chemical composition of these particles was identified as $Fe_3(OH)_5(SO_4)_2 \cdot 2H_2O$ (hexagonal) and $Fe_4(OH)_{10}(SO_4)$ (monoclinic), the former corresponding to the alunite group minerals (6).

The anions again have a great effect on the formation of ferric hydrous oxide sols. Solutions of $Fe(NO_3)_3$ yielded ellipsoidal particles whereas $FeCl_3$ solutions on heating gave rod-like particles. X-ray analysis showed the former to correspond to $\alpha-Fe_2O_3$, and the latter to $\beta-FeO(OH)$.

Very recent work showed that it is possible to prepare nearly spherical particles of narrow size distribution. The particles have been identified as $\alpha-Fe_2O_3$ and were obtained by aging acidified solutions of ferric chloride over long periods of time. Figure 5 shows a scanning electron micrograph of such a dispersion.

The presence of cations other than alkali metals also influenced the nature of the sols formed. Aging experiments were performed on a series of $Fe(NO_3)_3$ + Na_2SO_4 solutions in which a mole fraction of Na_2SO_4 was substituted by MSO_4 ($M = Cu^{2+}$, Ni^{2+}, Mg^{2+}, or Cd^{2+}). In all cases the concentration of SO_4^{2-} was kept constant in order not to affect the particle composition and shape by changing the content of sulfate ions in the growth medium. It was observed that most divalent ions (Cu^{2+}, Ni^{2+}, Mg^{2+}) caused an increase in modal particle size as their mole fraction was increased up to ~ 0.5 (relative to the content of Na_2SO_4). When MSO_4 represented more than 50% of the total sulfate content the particle size became smaller; complete inhibition of solid phase formation took

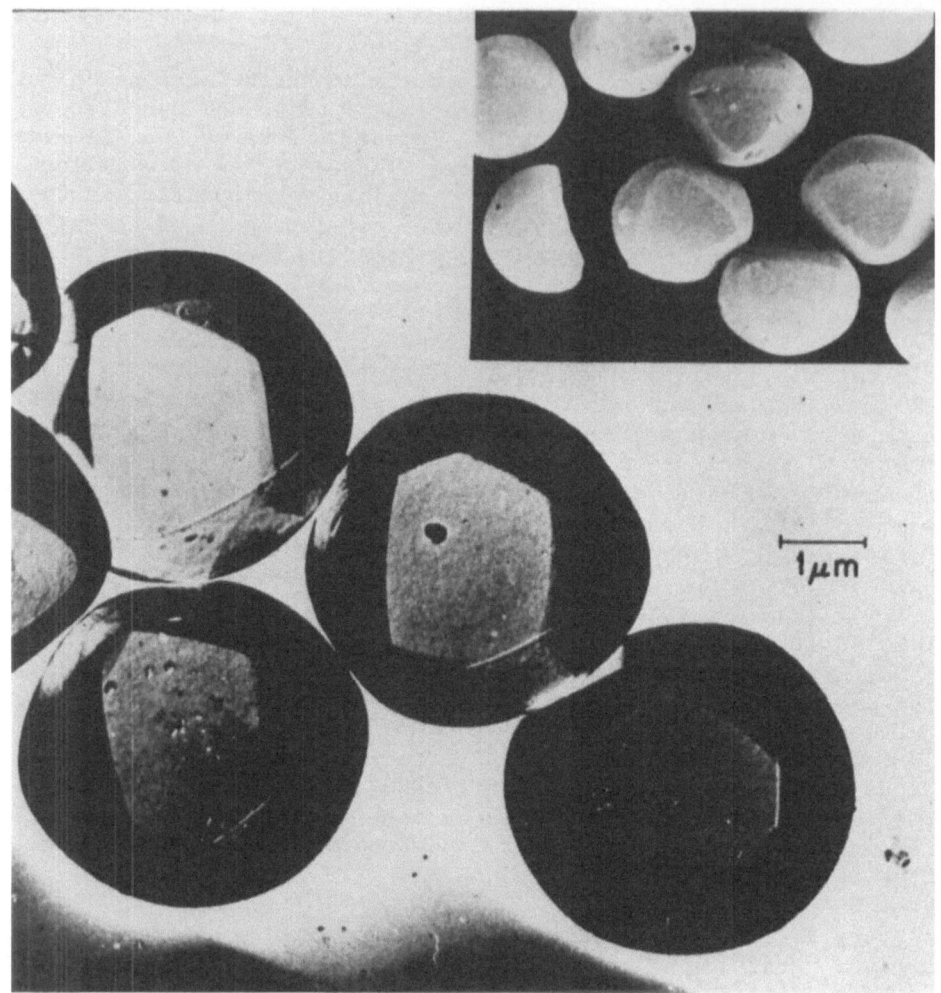

Figure 4.   Electron micrograph of a carbon replica of basic iron
(III) sulfate sol particles prepared from a solution 0.18M in
$Fe(NO_3)_3$ and 0.53M Na_2SO_4 aged in an oil, both heated from room
temperature to $80^{\circ}C$ at a rate of $1.5^{\circ}C$/min and then maintained at
$80^{\circ}C$ for 1.5 hours.   Insert shows a scanning electron micrograph
of the same particles.

place when a MSO_4 salt was used alone.   The latter effect was
found to be due to the change in pH.   As the fractional concentration of MSO_4 was increased the pH became lower, for example, in

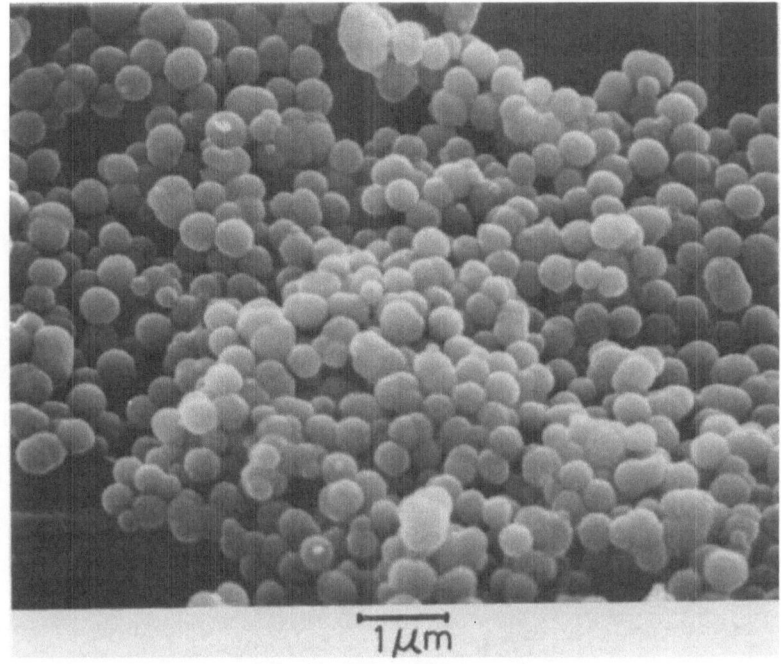

Figure 5.   Scanning electron micrograph of a sol obtained by aging a solution 0.032M in FeCl$_3$ and 0.0005 M in HCl for 2 weeks at 100°C. X-ray analysis showed the particles to be α-Fe$_2$O$_3$.

the case of gradual substitution of Na$_2$SO$_4$ with NiSO$_4$, the pH dropped from 1.78 to 1.45 as the mole fraction of NiSO$_4$ increased from 0 to 1.   If the same experiments were carried out at constant pH, particles formed over the entire concentration range of NiSO$_4$ although there was a variation in the average particle size.   These data point out the extreme sensitivity to pH of the processes involved in the formation of ferric hydrous oxides. Obviously, the change in pH affects the complexes in solution which act as the precursors to precipitation.   The nature of these complexes at various temperatures is now being investigated.

(e) Copper

Monodispersed sols were also obtained by aging copper salt solutions (similar to Fehling's solutions, but of different concentrations of components) in the presence of glucose (7).

All particles were of cubic crystal symmetry characteristic of
copper(I) oxide, but their habitus differed depending on the
concentration of the reacting components.

(f) Titanium

The size and shape of titanium dioxide particles are of
particular importance when this material is used as a pigment.
By extensive aging of titanium chloride solutions in the presence
of sulfate ions, it was possible to produce uniform titanium
dioxide sols consisting of spherical particles of rather striking
uniformity.   Their size could be altered by the length of aging.
A somewhat puzzling aspect of this system is that despite the
apparently spherical habitus (just as in the case of α-Fe_2O_3, Fig
5), the particles exhibit a definite x-ray pattern characteristic
of rutile.   This latter composition is further substantiated by
the finding that the refractive index which best fits the light
scattering data taken with the same sol is that of rutile (8).

IV. STUDIES WITH MONODISPERSED METAL HYDROUS OXIDE SOLS

The availability of inorganic latexes described in the previous
section has made it possible to carry out a number of investigations
on various aspects of the properties of colloidal dispersions.
Only a few examples will be offered here.

1. Mechanism of formation of the chromium hydroxide sol

Hardly any documented quantitative information is available on
the chemical processes involved in the precipitation of metal
hydroxides.   Recent work carried out with monodispersed chromium
hydroxide (4) represents an effort in this respect, and it has
produced evidence on the role of the solute complexes in the solid
phase formation.   Various species formed on aging of chromium
salt solutions in the presence of sulfate ions were separated by
paper electrophoresis.   They were further characterized by means
of radioactive tracers, primarily [57]Cr and [35]S.   Four species,
differing in ionic charge, were found in all aged solutions of
chrome alum.   Radio-paper electrophoretic chromatograms were
obtained at different times of aging and for different electrolyte
solutions and a quantitative analysis of the change in the concentration of various species has indicated that complexes of zero
charge and complexes of 2+ charge play the dominant role in the
chromium hydroxide formation.   All of the information points to a
mechanism   involving the formation of a complex chromium basic
sulfate precursor, upon which the neutral solute hydrolysis product,

$Cr(OH)_3$, nucleates and grows in a sequence of reactions as follows:

$$m\left[Cr_2(OH)_2SO_4\right]^{+2} + m\ SO_4^{-2}\ \text{and}\ n\ Cr(OH)_3^{\cdot}\ aq$$

$$\left[Cr(OH)SO_4\right]_m + n\ Cr(OH)_3^{\cdot}\ aq$$

$$\left[Cr(OH)_3\right]_{m+n}\ (solid) + m\ SO_4^{-2} + 2m\ H^+$$

This mechanism explains the role of sulfate ions and also the finding that the final product consists of spherical particles of chromium hydroxide which contain no sulfate.   For a detailed account of the results which lead to the described mechanism the reader is referred to a recently published paper (4).

Monodispersed chromium hydroxide sols have also been utilized to determine the mechanism of the particle growth (10).   For this purpose the Nielsen's chronomal anaylsis was applied (11), which indicated that the growth occurs via surface reaction.   The corresponding rate constants and energies of activation could also be evaluated.

2. Effects of polymers on particle growth

Numerous studies on the effects of surfactants on particle formation and growth have been reported with the generally accepted conclusion that solids precipitate in the presence of surface active agents and the particle number concentration is higher than in the absence of these additives;   consequently, the average particle size is smaller.   Much less work has been reported on the influence of polymers on particle formation and growth and the results are by no means conclusive (12).

The availability of monodispersed inorganic latexes and the ease with which they can be prepared makes it now possible to systematically investigate the influence of polymers on particle formation and growth.   Such a study has been undertaken with chromium hydroxide sols in the presence of various dextrans.   These polymers are particularly convenient because they can be obtained in rather narrow molecular weight fractions in nonionic, anionic, and cationic modifications.

One example of this type of investigation is given in Fig. 6 which shows the effect of dextran sulfate 2000 (molecular weight $\sim 4 \times 10^6$) on the particle size of chromium hydroxide.   The four electron micrographs are for sols obtained in the absence and in the presence of three different concentrations of the polyelectrolyte. Obviously, an addition of only 0.001% (by wt) of dextran sulfate

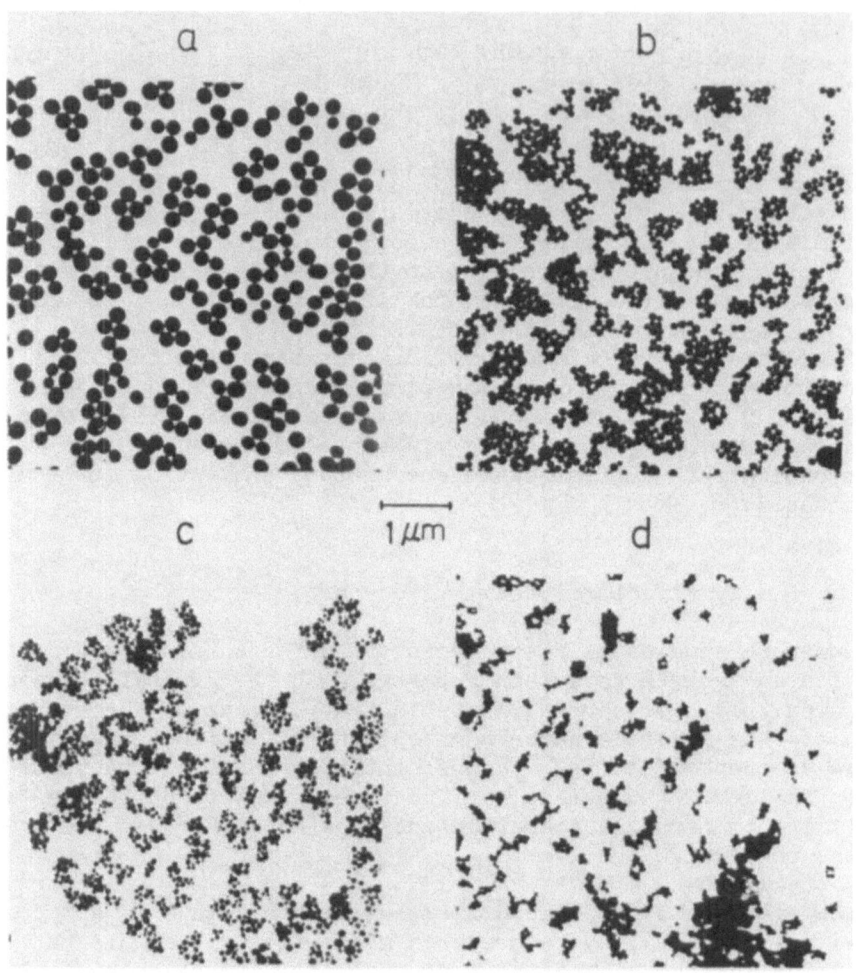

Figure 6.   The effect of dextran sulfate 2000 on the particle size
of a chromium hydrous oxide sol prepared by aging 4×10^{-4}M chrome
alum at 75°C for 26 hours.   The electron micrographs are for sol
particles in the presence of (a) 0%, (b) 0.001%, (c) 0.005% and
(d) 0.007% (by wt.) dextran sulfate respectively.

2000 has a considerable effect on the particle size; as the
concentration of the polymer increases, the average particle size
becomes smaller.   A similar observation was made with the cationic
dextran (DEAE) of the same chain length.   However, the concentration
of this polyelectrolyte necessary to cause the same reduction of
particle size was nearly two orders of magnitude higher than with
dextran sulfate.   Except in very high concentrations, nonionic
dextrans showed little influence on the formation of chromium
hydroxide sols.

It was assumed that the significant effect of dextran sulfate
on the particle size was due to ion binding of chromium ions by
the polyelectrolyte.   Dialysis equilibration studies on the same
system showed a considerable depletion of chromium solute species
on addition of the anionic polyelectrolyte.   This change in
solution concentration explains the difference in the size of the
particles, since it was shown that the dispersity of the sol depends
on the concentration of the metal salt in solution (1).

3. Heterocoagulation

Another application of the monodispersed inorganic latexes
has been found in the study of heterocoagulation.   According to
the current theories of coagulation of unlike sols, two parameters
are of primary importance, namely, the ratio of the sizes and the
ratio of the surface potentials of the interacting particles.
While the effect of varying sizes is easily explored with conventional
organic latexes, variation of surface potential is more difficult
to accomplish.   The latter problem can be overcome if the system
used to study heterocoagulation consists of an organic and an
inorganic latex.   Indeed, an investigation has been conducted on a
number of mixed sols consisting of monodispersed polyvinyl chloride
(PVC) latexes stabilized by fixed sulfate groups and monodispersed
chromium hydroxide sols.   The fact that both systems can be
obtained in a variety of modal diameters, allows for the elucidation
of the particle size effect.   Furthermore, a change in pH has
relatively little influence on the charge of the PVC latex; its
sign remains the same and the value changes slightly, particularly
over the higher pH range.   On the other hand, the charge of the
chromium hydroxide particles goes from positive to negative as the
pH rises and the magnitude of the charge itself varies considerably.
Thus, one can investigate the effects of surface potentials of
unlike particles over a wide range of conditions.   The results on
the kinetics of heterocoagulation of the described systems have
been compared with the existing theories and found in excellent
qualitative agreement, although quantitative correlation is still
lacking (13).

By using different inorganic latexes, such studies can now be extended to include the effects of particle shape and other parameters of interest in the behavior of mixed colloidal dispersions.

V. CONCLUDING REMARKS

The feasibility of the preparation of monodispersed metal hydrous oxide sols has now been amply demonstrated. It is recognized that this is only the first step in the elucidation of the processes involved in the formation of metal hydroxides and in the characterization of their various properties. There is obviously one common underlying principle involved in the precipitation of metal hydroxides: metal ion hydrolysis. Since the latter process varies with each metal, one would expect the conditions for the precipitation of the corresponding hydroxides to be different. The fact that ions other than hydroxyl may coordinate with the hydrolyzable metal species further complicates the picture. Thus, we would expect different specific chemical reactions to take place in solutions from which the metal hydroxides precipitate. Indeed, the experience so far has shown that the monodispersed systems vary considerably in the chemical composition of the final products, as well as in the particle size, shape, charge, and crystallinity. One of the intriguing questions is why certain systems appear in the form of amorphous particles, whereas some others give well defined crystalline materials. As a first approximation, it would appear that metal ions known for their tendency to continue to polymerize on heating or aging (such as aluminum and chromium) finally give amorphous particles, whereas metal ions which hydrolyze to well defined mono- and polynuclear species yield crystalline solids. Anions which promote polymerization of metal hydroxylated species (e.g., sulfate) would be expected to provide the formation of amorphous particles, whereas anions which give well defined complexes (e.g., chloride) should produce crystalline solids. This has indeed been found to be the case with aged aluminum salt solutions.

It is our belief that the new inorganic latexes described in this work will be employed in many fundamental investigations and that they will find numerous useful applications. For the latter purpose, procedures will have to be developed for their preparation economically in large quantities.

ACKNOWLEDGEMENTS

This presentation has been based on the work of a number of my associates to whom I am indebted for their contributions. Specifically, I would like to express my appreciation to Dr. A. Bell,

Mr. A. Bleier, Mr. R. Brace, Mr. M. Budnik, Mr. R. Cataldi, Dr. D. L. Catone, Dr. R. Demchak, Mr. M. Dennen, Dr. S. Kratohvil, Dr. A. D. Lindsay, Dr. P. McFadyen, Dr. J. B. Melville, Mr. M. Onofusa, Mr. R. Sapieszko, Mr. P. Scheiner, and Mr. W. Scott.   The work was supported by the NSF Grant GP 42331X and by the American Iron and Steel Institute project No. 63-269.

REFERENCES

1. R. Demchak and E. Matijević, J. Colloid Interface Sci. 31 (1969) 257.
2. E. Matijević, A. Bell, R. Brace and P. McFadyen, J. Electrochem. Soc. 120 (1973) 893.
3. E. Matijević, A. D. Lindsay, S. Kratohvil, M. E. Jones, R. I. Larson and N. W. Cayey, J. Colloid Interface Sci. 36 (1971) 273.
4. A. Bell and E. Matijević, J. Inorg. Nucl. Chem. 37 (1975) 907.
5. R. Brace and E. Matijević, J. Inorg. Nucl. Chem. 35 (1973) 3691.
6. E. Matijević, R. S. Sapieszko and J. B. Melville, J. Colloid Interface Sci. 50 (1975) 567.
7. P. McFadyen and E. Matijević, J. Colloid Interface Sci. 44 (1973) 95.
8. M. Budnik, M. S. Thesis, Clarkson College of Technology, 1975.
9. G. A. Parks, Chem. Rev. 65 (1965) 177.
10. A. Bell and E. Matijević, J. Phys. Chem. 78 (1974) 2621.
11. A. E. Nielsen, "Kinetics of Precipitation", Pergamon Press, Oxford, 1964.
12. A. D. Lindsay, E. Matijević and J. P. Kratohvil, Colloid Polymer Sci. 253 (1975) 581.
13. A. Bleier and E. Matijević, J. Colloid Interface Sci. (in press).

ELECTROCHEMICAL CONTROL OF THE FLOW BEHAVIOUR OF COAGULATED COLLOIDAL SOLS

ROBERT J.HUNTER   and   BRUCE A.FIRTH

Department of Physical Chemistry

University of Sydney

1.   INTRODUCTION

The flow properties of colloidal suspensions are of paramount importance in many of their technological applications and the control of those flow properties has been known for a long time to depend upon control of the interaction forces between the particles. Addition of strongly adsorbed ions (like H^+, OH^- and ionic surfactants) has often been used in an empirical way to alter dispersity and "viscosity" but only now are we beginning to obtain a more definite idea of how these ions work. In what follows we will concentrate attention for the most part on the behaviour of monodisperse suspensions of rigid spherical particles involved in attractive interactions with only brief preliminary reference to other types of interaction. The ions involved are in electro-chemical equilibrium with the bulk suspension medium and on adsorption they modify the particle charge and hence the energy of interaction between particles. It is the quantitative connection between the solution electrochemistry and the ultimate flow behaviour, that we will explore in this paper.

II. THEORY

Considerable progress has been made in recent years in the understanding of the flow behaviour of suspensions of spherical particles which are not undergoing interaction.[1]   A single"equation of state" can be given to describe the viscosity as a function of shear rate for particle concentrations up to, and even above 55% by volume.

When double layer effects are important and the particles are
repelling one another (stable systems) the elementary formulation
must be corrected for the primary[2] and secondary[3] electroviscous
effects and reasonable descriptions are now available for these[4]
though they are difficult to quantify exactly.

For systems involving attractive interactions between the
particles the colloidal forces are very large compared to the hydro-
dynamic forces acting between individual particles and quite new
types of behaviour become apparent. Such systems show non-Newtonian
behaviour even at very low particle concentrations when particle
self-crowding is unimportant and so too are the normal electroviscous effects. These unstable (coagulating) systems require a
distinctly different model for their description; that model must
be able to predict the dependence of viscosity, η, on shear rate D,
(or the shear stress (τ) - shear rate relation) in terms of the
colloidal characteristics of the system (particle size, concentration and interaction energy).

In the controlled studies to which we will refer below, the
system is always subjected to an initial very high shear rate and
measurements are made of the shear stress at successively lower
shear rates. We will assume that in that process the particles
are formed into small, compact floccules consisting of several
hundreds of individual particles whose size and packing density is
determined by the maximum stress to which they have been subjected.
We will then calculate the flow behaviour by assessing the energy
dissipation during flow assuming that these floccules are the
primary flow units.

The energy dissipation (per unit volume per second) in a flowing sol exhibiting plastic behaviour can be broken down into a
number of contributions:

$$E_{TOTAL} = \tau D = E_V + E_R + E_{FR} + E_{DR} \qquad (1)$$

where E_V is the viscous energy dissipated by the suspension medium
flowing around the flocs. We will assume it is given by the
Einstein relation:

$$E_V = D^2 \eta_{PL} = D^2 \eta_S (1 + 2.5 \ \phi_F \ \ldots \ldots)$$

$$= D^2 \ \eta_S (1 + 2.5 \ C_{FP} \phi_P + \ldots) \qquad (2)$$

where $C_{FP} = \phi_F/\phi_P$ is the floc volume ratio which should be predictable from the high shear history of the suspension.

E_R is the energy dissipated during rotation of the floc; we
will assume it to be negligible since collisions between flocs occur
more frequently than rotations at the volume concentrations used
here.[5]

The remaining energy terms relate to the effects of collisions

between flocs. Assuming a steady state size distribution at any
given D, we know that if two colliding flocs coalesce then some
other floc must be ruptured and the energy to do this is E_{FR}. If
the flocs merely form a doublet, which later separates, the energy
to separate it is E_{DR}. We postulate that $E_{DR} \gg E_{FR}$ i.e., although
floc coalescence and rupture are probably very important in establishing the initial floc structure and packing, they do not occur to any
great extent at the more modest shear rates (100-2000 s^{-1}) with
which we are most concerned. The energy E_{DR} consists of two parts:
E_{DL}, the energy required to rupture the links which form between
two flocs when the doublet forms, and E_{DE}, the energy needed to
stretch (but not break) the elastic links between particles within
the flocs as the tension is transmitted from the shear field to the
doublet contact area (Fig. 1b).

To calculate these energies we need (i) a satisfactory method
of describing the structure of the floc and hence the number of
links within it and (ii) a description of the interaction energy
(and hence, the force) between the particles as a function of
separation.

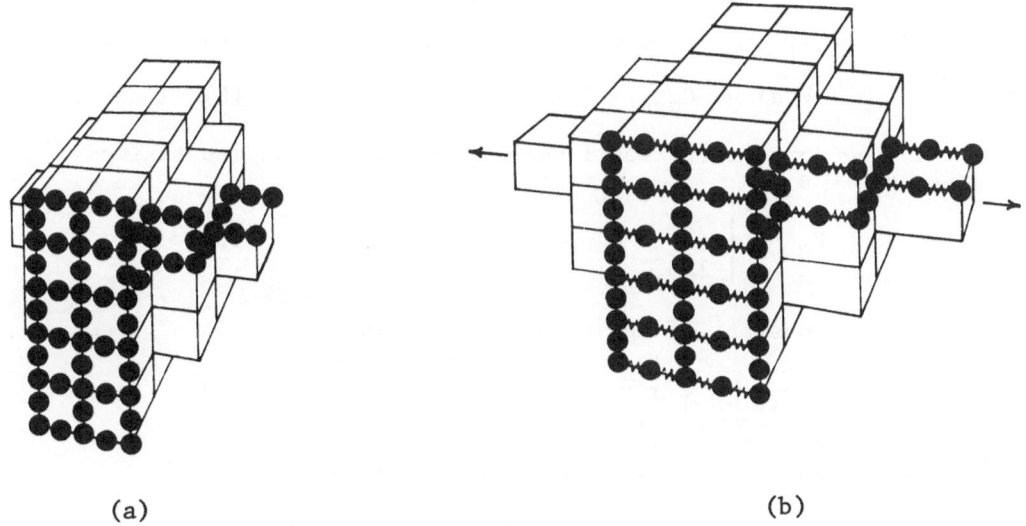

(a)   (b)

Fig. 1.   (a) Part of a floc with no applied stress:
          (b) The same with stress applied. (The relative particle
              movement is actually very small - only about 0.1%
              of a particle diameter.)

(i) <u>Floc structure</u> :   We assume the floc consists of a network of "girders" made up of chains of particles (Fig. 1a).   The least open girder structure is a cubic close packed array which has the minimum C_{FP} value (1.35) and for which the number of particle chains, n_c, crossing unit area of the floc in any given direction is a maximum.

Flocs with higher C_{FP} values can be constructed by removing particles from the close-packed structure in a regular way to create more open structures.   Examination of these structures[5] shows that n_c and C_{FP} are linked by the following semi-empirical relation:

$$n_c = (4C_{FP}r^2)^{-1} \qquad (3)$$

(ii) <u>Interaction Energy and Force</u> :   We assume that the interaction between particles is described by the DLVO theory[6] with the addition of a close range strong repulsion when the particles, or their strongly adsorbed solvent layers, make contact. The interaction is attractive at all separations $d > d_o$, and the maximum force required in the separation occurs at $d = d_1$

$$\text{For } d > d_1 : \quad V(d) = \frac{-Ar}{12d} + rG(\epsilon, \kappa d)\zeta^2 \qquad (4)$$

where A is the van der Waal's constant and

$$G(\epsilon, \kappa d) = \frac{\epsilon}{2} \ln \left[1 + \exp(-\kappa d) \right] \qquad (5)$$

For $d_o \leqslant d \leqslant d_1$ we assume a parabolic expression for the energy:

$$V(d) = -E_{SEP} + \frac{L}{2} (d-d_o)^2 \qquad (6)$$

so that the force increases linearly from $d = d_o$ to its maximum value at $d = d_1$:

$$F_M = r \left[\frac{A}{12d_1^2} + G'(\epsilon, \kappa d_1)\zeta^2 \right] \qquad (7)$$

The elastic constant of the bond, L, is then

$$L = \frac{r}{(d_1-d_o)} \left[\frac{A}{12d_1^2} + G'\zeta^2 \right] \qquad (8)$$

and the spring energy involved in stretching is

$$E_{SEP} + V(d) = \frac{r(d-d_o)^2}{2(d_1-d_o)} \left[\frac{A}{12d_1^2} + G'\zeta^2 \right] \qquad (9)$$

In addition to this energy we must add the energy E_L dissipated by the liquid as it flows into the narrow gap between the particles.

The number of links between flocs per unit area (n_F) is much smaller than n_c and even though a higher energy is involved in breaking them, (E_L + E_{SEP} + V(d)), where the last term, usually negative, is now zero). We still find that the main contribution from the energy dissipation comes from E_{DE} rather than E_{DL}.

Calculation of E_{DR} ($\simeq E_{DE}$)

At high shear rates, every doublet of flocs formed by collision must be broken up. The frequency of collisions per unit volume is[7]: $3\phi_F^2 D/\Pi^2 a^3$ where a is the floc radius. To break a doublet the stress must be transmitted through both flocs and the force applied to each bond is $F = \dfrac{n_F}{n_c} F_M$, which stretches them to a separation:

$$d = d_o + \frac{n_F}{n_c} (d_1 - d_o) \qquad (10)$$

Using equation (9) and adding in E_L to obtain the energy for each bond and noting that there are $n_c/2r$ bonds stretched in the process gives

$$E_{DR} \doteq E_{DE} = \frac{3\phi_F^2 D}{\Pi^2 a^3} \left\{ \frac{8}{3} \frac{\Pi a^3 n_c}{2r} \left[E_L + \frac{r n_F^2}{2 n_c^2} (d_1 - d_o)Q \right] \right\}$$

where $Q = \dfrac{A}{12 d_1^2} - \dfrac{\varepsilon \kappa \zeta^2}{2(\exp(\kappa d_1)+1)}$ \qquad (11)

Calculation of Flow Characteristics

(a) <u>Floc volume ratio, C_{FP}</u> : The maximum stress per unit area to which a floc is subjected during flow is[8]: $\sigma_T = 5\eta_S D$ and if the floc is to just withstand this stress we have $\sigma_T = n_c F_M$ and so from (3) and (7)

$$C_{FP} = \frac{1}{20 \, r\eta_S D C_{FP}} \left\{ \frac{A}{12 d_1^2} + G'(d_1)\zeta^2 \right\} \qquad (12)$$

where $D_{C_{FP}}$ is the maximum shear stress to which the system has been subjected.

(b) <u>Critical shear stress D_o</u> : At this point the shear stress-shear rate curve becomes linear and we infer that the shear field is then just capable of breaking down every floc, doublet which occurs by collision. The tensile stress applied at this shear rate must therefore be $\sigma_{TFD} = n_F \cdot F_M$. If we assume that $n_F \ll n_c$ and independent of ϕ_p, r and ζ (since it presumably depends only on the detailed conformation of the area of contact) then:

$$D_o = \frac{\sigma_{TFD}}{5\eta_S} = \frac{n_F r}{5\eta_S} \left\{ \frac{A}{12d_1^2} + G'(d_1)\zeta^2 \right\} \qquad (13)$$

(c) <u>Determination of τ_B</u> : The total energy dissipation for $D > D_o$ is given by eq. (1) which we can now simplify to $E = E_V^o + E_{DR} = \tau D$ and since $\tau = \tau_B + \eta_{PL} \cdot D$ at the higher shear rates we have: $\tau_B = E_{DR}/D$. Two extreme possibilities may be identified :

(i) For $E_{SEP} + V(d) \gg E_L$ we have:

$$\tau_B = \left[\frac{n_F^2 (d_1 - d_o)}{1000 \Pi \eta_S^3 D_{C_{FP}}^3} \right] \cdot \frac{\phi_P^2}{r} \cdot Q^4 \qquad (14)$$

and (ii) for $E_L \gg E_{SEP} + V(d)$:

$$\tau_B = \phi_P^2 \cdot E_L \, Q \Big/ (20 \, \Pi \eta_S D_{C_{FP}} \, r^4) \qquad (15)$$

III   EXPERIMENTAL

Equations (12), (13) and (15) have been tested against 12 different systems culled from the literature and one (TiO$_2$) which was studied by B.A.F.[5]   The details of the systems and the results are reported elsewhere.[5,12]   The systems referred to explicitly below are :

System 1 :   Poly-(methylmethacrylate) latices, $r = 0.106$ μm, Ionic strength (NaCl) = 2×10^{-2} g ion l^{-1}; $t = 25^\circ C$, ζ-potential controlled by quaternary ammonium soaps $(\zeta \geqslant 0)$.[11]

System 2a: As for 1 but with $r = 0.22$ μm $(\zeta \geqslant 0)$.

System 3:   TiO$_2$; $r = 0.07$ μm, ionic strength (KCl) = 10^{-2} g ion l^{-1}, $\phi_P = 0.02$; $t = 25^\circ C$; ζ-potential controlled with pH $(\zeta \leqslant 0)$.

System 4:   ZnO in oil[13], $\eta_S = 3.9$p $t = 30^\circ C$; $\phi_P = 0.10$.

System 5:   Carbon-black suspended in petroleum oil, $\eta_S = 0.63$ poise, $t = 30^\circ C$.

IV   RESULTS

The Flow Parameters

(a) The C_{FP} value   may be obtained by measuring the plastic viscosity η_{PL} and using equation (2) with the known volume concentration of particles, ϕ_p. It can also be obtained from sedimentation data[9] and the results are in general concordance[5]. It is found to be independent of particle concentration,[5] to be proportional to $1/r$ (Fig.2) (for $r \leqslant 1$ μm) and to decrease with decreasing attraction (Fig. 3). All of these results are consistent with equation (12).

(b) The critical shear stress, D_0 does not depend on particle concentration[5] but it does decrease with decreasing attractive interaction (Fig. 4) as equation (13) requires. We have insufficient data to test the radius dependence.

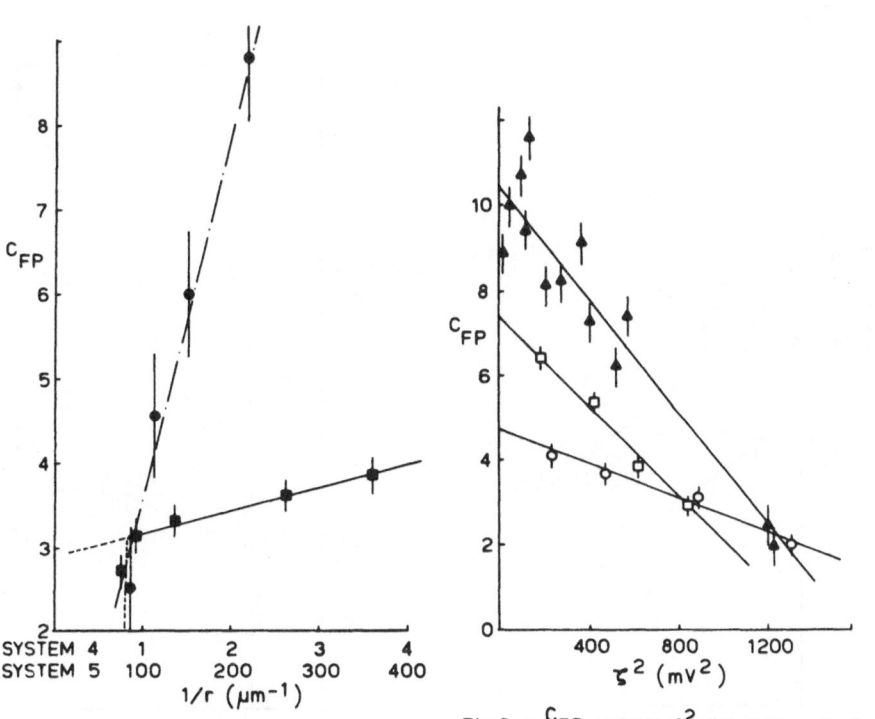

Fig. 2 C_{FP} versus $1/r$

Fig.3 : C_{FP} versus ζ^2 for system 1 - □; system 2a - ○; and system 3 - ▲.

(c)   The Bingham yield value, τ_B has often been shown to
depend on ϕ^2 and also to decrease with decreasing attraction[10,11]
as equation (15) predicts.   The alternative, equation (14)
would require a linear relation between $\tau_B^{1/4}$ and ζ^2 and although
this sometimes occurs, the $\tau_B - \zeta^2$ relation is always found to be
linear and the energy calculation using equation (15) is more
self-consistent[5].   We infer, therefore, that in these systems
$E_L \gg E_{SEP} + V(d)$.

We also find that for small particles $(r \leqslant 1\mu m)\tau_B$ depends
on $1/r$ (Fig. 5) which suggests, from equation (15), that
$E_L \propto \eta_S r^3$.   For non-deformable, spherical particles we find that
the ratio $E_L/\eta_S r^3$ is constant, within experimental error, and equal
to about $1-2 \times 10^5$ s^{-1}.

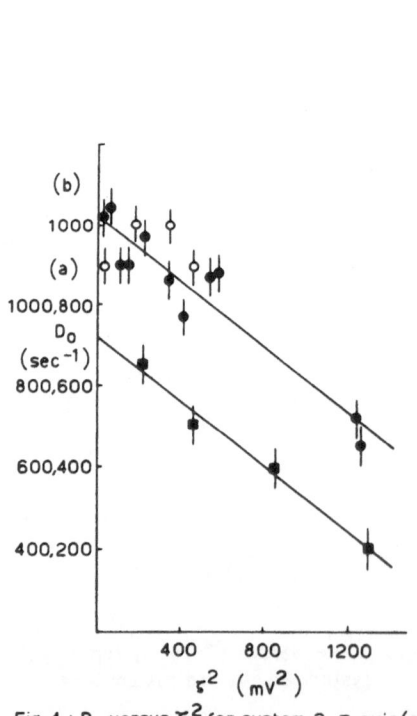

Fig. 4 : D_0 versus ζ^2 for system 2, ■ axis(a);
and system 3,● for ⧫$_P$ = 0.02, o for ⧫$_P$ = 0.01,
axis (b)

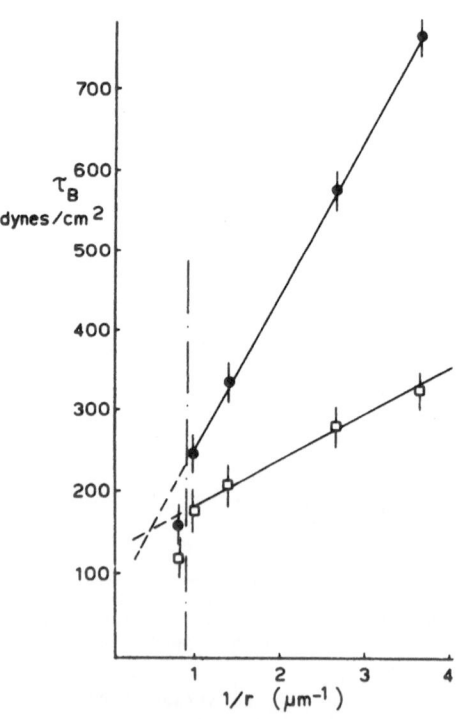

Fig. 5 : τ_B versus $1/r$ for system 4;
⧫$_P$ = 0.10,□ , 0.20,● .

The Microscopic Parameters of the Model

(i) <u>Distance of particle approach</u> - The ratio of slope to intercept in the plot of C_{FP} against ζ^2 gives a value for

$$T(d_1) = \frac{12d_1^2 G'(d_1)}{A} \qquad (16)$$

from which d_1 can be calculated using the now fairly reliable estimates of A. The values obtained for systems 1 and 2a are 8 ± 1 Å and 6 ± 1 Å respectively. For system 3, a similar value is obtained if it is assumed that in this case the constant charge approximation is more appropriate.[11] The same procedure applied to the D_o curves (Fig. 4) gives $d_1 = 7 \pm 3$ Å for system 2a and 6 ± 3 Å for system 3 at constant charge. The $\tau_B - \zeta^2$ data also give values around 8 Å for these systems.

(ii) <u>Elastic constant of the interparticle bond</u> - From equation (8) one can calculate L using reasonable values of the parameters : $r = 10^{-5}$ cm, $A = 10^{-13}$ erg; $d_1 = 10^{-7}$ cm ; (d_1-d_o) = 5×10^{-8} cm; $\zeta = 0$ gives L = 166 dyne cm^{-1} which compares favorably with van den Tempel's estimates[12] from the behaviour at small deformations (which would correspond to ~10^2-10^3 dyne cm^{-1} for our systems).

(iii) <u>n_c, n_F and D_c</u> - For a floc of radius 20 μm the maximum cross-sectional C_{FP} area is about 12×10^{-6} cm^2. The number of chains of particles in each direction in this central plane is 1320 ± 240 and the number of links between two flocs of this size is 20 ± 12 (for system 2a). Hence $n_F \ll n_C$ as we assumed. The values of $D_{C_{FP}}$ calculated for these systems are of order 10^5 s^{-1} which is quite conceivable considering that they are usually injected into the measuring apparatus through a micrometer syringe.

V. CONCLUSIONS

It would seem that the main features of the flow behaviour of coagulated sols, at least at high shear rates, can be accounted for by a model of the type considered above. When the energy dissipation is broken down into its component parts it appears that only two contributions are significant : (i) the viscous energy dissipated by the flow of the suspension medium around the floccules, and (ii) the energy involved in stretching the bonds *within a floc* in order to transfer enough tension to break the bonds *between* two colliding flocs. The fact that many intra-floc bonds must be stretched to break a few *inter*-floc bonds ensures that most of the energy is dissipated inside the flocs. Most of that energy dissipation (E_L) seems to be due to the movement

of liquid into the space between two separating particles even
though the movement is very slight (only a few $\overset{o}{A}$'s).

Now that we have a working model it must be tested
exhaustively at high shear rates and extended if possible, to low
shear rates. Most importantly we must find a suitable
explanation for the magnitude of the energy, E_L, which is
determined by the viscosity of the suspension medium and the
volume of liquid trapped between two particles.

The very significant conclusion from the electrochemical
viewpoint is that all three of the parameters which we have studied
(C_{FP}, D_o and τ_B) turn out to have an explicit and strong

dependence on the ζ-potential of the sol and this of course can be
controlled by controlling the electrochemical conditions in the
suspension medium. It is this aspect of the study which shows
most promise for the future quantitative description of these
systems.

REFERENCES

1. Krieger, I.M., Adv.Colloid Interface Sci., 3, 111–136 (1972).
2. Booth, F., Proc.Royal Soc. (London) 203A, 533–551 (1950).
3. Chan, F.S., Blachford, J. and Goring, D.A.I., J.Colloid
    Interface Sci., 22, 378–385 (1966).
4. Stone-Masui, J. and Watillon, A., J.Colloid Interface Sci.,
    28, 187–202 (1968): 34, 327–9 (1970).
5. Firth, B.A., Ph.D. Thesis, University of Sydney, 1975.
6. Verwey, E.J.W. and Overbeek, J.Th.G., *Theory of Stability of
    Lyophobic Colloids*, Elsevier, Amsterdam, 1948.
7. Smoluchowski, M. von, Physik.-Z, 17, 557, 583 (1916).
8. Bagster, D.F. and Tomi, D., Chem.Eng.Sci., 29, 1773 (1974).
9. Michaels, A.S. and Bolger, J.C., Ind.Eng.Chem.(Fundamentals)
    1, 24, 153 (1962).
10. Hunter, R.J. and Nicol, S.K., J.Colloid Interface Sci., 28,
    250 (1968).
11. Friend, J.P. and Hunter, R.J., J.Colloid Interface Sci.,
    37, 548 (1971).
12. Firth, B.A., and Hunter, R.J., Submitted to J.Colloid
    Interface Sci.
13. Weltmann, R.N. and Green, H., J.Applied Phys., 14, 569
    (1943).

NEW SPECTROSCOPIC METHODS FOR ELECTROCHEMICAL RESEARCH

J.D.E. McIntyre

Bell Laboratories, Murray Hill
New Jersey 07974
U.S.A.

INTRODUCTION

Electrochemists have long sought after experimental techniques which can provide a detailed picture on a molecular scale of chemical and physical processes occurring at the electrode-solution interface.   Such insight would enable us to gain a much more complete understanding of the many and varied factors which determine the rate and course of electrochemical reactions at solid surfaces.   In order to acquire such perception we must first ascertain empirically:

(i)   the chemical identity and concentration of ionic, atomic and molecular species (including dissociation fragments) at the electrode surface, either adsorbed or at the outer Helmholtz plane;

(ii)   the geometric or structural arrangement of these species and their positions relative to the substrate surface atoms;

(iii)   their vibrational, rotational, and translational motions on the surface;

(iv)   the charge distribution and energy level structure of the valence electrons in both adsorbate and substrate;

(v)   the effects of external perturbations such as electric and magnetic fields, light, pressure and temperature on factors (i)-(iv).

Owing to the recent rapid development of spectroscopic techniques for surface analysis, attainment of this goal is coming increasingly closer to hand.   The purpose of this paper is to illustrate and discuss how some of these new methods can be applied for the study of technologically important electrochemical processes such as corrosion, metal deposition, and electrocatalysis.

SURVEY OF SPECTROSCOPIC METHODS

Spectroscopic methods for electrochemical studies can be conveniently divided into two classes: *ex situ* and *in situ*. A number of the more important techniques in each class are listed in Table 1. This is not an exhaustive list of presently available techniques for surface characterization but rather is representative of those which have been, or could now be, profitably employed in conjunction with conventional electrochemical voltammetric techniques for the study of electrode processes. A detailed description of these spectroscopic methods is not possible within the content of this paper. I shall outline only their essential features, placing emphasis on those less likely to be familiar to electrochemists. More complete descriptions of the techniques and their applications in non-electrochemical surface studies may be found in a number of recent publications (1-9).

TABLE 1

Spectroscopic Methods for Electrochemical Research

Ex Situ

---

1. Electron Diffraction (LEED and RHEED)
2. Auger Electron Spectroscopy (AES)
3. Ultraviolet Photoelectron Spectroscopy (UPS)
4. X-ray Photoelectron Spectroscopy (XPS or ESCA)
5. Electron Energy-Loss Spectroscopy (ELS)
6. Ion Scattering Spectroscopy (ISS)
7. Secondary Ion Mass Spectrometry (SIMS)
8. Spark Source Mass Spectrometry
9. Mössbauer Spectroscopy

In Situ

---

1. Ellipsometric Spectroscopy
2. Infrared and Ultraviolet-Visible Spectroreflectometry
3. Raman Spectroscopy (surface and bulk solution)
4. Electron Spin Resonance Spectroscopy

Ex Situ Spectroscopy

Electron diffraction. In order to correlate the catalytic activity of a surface with its geometric and electronic structure, well-defined single- or stepped-crystal planes must be employed. At present, structural characterization of surfaces is carried out almost exclusively by low-energy electron diffraction (LEED). It should be remembered, however, that while LEED may show a surface to be well-ordered on an atomic scale, typical *real* crystal surfaces have dislocation densities of the order of 10^4-10^6 cm^{-2} (semi-conductors) or 10^6-10^8 cm^{-2} (metals) and are heterogeneous on both a microscopic and submicroscopic scale (1). The presence of atomic steps and ledges, single atoms and vacancies in crystal planes, which appear well-ordered atomically, can have a profound influence on such processes as heterogeneous catalysis and crystal growth. At present LEED is capable of revealing the two-dimensional trans-lational periodicity of a substrate surface or of a substrate covered by an ordered overlayer, but generally *not* the actual positions of the atoms in the surface layer. Recent progress in theoretical and computational analysis of LEED spectra (diffracted peak intensity *vs* incident electron energy in the range 10-200 eV) now enables surface structures to be determined with some certainty in a limited number of *simple* cases (10).

Reflection high-energy electron diffraction (RHEED) probes the surface structure by employing higher energy electrons (up to 100 keV) near grazing incidence. The grazing angle ensures small momentum transfer and restricts the mean free path of the elastically-scattered electrons to a few atomic layers. While LEED patterns yield essentially a plan view of the reciprocal net, RHEED patterns provide a side-view of the reciprocal-net rods. By combining LEED and RHEED on the same apparatus, information can be obtained about surface topography, the degree of ordered surface coverage, and the size and symmetry of the surface mesh (1). RHEED is extremely use-ful for crystal-growth studies since both reflection and transmission patterns can be observed (11).

Electron spectroscopies. The basic features of the ultra-high vacuum electron spectroscopies UPS, XPS and AES are illustrated in Fig. 1. UPS usually employs a helium resonance lamp as a source of 21.2 or 40.8 eV photons and hence is primarily employed for investigations of electronic surface structure at energies in the valence-level region. XPS and AES employ monochromatic X-rays of 1.2-1.5 keV energy (XPS) or variable-energy electrons of 0.05-5 keV energy (AES) for electronic excitation of the solid and thus can probe both core- and valence-level regions. All three techniques owe their high surface sensitivity to the small mean free path (5-50 Å) of the excited electrons prior to inelastic scattering. The characteristic binding energies of those electrons which escape from the solid surface are determined by measuring their kinetic

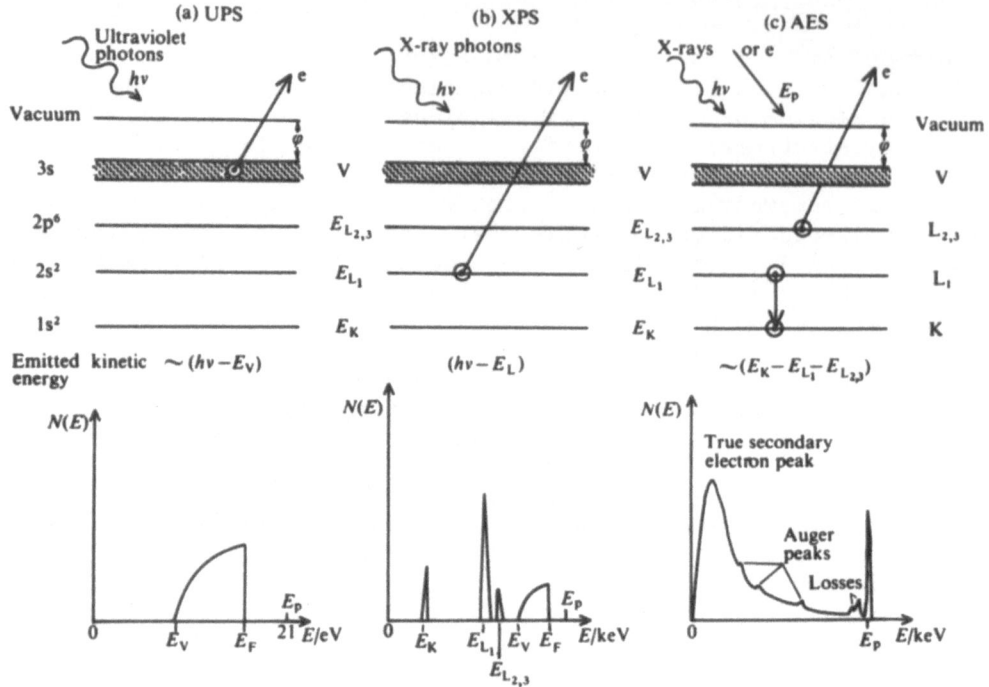

Fig 1   Schematic comparison of ultra-violet photoemission, X-ray
         photoemission and Auger electron spectroscopies. At the
         top are shown energy-level diagrams for a hypothetical free-
         electron metal (*not* to scale). Below are typical energy
         distribution curves for electrons emitted with kinetic
         energies between E and E + dE (Prutton (5)).

energies with retarding-field or electrostatic-deflection energy
analyzers. AES is usually the most suitable technique for
chemical identification of surface atoms because of the relatively
short scan times required. XPS, however, reveals 'chemical' shifts
of atomic core levels which result from the redistribution of
electronic density associated with chemical bonding (12). Such
shifts are characteristic of the degree of charge transfer in
surface chemical bonding and have been employed in solid-gas interfacial studies to distinguish different adsorption states of
chemisorbed molecules, e.g. CO on Mo and W films (13). Angular-
dependent XPS can be used to enhance selectively either surface or
bulk sensitivity, to investigate surface roughness, and to distinguish continuous from patched overlayers (14).

    The recent availability of tunable, high-energy synchrotron
radiation (h$\nu\sim$4-120 eV) has permitted the extension of UPS to

investigations of deep-lying d-band states of metals (15), bonding
states of adsorbed molecules (16), *empty* surface and conduction-
band states of semiconductors (17), and the energy dependence of
matrix element effects (15). The ability to vary the synchrotron
source energy continuously over a wide energy range enables determination of the relative contributions of photoionization cross-
sections and electron escape depth to the surface sensitivity of
photoelectron spectra (16). The wide range of energy states that
can be probed by UPS makes it useful as a 'fingerprinting' technique for identifying adsorbed molecular species (16, 18). Angular-
resolved UPS measurements now offer considerable promise of
revealing details of chemical bonding at solid surfaces (19, 20).
In addition, the ability of photoemission spectra to reveal *absolute*
energy levels of *initial* electronic states makes it particularly
useful for correlation with *in situ* optical surface spectroscopy
which measures the energy distribution of the joint density of
initial and final states (21).

   Electron energy-loss spectroscopy (ELS) analyzes the energies
of primary electrons both elastically and inelastically scattered
at a solid surface. As the energy of the incident beam is varied,
spectral features produced by inelastically-scattered electrons,
which have lost energy through excitation of the solid, are shifted
along the energy scale at a fixed energy separation from the elastic
peak (Fig. 1). At high primary energies ($E \sim 100$–2000 eV), losses
due to excitation of surface or bulk plasmons are observed together
with losses due to excitation of electronic transitions from core
or valence levels to unfilled bulk or surface states (22–28). This
technique may be used to investigate the energy distribution of
unoccupied electronic states *below* the vacuum level, a region
inaccessible to conventional UPS. ELS has also been used with low-
energy primary electrons ($E \sim 5$ eV) and a high resolution spectrometer
($\Delta E \sim 10$ meV) to detect vibrational frequencies of surface complexes
formed by adsorption, as well as surface phonons in the substrate
(29–31).

Ion spectroscopies. The basic principles of ion-scattering spectroscopy (ISS) are illustrated in Fig. 2. A beam of rare gas ions
(He^+, Ne^+, Ar^+) of mass M_1 and energy E_0 (0.5–10 keV) is directed
at the target surface consisting of atoms of mass M_2. At these low
energies the primary ions are *elastically* scattered over a range of
angles θ_L (laboratory coordinates) such that (33, 34):

$$\frac{E_1}{E_0} = \frac{1}{(1 + \mu)^2} \left[\cos\theta_L + (\mu^2 - \sin^2\theta_L)^{\frac{1}{2}} \right]^2 \tag{1}$$

where $\mu = M_2/M_1$. For ions scattered at $90°$:

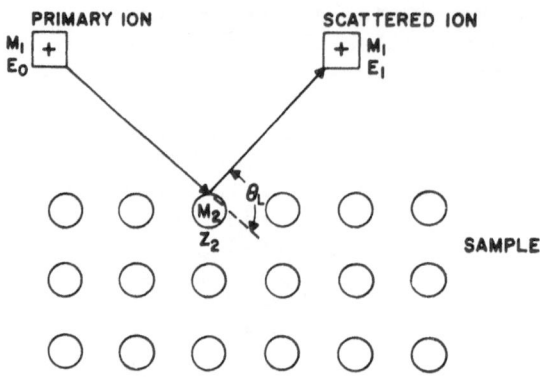

Fig. 2   Schematic diagram of low-energy ion scattering at a solid
         surface (Honig (32)).

$$\frac{E_1}{E_0} = \frac{M_2 - M_1}{M_2 + M_1} \tag{2}$$

The energies of the scattered ions are determined with an electrostatic analyzer combined with an electron multiplier. Owing to
the high probability of neutralization of ions which penetrate
beyond the first atomic layer, this technique is extremely surface
sensitive. Thus, the energy distribution curve of the elastically
scattered ions essentially yields an elemental analysis of the
surface atom layer *per se*, the peak energies E_i being proportional
to the masses M_i of the constituent surface atoms i and the peak
areas to their surface concentration.

From Eq. (2) it is evident that resolution of ISS is very
limited for heavy ions of similar masses. It has recently been
observed, however, that anomalous surface neutralization occurs
when He ions are scattered from the elements: Pb, Bi, In, Ge, Sb,
Ga and Sb (35). Pronounced oscillations are observed in the
scattered-ion yield due to oscillations in the ion-neutralization
probability. The above target atoms have d-state energy levels
lying within 10 eV of the He ionization potential (24.6 eV). It
has been proposed (36) that this phenomenon arises from a quantum
mechanical phase interference associated with near-resonant ion-
atom charge exchange. This oscillatory effect may provide a useful
method for investigating the surface composition and structure of
heavy metal surface alloys and adsorbed metal layers. A number of
such systems (e.g. Pb/Pt, Pb/Au) are reported to show electrocatalytic activity (37-40).

ISS may also be employed to investigate the atomic structure

of surface layers with two or more constituents (e.g. GaP, NiO).
The scattered ion spectrum of such a surface contains peaks corresponding to single scattering events at the individual atoms. If
the azimuthal angle of the incident ion beam at the surface is
varied, atoms lying below the outermost surface plane may be geo-
metrically shadowed, e.g. the Ni atoms when a (2x1)O layer is
adsorbed on a Ni(110) surface (41). This causes a decrease in the
scattered ion peak ratio which reflects the local surface topography.

In surface analysis by secondary ion mass spectrometry (SIMS),
the energy analyzer used in ISS is replaced by a quadrupole mass
spectrometer. Unlike ISS, this technique is not restricted to
irradiation of a small area to obtain good mass resolution. By
employing low primary ion flux rates, the sample sputtering rate
can be made very slow. Sensitivities of 10^{-6} of a monolayer can
be achieved with this technique (42, 43). It is difficult to use
SIMS for quantitative analysis, however, because the emission
probability of a secondary ion depends critically on its local
environment in the solid and may vary by orders of magnitude among
samples of different composition and structure.

Spark source mass spectrometry can conveniently be adapted
for surface studies and thin film analysis (44, 45) and is capable
of detecting relative concentrations of elemental impurities in the
mass range 10-200 amu, both rapidly and simultaneously. The
sensitivity is such that a fractional monolayer of a surface impurity
can be detected. With this technique a craterful of material at the
sample surface, 0.2-5 μm deep, is volatilized by generating a high-
voltage RF spark between a fine-tipped probe electrode and the
sample. The resulting ions of widely differing kinetic energy and
charge are consecutively focussed by electric and magnetic fields
to a planar focal locus and detected with an Ilford-type ion
sensitive AgBr plate. Surface species can be distinguished from
those in the bulk by repetitively sparking the same sample spot.
While not possessing the extreme surface selectivity or yielding
the same types of 'chemical' information as some of the ex situ
methods discussed above, the technique is semi-quantitative and
valuable for analysis of specimen stains, inclusions or composition
gradients (45). It has been employed to advantage in analysis of
impurities in electroplates in both the surface and subsurface
regions (46-48).

An elemental depth composition profile of a solid sample or
surface film can be obtained by applying a surface analytical
technique such as AES, XPS, ISS or SIMS in combination with ion
sputter etching. The elemental composition at the bottom of the
crater formed by the ion beam is monitored continuously during
the erosion process (49, 50). AES and SIMS can also be applied to
obtain an image of the lateral distribution of elements across a
surface with a resolution of 1-5 μm. By combining depth composition

analysis with lateral scanning of the ion or electron-probe beam,
it is possible to obtain a three-dimensional elemental map of the
sample. Such analyses are of particular value for composite thin
film structure determination and mapping the spatial distribution
of unwanted impurities in microelectronic circuits (50).

Mössbauer spectroscopy. Mössbauer spectroscopy involves measurement of the recoil-free γ-ray emission or resonance absorption
spectra of nuclei bound in solids (51). The free-atom recoil
energy of an atom of mass M emitting a γ-ray of energy E_γ is given by:

$$E_R = E_\gamma^2 / 2Mc^2 \qquad (3)$$

where c is the velocity of light. If E_R is smaller than the
binding energy of the atom in the solid (1-10 eV), but greater than
the characteristic energy of the lattice vibrations (0.01-0.1 eV),
the recoil energy will be primarily dissipated by excitation of
phonons. However, if $E_R << \theta_D$, the Debye temperature of the solid,
no phonons can be excited and the momentum is transmitted to the
atoms ($> 10^8$) of the surrounding lattice. As a consequence,
negligible energy is lost to recoil in the emission process and
the linewidths of the emitted γ-rays are extremely narrow, corresponding closely to the natural lifetimes of the excited nuclear
states. It thus becomes possible to detect the very small changes
in nuclear energy levels (10^{-7} eV) associated with the electrostatic
interaction of the nucleus with the surrounding electrons of the
molecule or complex in which it is chemically bound. The recoilless
γ-ray absorption spectra reveal isomer shifts, quadrupole splittings
and magnetic hyperfine splittings and thus provide information
about the symmetry and structure of the molecule, as well as the
oxidation state of the absorbing atom. Among those isotopes for
which a Mössbauer effect has been observed and which also are of
interest in electrochemical studies are: ^{57}Fe, ^{57}Co, ^{61}Ni, ^{67}Zn,
^{99}Ru, ^{117}Sn, ^{127}I, ^{129}I, ^{131}Xe, ^{181}Ta, ^{182}Ta, ^{182}W, ^{187}Re, ^{191}Ir,
^{193}Ir, ^{195}Pt and ^{197}Au.

The fraction of γ-rays emitted without energy loss to the
lattice is given by the Debye-Waller factor:

$$f = \exp\left[-\frac{E_0}{k\theta_D} \left(\frac{3}{2} + \frac{\pi^2 T^2}{\theta_D} \right) \right] \qquad (4)$$

where E_0 is the energy of the excited nuclear state above the ground
state, T is the absolute temperature, and k is Boltzmann's constant.
For isotopes with high values of E_0 (e.g. for ^{197}Au, E_0 = 77 keV),
Mössbauer spectra must usually be measured at cryogenic temperatures,
ex situ. However, for isotopes with low values of E_0 (e.g. for ^{57}Fe,
E_0 = 14.4 keV) the Mössbauer effect may be observed at temperatures
over 1000°C. It is then feasible to use Mössbauer spectroscopy as

an *in situ* method. This fact has been exploited in a number of *in situ* corrosion studies of Fe, Co and Sn and their alloys (for a recent review, see Ref. (52)). The technique has sensitivity adequate for the study of corrosion films of only 10-50 Å thickness (52-55).

Coupling of surface analysis and electrochemical systems. Few of the *ex situ* ultra-high vacuum spectroscopies described above have been applied for the study of electrochemical systems owing to the requirement that the electrochemical procedures be carried out without contamination of either the vacuum system or the electrode surface (56). Two approaches to the solution of this problem are currently being investigated. In the first (56, 57), the surface of a metal single-crystal electrode (e.g. Pt) is cleaned by Ar^+ ion bombardment and its surface composition and structure monitored by AES and LEED. The sample is then positioned in contact with the end of a dual capillary to form a thin-layer electrochemical cell (TLEC). After back-filling the chamber with purified argon, electrolyte is admitted into the TLEC and the desired electrochemical experiments are carried out. The electrode is then removed from potential control and rinsed; excess solvent is evaporated. Subsequently the surface is re-examined by various electron spectroscopies without ever having been removed from the vacuum chamber. An advantage of the TLEC configuration is that surface contamination by adsorbable impurities is minimized owing to the very small volume of electrolyte employed ($\sim 10^{-3}$ cm^3).

In the second approach (58-61), the sample preparation, electrochemistry and spectroscopic analysis are carried out in separate evacuable chambers. Provision can be made to freeze the electrode and surrounding electrolyte very rapidly, with potential control still applied. Excess solvent and electrolyte are sublimed away under vacuum and the electrode, together with chemisorbed water and other species, is then transferred back to the electron spectrometer at liquid nitrogen temperature for analysis of surface composition and structure. In this way it is hoped to avoid changes in the structure of the compact water layer and chemisorbed overlayer(s) and to minimize decomposition of surface films by high-energy X-ray or electron beams. These problems are of particular concern with the first method described above.

Whether these techniques will yield spectroscopic information representative of the actual interfacial conditions existent at a working electrode remains an open question. Until more definitive results become available, corroborated where possible by *in situ* spectroscopy, the interpretation of the results of *ex situ* experiments involving electrode surface films of only monolayer thickness must be treated with considerable caution.

In Situ Spectroscopy

In situ spectroscopic studies of chemical and physical processes occurring at an electrode-electrolyte interface are primarily restricted to the use of optical methods. The sensitivities of modern surface reflection spectroscopy techniques are now sufficiently high to permit their use in studies of adsorption and catalysis on well-defined single- or stepped-crystal surfaces. Ultraviolet-visible spectroscopy can detect the formation of various adsorption states of 'known' species and provide information about the electronic-charge and energy-level structure of the surface. Infrared and Raman spectroscopy can provide detailed information about the chemical identity of molecular species, as well as their bonding configurations and interactions. Owing to their rapid response, these techniques are also useful for kinetic measurements and spectroscopic detection of dissolved reaction intermediates and products in the electrochemical diffusion layer.

Ultraviolet-visible reflection spectroscopy. With the advent of minicomputer-controlled, automatic photometric reflectometers and ellipsometers (21, 62), it is now possible to measure the continuous optical adsorption spectrum of an electrode-electrolyte interphase over a wide photon-energy range (1-6 eV). Alteration of the optical properties of the interfacial or 'surface' region (which in electrochemical systems includes the ionic double-layer, adsorbate and perturbed substrate layers) modulates the absorbances of both surface (s) and bulk substrate (b) such that the net reflectance change is given by:

$$-\Delta R = \Delta A_s + \Delta A_b \tag{5}$$

For a very thin surface region whose thickness d is much less than the wavelength λ of the incident radiation, the fractional reflectance changes $\delta R/R|_\perp$ and $\delta R/R|_\shortparallel$ measured by reflectometry are closely related to the angular quantities $\delta\psi$ and $\delta\Delta$ measured by ellipsometry. If both surface and bulk regions are strongly absorbing and the angle of incidence $\phi \leq 45°$, then to first order in d/λ (21):

$$\left.\frac{\delta R}{R}\right|_\perp = -\frac{8\pi d n_a \cos\phi}{\lambda} \mathrm{Im}\left[\frac{\hat{\epsilon}_s - \hat{\epsilon}_b}{\epsilon_a - \hat{\epsilon}_b}\right] \tag{6}$$

$$\left.\frac{\delta R}{R}\right|_\shortparallel = \frac{1}{\cos^2\phi}\left.\frac{\delta R}{R}\right|_\perp \tag{7}$$

$$\delta\psi = \frac{\sin 2\psi \tan^2\phi}{4}\left.\frac{\delta R}{R}\right|_\perp \tag{8}$$

$$\delta\Delta = \frac{4\pi d n_a \cos\phi \tan^2\phi}{\lambda}\mathrm{Re}\left[\frac{\hat{\epsilon}_s - \hat{\epsilon}_b}{\epsilon_a - \hat{\epsilon}_b}\right] \tag{9}$$

Here $\hat{\varepsilon}_s$ and $\hat{\varepsilon}_b$ represent the complex optical-frequency dielectric constants of the absorbing surface and bulk regions, respectively; $n_a = \sqrt{\varepsilon_a}$ is the pure real refractive index of the transparent ambient; and ψ is the azimuthal angle measured by ellipsometry for the 'bare' (or unperturbed) surface.

In certain instances, e.g. when the substrate is highly reflecting, $\Delta A_s >> \Delta A_b$ and the absorption spectrum of the surface region can be measured *directly* by reflectance spectroscopy (*cf.* Eq. (5)). Usually, however, the ΔA_b term is not negligible and may even be dominant for substrates of low reflectivity. It is then necessary to apply a deconvolution procedure (63), analogous to that employed in ellipsometric spectroscopy, to determine Im $[\hat{\varepsilon}_s (\omega)]$. The latter quantity is proportional to the joint density of initial and final electronic states (JDOS) in the surface region (21).

For very thin surface regions ($d << \lambda$), the relative contributions of the normal and tangential components of A_s and ΔA_b to the ΔR spectrum can be calculated using an idealized stratified-layer model (21). This model does not attempt to account for strong interactions between adjacent layers of the type involved in chemisorption. Nonetheless, such calculations are valuable for distinguishing the origins of many optical effects observed in experimental differential reflection spectra. Figs 3 and 4 illustrate the results of model spectra calculations for two systems of interest in 'underpotential' metal deposition studies. Here the optical properties of the metal overlayers were modelled using *bulk* metal optical constants. Several features of the calculated spectra are noteworthy. For the case of Pb/Ag (Fig. 3): (i) at low photon energies the primary contribution to ΔR is A_s; (ii) at high photon energies ΔA_b becomes dominant (owing to the onset of a d-band $\rightarrow$ Fermi surface transition in the Ag substrate) and causes a sign reversal in $\Delta R/R$; (iii) the true ΔR spectra are highly distorted in experimental measurements of $\Delta R/R$ in energy regions where the substrate reflectance falls to low values. It is possible, however, to measure the *absolute* reflectance spectrum of an electrode *in situ* at $\phi = 45°$ by use of the Abelès relation: $R_\perp = R_\parallel / R_\perp$ (64). Conjugate measurement of $R_\perp(\omega)$ and $R_\parallel(\omega)$ then enables experimental $\Delta R/R$ spectra to be converted to ΔR spectra. Finally, for the case of Ag/Au (Fig. 4), it is evident that there is no energy region in which the true form of the surface absorption spectrum can be discerned directly from measurements of $\Delta R(\omega)$.

Interpretation of ΔR spectra of metal electrodes is further complicated by the fact that in its surface region, the intrinsic energy distribution of the JDOS of the highly-absorbing substrate may be strongly perturbed by the formation of chemisorptive bonds. This effect is in large part responsible for the anomalously high absorption coeffcients (10^5-10^6 cm^{-1}) measured by reflectometry or

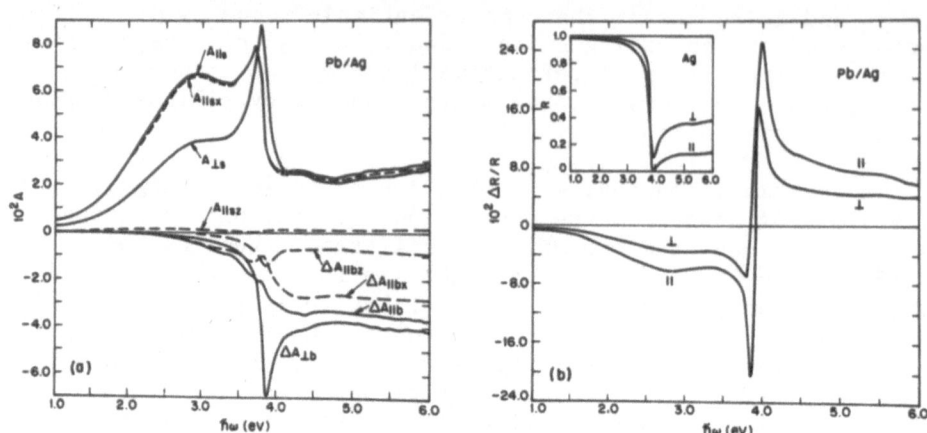

Fig. 3 (A) Calculated frequency dependence of the surface (s) and
        bulk substrate (b) absorbance components produced by
        deposition of a monolayer of Pb (d = 3.50 Å) on an Ag
        electrode in aqueous electrolyte. The optical properties
        of the Pb overlayer were modelled using the optical
        constants of bulk Pb. (B) Normalized differential reflect-
        ion spectra produced by surface monolayer deposition as in
        (A). ϕ = 45.0°.

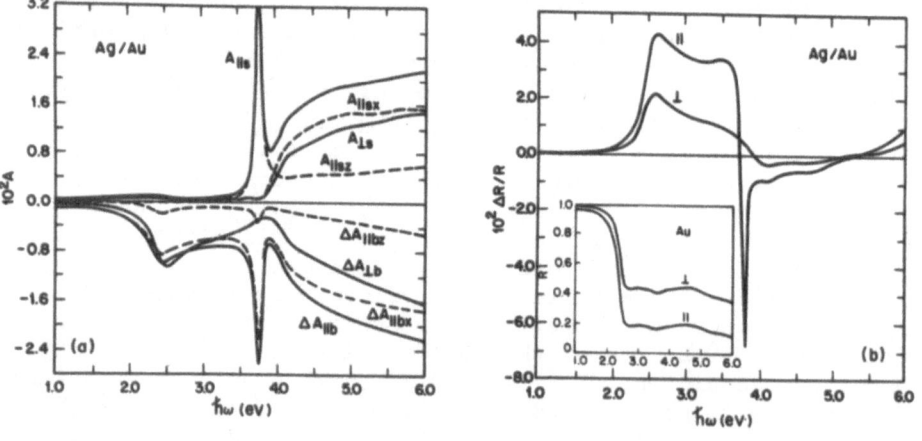

Fig. 4 (A) Calculated frequency dependence of the surface and
        bulk substrate absorbance components produced by deposition
        of a monolayer of Ag (d = 2.88 Å) on an Au electrode in
        aqueous electrolyte. The optical properties of the Ag
        overlayer were modelled using the optical constants of
        bulk Ag. (B) The normalized differential reflection
        spectra from (A). ϕ = 45.0°.

ellipsometry for adsorbed monolayers (65-67). In addition, while UPS yields information on the probability of optically exciting an electron to a given final energy E_f, the parameter Im $\varepsilon_s(\omega)$ deduced from reflection spectroscopy is proportional to the sum of electronic transition probabilities to *all* accessible final states for which $\hbar\omega = E_f - E_i$ (E_i = initial energy). The integral nature of the absorption process tends to obscure spectroscopic structural details. Owing to the combination of these effects, chemical identification of adsorbed reaction products by UV-visible reflection spectroscopy is usually precluded. The technique is extremely useful, however, for detecting changes in the electronic interactions of chemisorbed species with and through the substrate as a function of surface coverage.

Infrared and Raman spectroscopy. Electronic absorption spectroscopy in the UV-visible region has not proved generally useful for chemical identification of either dissolved or adsorbed species generated electrochemically, owing to the relatively low information content of the spectra observed. In contrast, vibrational spectroscopy has a much greater molecular specificity, yielding information about chemical structure and bonding, the nature of adsorption sites on solids, and interactions with the ambient environment. Coupling infrared (IR) spectroscopy with electrochemical methods is very difficult, however, owing to: (i) strong competitive light absorption by the solvent-electrolyte system; (ii) the low extinction coefficients of molecular species in the IR region; (iii) the low detection sensitivity for species adsorbed at highly-reflecting metal surfaces, and (iv) a low signal-to-noise ratio which precludes fast kinetic studies of transient species at low-level concentrations. Some limited success has been achieved, however, by use of thin-layer (68, 69) and internal-reflection (70, 71) spectroelectrochemical cells.

Conventional Raman spectroscopy is in general much too insensitive for spectroelectrochemical studies of the low level of soluble electrogenerated scatterers within an electrochemical diffusion layer. A very promising new approach to the problem of measuring vibrational spectra of such species, however, is the application of *resonance Raman spectroscopy* (72). Laser excitation within an electronic absorption band of a species of interest effectively enhances the Raman scattering cross-section by factors up to 10^6 for certain vibrational modes. As a result it becomes possible to detect species with extinction coefficients $>10^3$ M^{-1} cm^{-1} at millimolar concentration levels. Since the exciting line is in visible region, there is only low-level interference from the solvent or supporting electrolyte (unless they are colored).

The physical origins of the resonance Raman effect can be seen as follows (Fig. 5). The total scattering intensity I_s (photons molecule^{-1} sec^{-1}) for a transition from initial ground

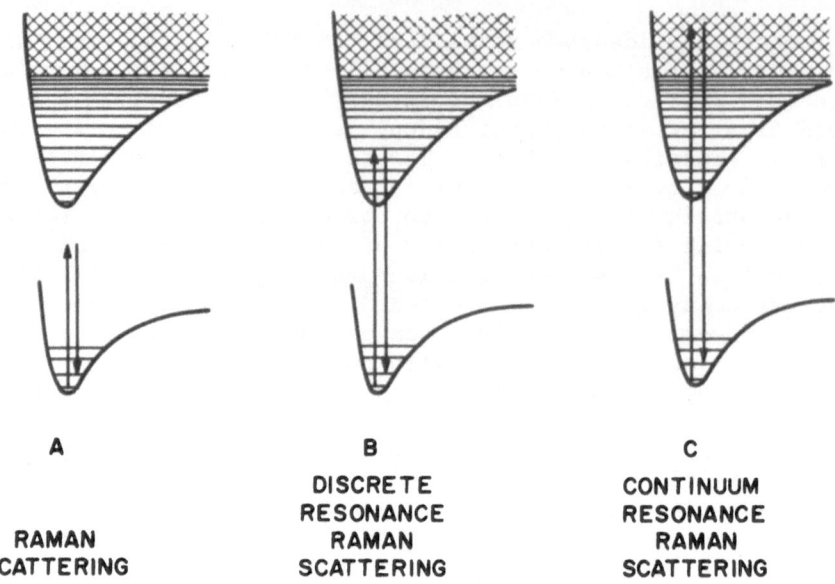

A

B
DISCRETE
RESONANCE
RAMAN
SCATTERING

C
CONTINUUM
RESONANCE
RAMAN
SCATTERING

**RAMAN
SCATTERING**

Fig. 5   Classification of Raman scattering according to laser
frequency.   (A) The incident laser frequency is far from
resonance with any real electronic transition; normal
Raman scattering is observed.   (B) Electronic transitions
are excited from the ground state to discrete levels of
an intermediate state.   (C) The incident frequency excites
transitions into the range of a dissociative continuum
(Rousseau and Williams (75)).

state $|G>$ to final state $|F>$ is given by (73, 74)

$$I_s = \frac{8\pi\omega_s^4 I_L}{9c^4} \sum_{\rho\sigma} \left| (\alpha_{\rho\sigma})_{GF} \right|^2 \tag{10}$$

where I_L is the incident intensity at frequency ω_L, ω_s is the
scattered frequency, and $(\alpha_{\rho\sigma})_{GF}$ is the polarizability tensor for
the transition $|G> \to |F>$ with incident and scattered polarizations
ρ and σ, respectively.   From second-order perturbation theory (75)

$$(\alpha_{\rho\sigma})_{GF} = \frac{1}{\hbar} \sum_I \left[\frac{<F|p_\rho|I><I|p_\sigma|G>}{\omega_{GI} - \omega_L} + \frac{<I|p_\rho|G><F|p_\sigma|I>}{\omega_{IF} + \omega_L} \right] \tag{11}$$

where ω_{GI} is the energy spacing between an excited rovibronic state
$|I>$ and ground state $|G>$, and p is the electron momentum operator.
Near resonance ($\omega_L \approx \omega_{GI}$), the first term in Eq. (11) is dominant.
By separating out the vibrational wave-function and expanding the

resultant expression for the electronic matrix element $M(\xi)$ to first
order in $\Delta\xi$, the normalized deviation from the equilibrium separation ξ_0, the following simplified expression is obtained for the
case of *discrete* resonance Raman scattering (75):

$$I_s = \frac{8\pi\omega_s^4 I_L}{9c^4\hbar^2} \left[M(\xi_0)\right]^2 \frac{|<f|i><i|g>|^2}{(\omega_{GI}-\omega_L)^2 + \Gamma^2} \tag{12}$$

Here states $|g>$ and $|f>$ correspond to vibrational wavefunctions
of the ground electonic states $|G>$ and $|F>$, while state $|i>$
corresponds to those of the excited electronic state $|I>$. Γ is
the linewidth of the discrete transition; the matrix elements
$<f|i>$, $<i|g>$ correspond to the Franck-Condon overlap amplitudes.
From Eq. (12) it is evident that the spectrum of the Raman-scattered
radiation depends on the specific vibrational and rotational quantum
numbers of the excited state selected by tuning ω_L. A small change
in ω_L will induce resonance with a different discrete state and
may produce marked differences in the appearance of the re-emission
spectrum. In contrast, for normal Raman scattering (Fig. 5A)
$\omega_L<<\omega_{gi}$ and the resonance denominator remains effectively constant,
with the result that all of the scattered radiation appears in the
ω_s^4 term (Eq. (12)). For the case of continuum resonance Raman
scattering, all populated vibrational-rotational levels are at
resonance with some continuum state since the energy separation of
the continuum levels is infinitesimally small. As a result, the
Raman scattering spectrum varies smoothly with ω_L since the re-
emission corresponds to the summation of Raman transitions from
all populated levels. Further characteristic differences among
the three types of Raman scattering are detailed in Ref.(75).

APPLICATIONS IN CATALYSIS AND ELECTROCHEMICAL RESEARCH

Applications of the methods discussed above may be illustrated
by a number of examples, where both *ex situ* and *in situ* spectroscopies of widely differing energy ranges have been employed as
interfacial probes.

Ex Situ

Surface segregation in binary alloys. Copper-nickel alloys are of
considerable interest as hydrogenolysis and dehydrogenation
catalysts (76, 77). In order to correlate their observed activity
with electronic and geometric factors, it is essential to know
the actual surface composition of the alloy since this may differ
markedly from that in its bulk. Fig. 6 illustrates the resolv-
ability of the neighbouring elements Cu and Ni by ISS and the

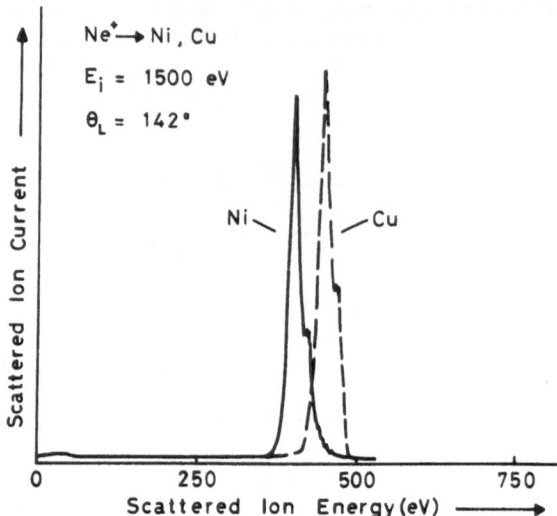

Fig. 6   Spectra for pure Ni and pure Cu as studied by Ne[+] ion
         scattering (Brongersma and Buck (78)).

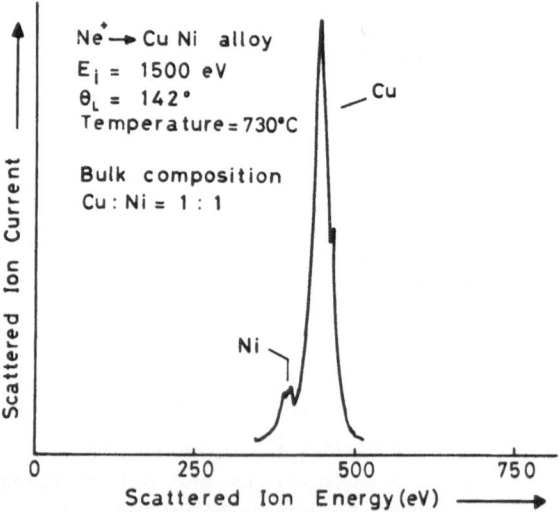

Fig. 7   Ion scattering spectrum of an annealed Cu-Ni alloy,
         illustrating the pronounced surface enrichment in Cu
         (Brongersma and Buck (78)).

nearly equal response observed in measurements on pure bulk
samples. Fig. 7 illustrates the almost complete suppression of
the Ni peak observed for the scattering of Ne ions from the surface
of a Cu-Ni alloy of 1:1 bulk composition (78). Here, the sample
was maintained at a temperature of 730°C to anneal sputtering
damage rapidly. The spectrum indicates that the composition of the
surface layer is almost pure Cu. This result cannot be attributed
to preferential sputtering since Cu is the more easily removed
component. Marked enrichment was also observed for a 10% Cu-90%
Ni alloy. Similar results have been obtained by low-energy
(∿100 eV) AES (79) and UPS (80). It should be noted that high-
energy (∿1000 eV) AES cannot detect this surface segregation
owing to the much larger electron escape depth (∿20 Å).

The observed surface enrichment in Cu is consistent with the
thermodynamic prediction that the component with the lowest latent
heat of sublimation will preferentially segregate to the surface
(81). This fact is apparently responsible for the rapid decrease
in activity of Cu-Ni alloys for ethane hydrogenolysis with their
rather constant activity for cyclohexane dehydrogenation (82).
Under actual catalytic conditions, the alloy's surface composition
may differ somewhat from that found in ultra-high vacuum since Ni
may be stabilized in the surface layer by forming chemisorptive
bonds with H atoms. Similar effects may occur in electrochemistry
where an oxide-forming component of an alloy electrocatalyst may
be segregated to the surface.

Corrosion and passivation. In order to understand the mechanism
of passivation of various stainless steel alloys, it is essential
to establish the composition and structure of the surface film.
Fig. 8 illustrates the ISS response observed (83) for a Fe-24 Cr
alloy sample as a function of the time of sputtering with Ne^+ ions.
In this system the Cr sensitivity is 1.5 times that of Fe (*cf.*
Fig. 6), but there is no preferential sputtering of the metallic
components in either the bulk alloy or surface oxide phases. The
variation of the Cr/Fe content through the oxide film is evident
from Fig. 8. In Fig. 9 is shown the corrected Cr/Fe atom composition
ratio as a function of depth from the surface for four different
Fe-Cr alloys. The oxide thicknesses were estimated to be ∿20 Å
for the high-Cr alloys (Cr>12%) and 30-40 Å for the lower Cr alloys.
The average Cr content of the oxide was found to vary linearly with
alloy composition, possibly as a result of the greater affinity
of Cr for O. Complementary ISS studies with He^+ ions showed the
film composition to be oxygen-rich (or metal-deficient) M_2O_3 at the
air/oxide interface and oxygen-deficient at the oxide/alloy interface.

From these results it was concluded (83) that the 'stainless'
behaviour of these alloys is not associated with the quantity of
Cr in the film, but rather with its distribution, the defect

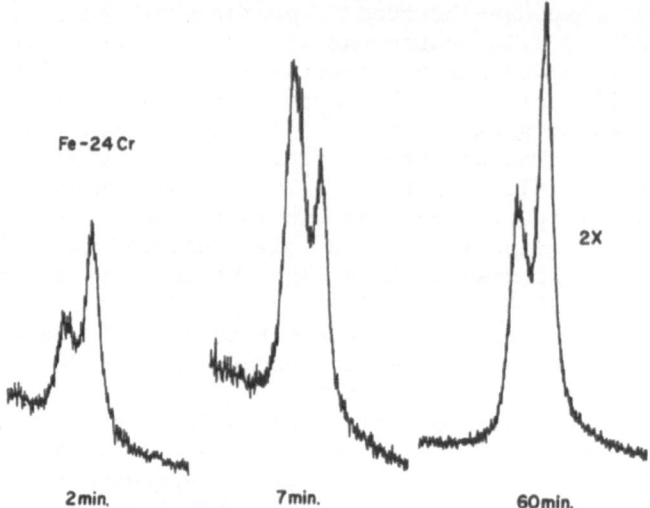

Fig. 8 ^{20}Ne$^+$ ion scattering spectrum of an Fe-24 Cr alloy at
different sputtering times. The Cr sensitivity is 1.5
times that of Fe. The two spectra on the left were
recorded in the surface oxide region; that on the right
is for the bulk alloy (Frankenthal and Malm (83)).

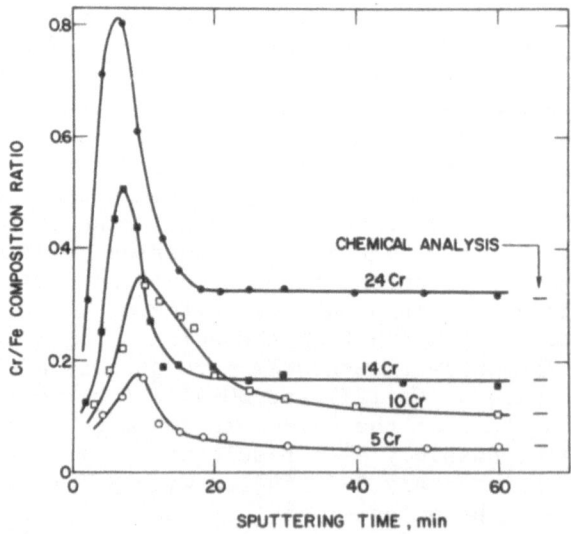

Fig. 9   Cr/Fe atomic composition ratio as a function of ^{20}Ne$^+$ ion
sputtering time for four Fe-Cr alloys. The curves are
corrected for the relative scattering sensitivities of
the two components (Frankenthal and Malm (83)).

structure of the oxide, and the bonding of the metal cations. ISS gives no information about the valence state(s) of these cations. It would seem, therefore, that complementary angular-dependent XPS and Mössbauer spectroscopy studies of such alloys could yield valuable additional information about the nature of such mixed-oxide surface films, whose passivating properties depend on the state of aggregation, particle size, and electronic and ionic conductivities.

In Situ

Electrocatalysis and metal deposition. Recent studies (37-40) have shown that underpotential deposition (UPD) of heavy metal atoms on a foreign metal substrate can significantly enhance the rates of many electrochemical reactions. This catalysis may result from: (i) increasing the surface electronic density of states at the Fermi level; (ii) decreasing the local work function and enhancing electron tunneling; (iii) shifting the point of zero charge (pzc) and hence altering the influence of the electrical double-layer on electrode processes where the electroactive species have weak interactions with the electrode (Frumkin theory); (iv) shifting the pzc and altering heats of adsorption, surface coverages or desorption rates in electrocatalytic processes where there is strong interaction with the electrode (volcano theory); (v) generating dislocations in the bulk or nucleation centers for crystal growth in electrodeposition processes; (vi) preventing the blockage of catalytically active centers or crystal growth sites by poisons; and (vii) accelerating the rates of slow electrode reactions through coupling with fast displacement reactions.

Classical electrochemical voltammetric measurements do not yield detailed information about the nature of the adsorption states of such chemisorbed species. When combined with an *in situ* probe such as optical reflection spectroscopy, however, the electronic structure of the electrode surface may be investigated. Fig.10 illustrates a series of $\Delta R/R$ spectra measured with a rapid-scanning spectroreflectometer (21, 63) in a study of the UPD of Pb on an evaporated Ag film electrode. The uppermost spectra arise solely from the electroreflectance (ER) effect (84) produced by charging the electrical double-layer from 0.400 to 0.050 V(NHE); UPD of Pb does not begin until 0.0 V. The small peaks at $\hbar\omega = 3.14$ eV originate from an optically-excited surface plasma resonance, while the large positive peaks at 3.76 eV result from the steep drop in reflectance of Ag at its bulk plasma frequency. At the more cathodic potentials of -0.175 and -0.225 V, Pb is adsorbed on the Ag surface to coverages of ∿0.8 and 1.0 monolayer, respectively. The $\Delta R/R$ spectra for these cases originate from a combination of ER effects, the optical properties of the adsorbate and a perturbation of the electronic density of states of the Ag substrate

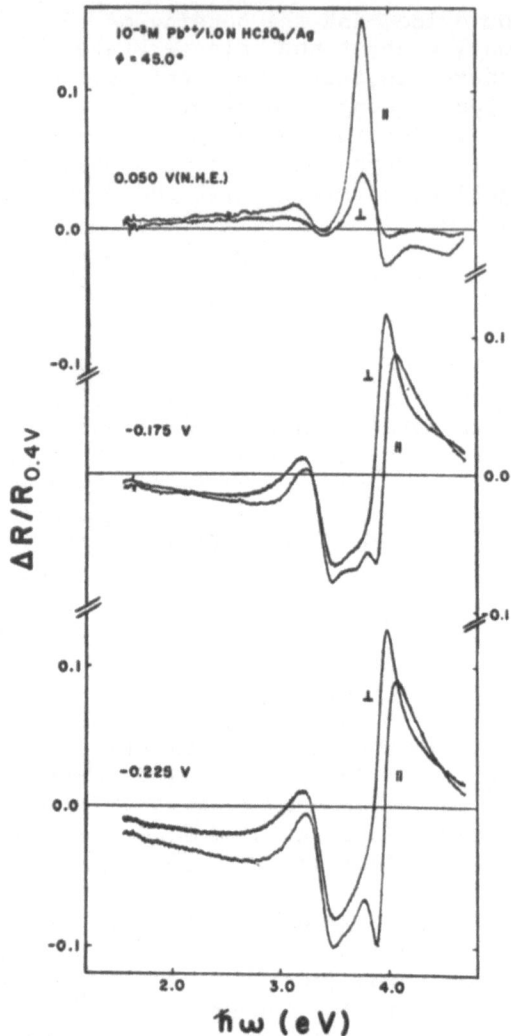

Fig. 10 Spectral dependence of the relative reflectance changes
         observed for various coverages (θ) of an evaporated Ag film
         electrode by adsorbed Pb. The uppermost curves (θ=0) show the
         electroreflectance effect produced by a potential change
         from E = 0.400 to 0.050 V(NHE). [Pb++] = 1.0 x 10⁻³ M;
         [HClO₄] = 1.0 M; φ = 45.0° (McIntyre (21)).

surface. Comparison of Fig. 10 with Fig. 3 reveals that the gross
features of the experimental spectra are well-predicted by the
model calculation - in particular, the broad reflectance decrease
over the region 1.5 to 3.0 eV, the shift of the principal positive

$\Delta R/R$ peaks from 3.76 to 4.0 eV, and the sharp dip in $\Delta R/R$ at 3.9 eV. It is clear that the primary spectroscopic effects in this example do not originate from an adsorption-induced ER effect, as has been proposed in other optical-studies of UPD (85), but rather from the optical properties of the strongly absorbing Pb overlayer *per se*.

The fine structure observed in the photon-energy range 3.5-3.9 eV is unique. This structure could conceivably result from the formation of resonant bound states of Pb adatoms, as observed in UPS studies (86) of metal overlayers on Ag, but the $\Delta R/R$ peaks are too narrow for this to be the case. Alternatively, an increase in free-electron concentration in the Ag surface due to charge transfer from divalent Pb atoms could cause shifts (in opposite directions) of the $L_2 \rightarrow L_1$ and $L_3 \rightarrow L_2'$ (Fermi surface) interband transitions. Such effects have been observed in a number of bulk alloys (*cf*. Ref. (87)) as a result of the expansion of the Fermi surface. In the present case, however, analogous effects were observed for the UPD of *monovalent* Tl overlayers. We conclude that the sharp spectral structure, which is made clearly visible by the adsorption-induced shift of the main $\Delta R/R$ peak to higher photon energies, originates from lattice strain or defects present in the nonannealed Ag film. Strain-induced variation of the lattice spacing is probably greatest in the highly disordered regions of the sample near grain boundaries and has its principal effect on the interconduction-band transition $L_2 \rightarrow L_1$, which is lowered in energy and broadened (88).

In contrast to the Pb/Ag system, Ag and Au are miscible in all proportions. The Ag/Au system is well-suited for optical studies because the threshold energies of interband transitions in the pure bulk metals are well-separated, as illustrated by the characteristic reflectance spectra shown in the insets in Figs 3 and 4. Such studies may reveal spectroscopic features characteristic of surface alloy formation when Ag is electrodeposited on Au. Further, since Ag and Au have significantly different electronegativities ($\chi_{Ag} = 1.9$; $\chi_{Au} = 2.4$), optical effects due to charge transfer may be investigated. Fig. 11 illustrates the differential reflection spectra measured at a series of surface coverages by rapid-scanning spectroreflectometry. Again note the fine spectral detail observed with this technique. A detailed discussion of these spectra must await an analysis for Re $\hat{\epsilon}_s$ (ω) and Im $\hat{\epsilon}_s$ (ω), but certain features are noteworthy. Comparison of Fig. 11 with Fig. 4 reveals that the model calculation does not correctly predict the form of the experimental $\Delta R/R$ spectra. At coverages of one monolayer and less, the $\Delta R/R$ spectra strongly resemble the forms of ER spectra produced by electric-field modulation (89). This could be indicative of charge-transfer effects but the evidence is not convincing. At these coverages extensive charge transfer due to electronegativity differences

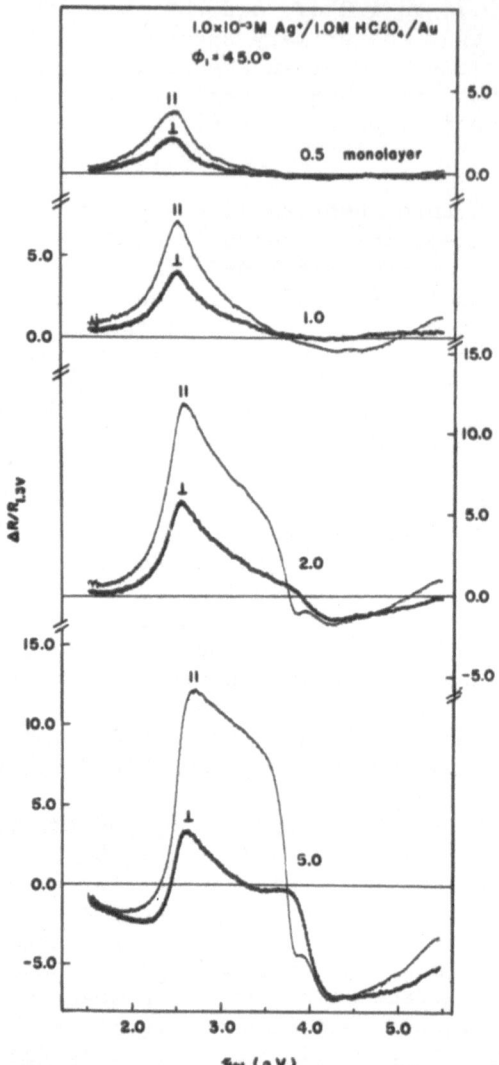

Fig. 11 Normalized differential reflectance spectra
         observed for various coverages of an evaporated Au film
         electrode by Ag. [Ag$^+$] = 1.0 x 10^{-3} M; [HClO$_4$] = 1.0 M;
         ϕ = 45.0°.

should be overridden by Coulombic repulsion. Further, a compen-
satory ER effect of opposite sign should be produced in the Ag
overlayer, but this is not observed. At higher coverages, the
spectral structure at 3.9 eV indicates the formation of Ag d-bands.
This may indicate that a separate Ag overlayer is formed rather

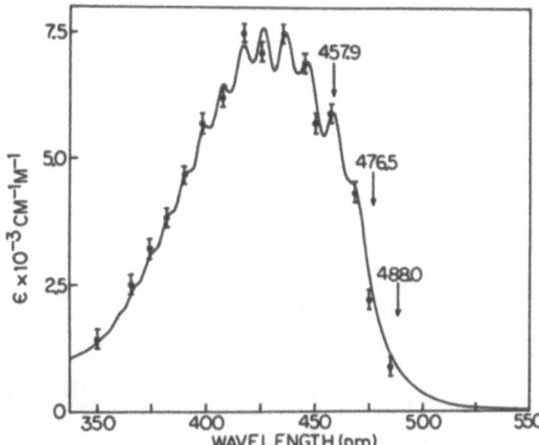

Fig. 12 Electronic absorption spectrum of TCNE⁻. in 0.1 M TBAP
-acetonitrile solution. Solid curve: spectrum of TCNE⁻.
prepared by vacuum electrolysis.Points:transient absorption
spectrum determined by potential-step chronoabsorptometry
(Jeanmaire, Suchanski and Van Duyne (72)).

than a surface alloy. As yet no explanation can be offered for
the decrease in reflectance below 2.0 eV at the highest coverage.
This is well below the threshold energies for d-band → Fermi
surface transitions near L and X in the energy bands of Au.

Spectroelectrochemistry. The anion radical generated by electro-
reduction of tetracyanoethylene (TCNE) has a strong absorption
band in the visible-near UV region (Fig. 12). Fig. 13(A) shows
the normal Raman scattering spectrum observed (72) for an aceto-
nitrile solution containing 2.1 mM TCNE and 0.1 M tetrabutylammonium
perchlorate (TBAP). All bands in this spectrum are due to CH_3CN
except for the 931 cm^{-1} band which is due to the $\nu_1(A_1)$ Cl-O
stretching frequency of ClO_4^-. No bands are observed for TCNE
owing to its low concentration and the lack of overlap of any
electronic absorption band with the exciting Ar^+ ion laser lines.
Thus, there is no resonance enhancement in this background spectrum.
Figures 13(B), (C) and (D) show Raman spectra obtained from TCNE
solutions electrolyzed at -0.20 V *vs* SCE with exciting lines of
different wavelengths. The principal new features in the spectra
of the electrolyzed solutions appear at 540, 1421, 1961 and 2194
cm^{-1}. Very weak new bands are also seen at 615, 870, 1306 and
1565 cm^{-1} which are attributed to TCNE⁻. decay products. The latter
are not observed in spectra obtained by controlled potential-step
modulation experiments (at 10 Hz) because in this case TCNE⁻. is
reoxidized before it has a chance to react.

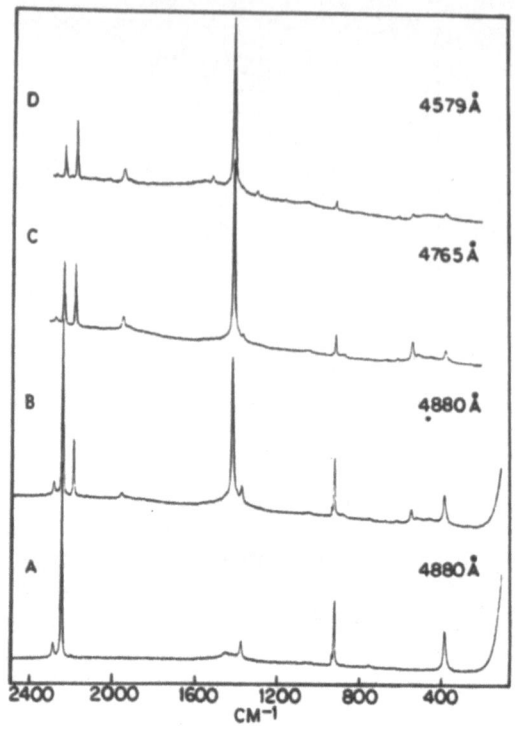

Fig. 13   Resonance Raman spectra of TCNE⁻· electrochemically
          generated by controlled-potential coulometry at -0.20
          V(SCE) in 0.1 M TBAP/CH$_3$CN.   (A) Background scan, [TCNE]
          = 2.1 mM;   (B) [TCNE⁻·] = 2.1 mM;   (C) [TCNE⁻·] = 3.3 mM;
          (D) [TCNE⁻·] = 3.3 mM.   The wavelengths of the exciting
          Ar⁺ ion laser lines are indicated (Jeanmaire, Suchanski
          and Duyne (72)).

The assignment of the main bands of the TCNE⁻· spectra (after
Ref. (90)) is as follows: 540 cm^{-1} - ν_3(A$_g$) C—C symmetric stretch;
1421 cm^{-1} - ν_2(A$_g$) C=C symmetric stretch; 1961 cm^{-1} - $\nu_2 + \nu_3$
combination band; and 2194 cm^{-1} - ν_1(A$_g$) C≡N symmetric stretch.
The frequency shifts observed are consistent with the concept that
addition of an extra electron to the lowest antibonding molecular
orbital of TCNE increases the C—C bond character. It may be
noted that the C≡N stretch frequency of the 3 mM TCNE product is
totally resolvable from that of the 19 M CH$_3$CN solvent. Such
resolution could not be obtained with conventional IR spectroscopy
owing to the breadth of the absorption bands.

Resonance Raman spectroscopy appears to offer great promise
for *in situ* studies of electrode reactions owing to its sensitivity,
high resolution, and molecularly specific character. Not only can

it be used for 'fingerprint' identification of electrogenerated species, but also for structural, kinetic and mechanistic investigations. An intriguing new aspect is the use of Raman spectroscopy for surface studies. It has recently been demonstrated (91-93) that adsorbed monolayers of strongly scattering molecules such as pyridine can be detected on roughened metal electrodes. The appearance of new vibration bands with changing potential provides previously inaccessible information about the orientation and bonding of molecules adsorbed at electrode surfaces. The origin of the high sensitivity observed for such monolayer films is not yet clear. It may be related to a resonance enhancement resulting from configuration interaction of the valence electrons of the adsorbate with the continuum of metallic substrate states, in a manner somewhat analogous to that observed for continuum resonance Raman spectra of 'free' scatterers (Fig. 5(C)). This phenomenon may be investigated further using a tunable dye laser as a source of exciting radiation and correlating the Raman scattering signal strength with the absorption spectra of both the adsorbate and the metal substrate.

CONCLUSIONS

Modern spectroscopic methods for surface research promise to provide electrochemists with the means for studying processes at the electrode-electrolyte interface at a level not considered possible just a few years ago. In this paper I have attempted to outline briefly the principles and utility of some of the more novel methods. Others, such as XPS (ESCA) and AES, have already proven their worth and are being widely applied. Application of these techniques in fundamental studies of simple, well-defined electrochemical systems should increase our ability to isolate and understand the factors that govern many complex technological processes. This knowledge should in turn enable us to upgrade existing electrochemical processes and to devise new and improved processes and devices to meet our future needs in a manner which will effectively conserve energy and material resources. Achievement of this goal will require close cooperation among scientists pursuing fundamental surface studies and those engaged in the development and application of practical catalytic and electrochemical systems.

ACKNOWLEDGMENTS

The author wishes to express his appreciation to colleagues at Bell Laboratories for helpful discussions during this work, and particularly to D.E. Aspnes, T.M. Buck, R.L. Cohen, R.P. Frankenthal, D.L. Malm, D.L. Rousseau and J.E. Rowe. The technical assistance of W.F. Peck, Jr. is also gratefully acknowledged.

REFERENCES

1. G.A. Somorjai, *Principles of Surface Chemistry*, Prentice-Hall, Englewood Cliffs, N.J., U.S.A., 1972.

2. D.A. Shirley (Ed.), *Electron Spectroscopy*, North-Holland, Amsterdam, 1972.

3. P.F. Kane and G.R. Larrabee (Eds.), *Characterization of Solid Surfaces*, Plenum, New York, 1974.

4. J.M. Blakely (Ed.), *Surface Physics of Materials*, Academic Press, New York, 1975.

5. M. Prutton, *Surface Physics*, Clarendon Press, Oxford, 1975.

6. A.W. Czanderana, *Methods of Surface Analysis, Vol. 1, Characterization of Surfaces and Adsorbed Species*, Elsevier, Amsterdam, 1975.

7. U. Gonser (Ed.), *Mössbauer Spectroscopy*, Springer-Verlag, Berlin, 1975.

8. R.B. Anderson and P.T. Dawson, *Experimental Methods in Catalytic Research*, Academic Press, New York, 1976.

9. E.G. McRae and H.D. Hagstrum in N.B. Hannay (Ed.), *Treatise on Solid-State Chemistry*, Vol. 6, Plenum, New York, in press.

10. J.A. Strozier, Jr., D.W. Jepsen and F. Jona in Ref. (4), p. 1.

11. G.H. Hill, J. Marklund, J. Martinson and B.J. Hopkins, Surface Sci., 24 (1-71) 435.

12. K. Siegbahn in Ref. (2), p. 15.

13. S.J. Atkinson, C.R. Brundle and M.W. Roberts, J. Electron Spectroscopy, 2 (1973) 105.

14. C.S. Fadley in G.A. Somorjai and J. McCaldin (Eds.), *Progress in Solid-State Chemistry*, Pergamon Press, New York, 1976.

15. J. Freeouf, M. Erbudak and D.E. Eastman, Solid State Commun., 13 (1973) 771.

16. T. Gustaffson, E.W. Plummer, D.E. Eastman and J.L. Freeouf, Solid State Commun., 17 (1975) 391.

17. D.E. Eastman and J.L. Freeouf, Phys. Rev. Letters, 33 (1974) 1501.

18. J.E. Demuth and D.E. Eastman, Phys. Rev. Letters, 32 (1974) 1123.

19. W.F. Egelhoff and D.L. Perry, Phys. Rev. Letters, 34 (1975) 93.

20. J. Anderson and G.J. Lapeyre, Phys. Rev. Letters, 36 (1976) 376.

21. J.D.E. McIntyre in B.O. Seraphin (Ed.), *Optical Properties of Solids - New Developments*, North-Holland, Amsterdam, 1976, Ch. 11.

22. R.L. Gerlach, J. Vac. Sci. Technol., 8 (1971) 599.

23. F. Steinrisser and E.N. Sickafus, Phys. Rev. Letters, 27 (1971) 992.

24. J. Küppers, Surface Sci., 36 (1973) 53.

25. S. Ohtani, K. Terada and Y. Murata, Phys. Rev. Letters, 32 (1973) 415.

26. J.E. Rowe and H. Ibach, Phys. Rev. Letters, 31 (1973) 102.

27. H. Ibach and J.E. Rowe, Phys. Rev. B, 9 (1974) 1951.

28. J.E. Rowe, Solid State Commun., 15 (1974) 1505.

29. F.M. Propst and T.C. Piper, J. Vac. Sci. Technol., 4 (1967) 53.
30. H. Ibach, J. Vac. Sci. Technol., 9 (1972) 713.
31. H. Ibach, K. Horn, R. Dorn and H. Lüth, Surface Sci., 38 (1973) 433.
32. R.E. Honig, Thin Solid Films, 31 (1976) 89.
33. T.M. Buck in Ref. (6), Ch. 3.
34. J.M. Poate and T.M. Buck in Ref. (8), Ch. 6.
35. R.L. Erickson and D.P. Smith, Phys. Rev. Letters, 34 (1975) 297.
36. N.H. Tolk, J.C. Tully, J. Kraus, C.W. White and S.H. Neff, Phys. Rev. Letters, 36 (1976) 747.
37. R.R. Adzić and A.M. Despić, J. Chem. Phys., 61 (1974) 3482.
38. R.R. Adzić, D.N. Simić, A.R. Despić and D.M. Drazić, J. Electroanal. Chem., 65 (1975) 587.
39. R.R. Adzić and A.R. Despić, Z. Phys. Chem. N.F., 98 (1975) 95.
40. J.D.E. McIntyre and W.F. Peck, Jr., J. Electrochem. Soc., in press.
41. W. Heiland and E. Taglauer, J. Vac. Sci. Technol., 9 (1972) 620.
42. A. Benninghoven and E. Loebach, Rev. Sci. Instrum., 42 (1971) 49.
43. A. Benninghoven, Surface Sci., 35 (1973) 427.
44. D.L. Malm in E.M. Hurt and W.G. Guldner (Eds.), *Physical Measurement and Analysis of Thin Films*, Plenum Press, New York, 1969, p. 148.
45. J.M. McCrea in Ref. (3), p. 577.
46. G. Vidal, P. Galmard and P. Lanusse, Int. J. Mass Spectry. Ion Phys., 2 (1969) 373.
47. M.J. Vasile and D.L. Malm, Anal. Chem., 44 (1972) 650.
48. D.L. Malm and M.J. Vasile, J. Electrochem. Soc., 120 (1973) 1484.
49. C.C. Chang in Ref. (3), p. 509.
50. N.C. MacDonald in D. Beaman and B. Siegel (Eds.), *Electron Microscopy: Physical Aspects*, Wiley, New York, in press.
51. G.K. Wertheim, *Mössbauer Effect: Principles and Applications*, Academic Press, New York, 1964.
52. G.W. Simmons and H. Leidheiser, Jr. in R.L. Cohen (Ed.), *Applications of Mössbauer Spectroscopy*, Vol. 1, Academic Press, New York, 1976, Ch. 3.
53. W.E. O'Grady, Chem. Phys. Letters, 5 (1970) 116.
54. J.O'M. Bockris, A. Damjanović and W.E. O'Grady, J. Coll. Interface Sci., 34 (1970) 387.
55. W.E. O'Grady and J.O'M. Bockris, Surface Sci., 38 (1973) 279.
56. A.T. Hubbard, Crit. Revs. Anal. Chem., 3 (1973) 201.
57. A.T. Hubbard, R.M. Ishikawa and J.A. Schoeffel in M.W. Breiter (Ed.), *Proc. Symp. Electrocatalysis*, The Electrochemical Society, Princeton, N.J., 1974, p. 258.
58. J.A. Joebstl and G.W. Walker, papers presented at the *145th National Meeting of The Electrochemical Society*, San Francisco, Calif., May 1974, Extended Abstract Nos. 322 and 327.
59. R.W. Revie, B.G. Baker and J.O'M. Bockris, J. Electrochem. Soc., 122 (1975) 1460.

60. J.A. Joebstl, paper presented at the *First Chemical Congress of the North American Continent*, Mexico City, Mexico, Dec. 1975, Abstract PHSC 18.

61. W.E. O'Grady, M.Y.C. Woo, P.L. Hagans and E. Yeager, paper presented at the *149th National Meeting of The Electrochemical Society*, Washington, D.C., May 1976, Extended Abstract No. 260.

62. D.E. Aspnes in B.O. Seraphin (Ed.), *Optical Properties of Solids - New Developments*, North-Holland, Amsterdam, 1976, Ch. 15.

63. J.D.E. McIntyre, M.F. Robbins and W.F. Peck, Jr., paper presented at the *First Chemical Congress of the North American Continent*, Mexico City, Mexico, Dec. 1975, Abstract PHSC 90.

64. J.D.E. McIntyre, M.F. Robbins and W.F. Peck, Jr., paper presented at the *149th National Meeting of The Electrochemical Society*, Washington, D.C., May 1976, Extended Abstract No. 363.

65. J.D.E. McIntyre and W.F. Peck, Jr. in M.W. Breiter (Ed.), *Proc. Symp. Electrocatalysis*, The Electrochemical Society, Princeton, N.J., 1974, p. 212.

66. J. Anderson, G.W. Rubloff, M.A. Passler and P.J. Stiles, Phys. Rev. B, 10 (1974) 2401.

67. J. Horkans, B.D. Cahan and E. Yeager, Surface Sci., 46 (1974) 1.

68. D.R. Tallant and D. Evans, Anal. Chem., 41 (1969) 835.

69. P.T. Kissinger and C.N. Reilley, Anal. Chem., 42 (1970) 12.

70. H.B. Mark, Jr. and B.S. Pons, Anal. Chem., 38 (1966) 119.

71. J. Mattson and C.A. Smith, Anal. Chem., 47 (1975) 736.

72. D.L. Jeanmaire, M.R. Suchanski and R.P. Van Duyne, J. Am. Chem. Soc., 97 (1975) 1699.

73. A.C. Albrecht, J. Chem. Phys., 34 (1961) 1476.

74. J. Tang and A.C. Albrecht in H.A. Szymanski (Ed.), *Raman Spectroscopy*, Vol. 2, Plenum Press, New York, 1970, p. 33.

75. D.L. Rousseau and P.F. Williams, J. Chem. Phys., 64 (1976) 3519.

76. J.H. Sinfelt, J. Am. Chem. Eng. 19 (1973) 673.

77. J.H. Sinfelt, Crit. Revs. Solid State Sci., 4 (1974) 311.

78. H.H. Brongersma and T.M. Buck, Surface Sci., 53 (1975) 649.

79. C.R. Helms and W.E. Spicer, Phys. Rev. Letters, 32 (1974) 228.

80. K.Y. Yu, C.R. Helms and W.E. Spicer, to be published.

81. F.L. Williams and D. Nason, Surface Sci., 45 (1974) 377.

82. J.H. Sinfelt, J.L. Carter and D.J.C. Yates, J. Catal., 24 (1972) 283.

83. R.P. Frankenthal and D.L. Malm, J. Electrochem. Soc., 123 (1976) 186.

84. J.D.E. McIntyre in R.H. Muller (Ed.), *Advances in Electrochemistry and Electrochemical Engineering*, Vol. 9, Wiley-Interscience, New York, 1973, p. 61.

85. D.M. Kolb, D. Leutloff and M. Prazsnyski, Surface Sci., 47 (1975) 622.

86. D.E. Eastman and W.D. Grobman, Phys. Rev. Letters, 30 (1973) 177.

87. C.J. Flaten and E.A. Stern, Phys. Rev. B, 11 (1975) 638, and references cited therein.

88. P. Minsemius, H.P. Lengkeek and F.F. Van Kampen, Physica, 79b (1975) 529.
89. J.D.E. McIntyre, Surface Sci., 37 (1973) 658.
90. J.J. Hickel and J.P. Devlin, J. Chem. Phys., 48 (1973) 4750.
91. M. Fleischmann, P.J. Hendra and A.J. McQuillan, Chem. Phys. Letters, 26 (1974) 163.
92. A.J. McQuillan, P.J. Hendra and M. Fleischmann, J. Electroanal. Chem., 65 (1975) 933.
93. R.L. Paul, A.J. McQuillan, P.J. Hendra and M. Fleischmann, J. Electroanal. Chem., 66 (1975) 248.

THE APPLICATION OF AUGER ELECTRON SPECTROSCOPY TO THE STUDY OF ELECTRODE SURFACES

B.G. Baker

School of Physical Sciences
The Flinders University of South Australia
Adelaide, South Australia

INTRODUCTION

Most of the techniques which have been developed for the study of the composition of solid surfaces cannot be directly applied to the solid-liquid interface. It is necessary to transfer the solid sample from solution into vacuum in order to have access to those methods which have been developed for the study of the solid-vacuum interface. The success of such experiments is obviously dependent on the conditions of the transfer and the subsequent treatment of the sample, since the surface might well undergo changes which destroy the relevant information. This paper is concerned with the design of experiments to determine the composition of electrode surfaces by surface analysis in vacuum.

SURFACE ANALYSIS

Auger electron spectroscopy (AES) has proved to be a most suitable technique for surface analysis. It provides a qualitative analysis for the elements at, or very close to the surface, with a detection limit of about 1% of an atomic layer. With appropriate calibration, it can give a quantitative estimate of the components near the surface. It is also possible to detect differing modes of chemical bonding by observing chemical shifts in the Auger energies. Detailed discussion of these aspects of the technique has appeared elsewhere (1).

The high degree of surface sensitivity of AES is readily understood from the dependence of mean free path on electron energy as shown in Figure 1. While this data should be regarded as only semiquantitative, since there is some dependence on the

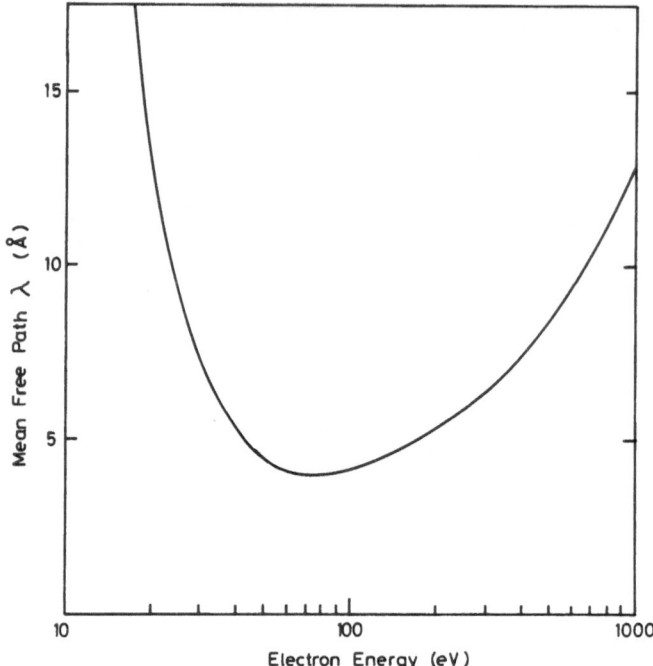

Fig. 1   Mean free path of electrons in the metals as a function of
         energy.

nature of the solid (2,3), the general observation is that Auger
electrons detected with energies in the range 20 - 1000 eV will
have come predominantly from the outmost 2 - 6 atomic layers of
the solid.   Fractions of a surface monolayer are readily detected.
While this is in accord with the objective of our experiment, it
is also apparent that the analysis will be likewise sensitive to
contaminants.   Failure to control the environment of the sample
can introduce gross inaccuracies and may result in the complete
obscuring of the sample by contaminant.

    An electrode surface may become contaminated by adsorption
from solution after the potential is removed, or by adsorption
from an ambient gas phase during transfer to the vacuum.   It may
also undergo decomposition due to the evaporation of constituents
into the vacuum, or by the effects of the primary electron beam
which is used to excite the Auger transitions (4).   The design of
the surface analysis experiment will clearly depend on the particular material under study.   In order to illustrate ways in
which these problems can be overcome, the experimental aspects of
two recent studies, successfully completed in this laboratory will
now be discussed.

AN AUGER ELECTRON ENERGY ANALYSER

The analysers most commonly used for AES are the cylindrical mirror and retarding grid analysers. For the present experiments a retarding grid analyser constructed in this laboratory was mounted in a Pyrex tube (5). This analyser has a post-monochromator after the design of Huchital and Rigden (6). The principle is shown in Figure 2. The grid nearest the sample is at ground potential so that all the electrons pass, but the second grid is swept increasingly negative to cut-off electrons of low energy. This grid also has a small modulating voltage to facilitate phase sensitive detection. The electrons with energy only just in excess of the cut-off then pass into the influence of the post-monochromator. A cylindrical mesh 5 V more negative than the second grid repels these electrons. They follow trajectories illustrated as A and arrive at the collector which is the most positive electrode. The electrons of higher energy which are not appreciably influenced by

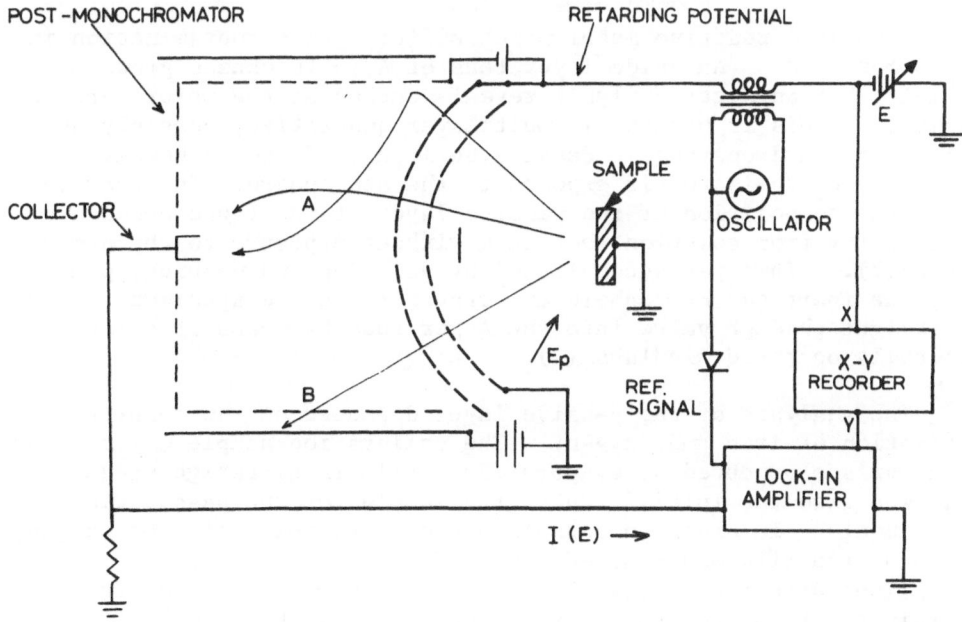

Fig. 2   Auger electron energy analyser: retarding grid type with post-monochromator.

these small fields, pass through the mesh and are collected by the outer positive screen (trajectory B).   The effect of the post-monochromator is to remove high energy background and provide a better signal/background ratio.   This is most important for analyses involving peaks at energies < 100 eV where the secondary electron background is large.

A particular advantage of this analyser for the study of electrodes is that the resolution is not critically dependent on the direction of the admitted electrons.   It is therefore possible to use a primary electron beam of broad focus and so have a small electron current density at the sample, thus minimizng the possible damaging effects.

To perform satisfactorily, the analyser tube must be initially baked in vacuum and the metal parts thoroughly outgassed.   The electrode sample should not be present during this system pre-paration but can be introduced later via a vacuum lock as described below.

THE PASSIVE LAYER ON IRON

Iron is a reactive metal which suffers gross contamination in the atmosphere.   An oxide layer tens of $\overset{o}{A}$ in thickness grows rapidly, but an Auger analysis reveals carbon as the major surface element.   This is present in multilayer quantities, possibly as a polymerized hydrocarbon.   Passivated iron is found to become similarly contaminated if exposed to the atmosphere.   In order to study the composition of the passive layer it was found necessary to transfer from solution to vacuum without exposure to the atmos-phere (7).   This was accomplished by draining and washing, pumping away the inert gas atmosphere and transferring the specimen   through a straight-through valve into the Auger tube by means of a mag-netically operated windlass (7).

The analysis of the passive layer depended on the quantitative estimation of iron and oxygen.   The calibration sample was a clean iron surface produced by evaporating a film in ultrahigh vacuum. This was also the initial state of a sample for the passivation experiment.   In order to establish a calibration basis for oxygen, a clean iron film was reacted with gaseous oxygen to form Fe_2O_3. Comparison with the passive layer spectrum showed that the latter contained more oxygen than Fe_2O_3, consistent with a composition of $Fe_2O_3.H_2O$.   Note that hydrogen is not detected by AES.

The sample taken from an aqueous environment presents a diffi-culty when introduced to the highly anhydrous conditions of a baked UHV system.   Traces of water vapour react with the walls of the vessel to displace CO and CO_2 which can then contaminate the sample. It was found possible to pre-condition a UHV system with small

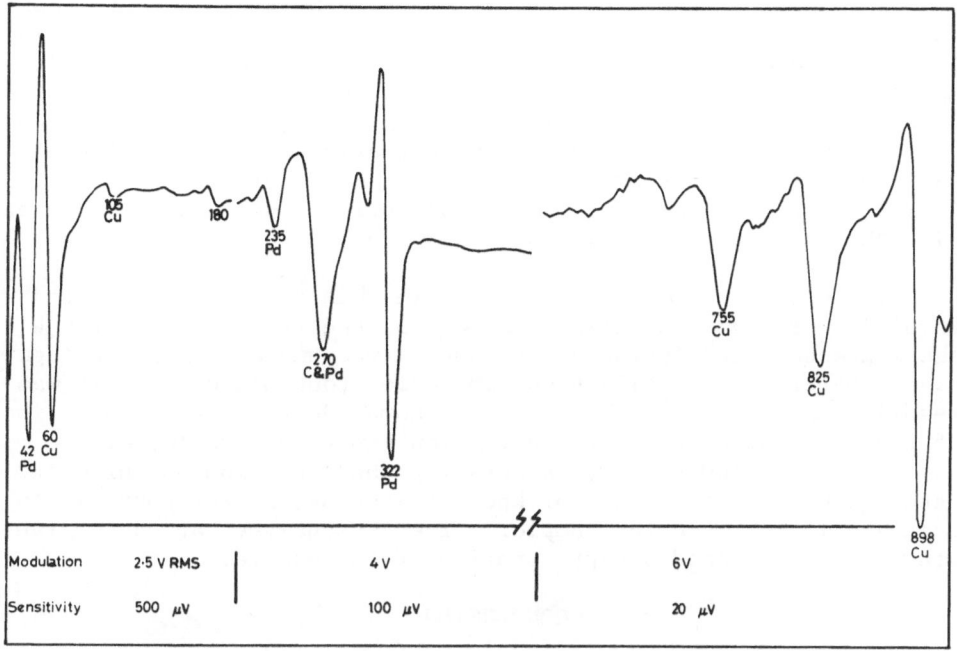

Fig. 3   Auger spectrum (second derivative of current) from a copper-
         palladium alloy. Peak energies in eV.

amounts of water vapour and to pump away the desorbed gases.
While the ultimate vacuum was inferior ($\sim$ 5 x 10^{-8} torr) the system
was then found to cause less contamination of the electrode sample.

CORROSION OF ALLOYS

The differing rates of dissolution at an alloy surface result
in a change in surface composition which is ideally suited to an
AES determination.   A series of copper-palladium alloys were the
subject of a recent study (8).   One reason for the choice of this
system is that copper and palladium have a non-interfering Auger
spectrum as shown in Figure 3.   Note that the two metals are
clearly resolved and can be measured by both low and high energy
peaks.   Since these electrons have different mean free paths this
can aid interpretation of a composition variation with depth from
the surface.   Note also that major contaminant peaks: S, 150 eV;
B, Cl 180; 0, 515, do not interfere.   The carbon peak at 270 eV
interferes with a Pd peak but the other Pd peak at 322 eV is used
for the analysis.

In this work the alloy was prepared by melting in argon, homo-
genized and cold-rolled.   A clean sample surface was produced in
the Auger tube by argon ion bombardment.   This removed contaminant
and exposed a fresh alloy surface.   This procedure introduces the
possibility that differential sputtering could change the surface
composition.   In fact Cu and Pd have nearly equal rates of sputtering.   By calibrating the Auger peaks with pure Cu and pure Pd,
it was shown that the sputtered alloy surface analysed in agreement
with the bulk composition.

The electrochemical experiments in this case were conducted
in a flow cell and the alloy sample was then transferred through
the atmosphere and introduced through a vacuum-lock into the Auger
spectrometer.   Only slight contamination took place and this was
readily removed by a light argon ion bombardment.   The composition
of the Cu-Pd alloy after the corrosion experiment varied over a
depth of up to 100 Å.   By calibrating the argon ion bombardment
rate against a copper film of known thickness, it was possible to
quantitatively erode the sample.   The AES analyses at successive
stages then provided a depth profile of the sample.

CONCLUSION

Electrode surfaces can be analysed by AES provided appropriate
care is taken to protect the sample during the transfer from
solution to vacuum.   Quantitative elemental analysis by AES is
facilitated by *in situ* calibration with samples of known surface
composition.

ACKNOWLEDGEMENTS

I am grateful to R.W. Revie, J. Gneiwek and J.O'M. Bockris
for discussion and for developing the techniques in practice.

This work was supported in part by the Australian Research
Grants Committee.

REFERENCES

1.   B.G. Baker, in J.O'M. Bockris and B.E. Conway (Eds.) *Modern
      Aspects of Electrochemistry*, Plenum Press, New York, Volume 10,
      1975, p. 93.

2.   P.W. Palmberg, Anal. Chem.,45 (1973) 549A.

3.   C.J. Powell, Surface Science,44 (1974) 29.

4.   B.G. Baker and B.A. Sexton, Surface Science,52 (1975) 353.

5. B.A. Sexton, B.G. Baker and R.A. Armstrong, submitted to J. Vac.
    Sci. and Technol. (1976).

6. D.A. Huchital and J.D. Rigden, Appl. Phys. Letters,16 (1970)
    348; J. Appl. Phys.,43 (1972) 2291.

7. R.W. Revie, B.G. Baker and J.O'M. Bockris, J. Electrochem. Soc.,
    122 (1975) 1460.

8. J. Gniewek, J. Pezy, B.G. Baker and J.O'M. Bockris, to be
    submitted to J. Electrochem. Soc. (1976).

APPLICATION OF ANODIC STRIPPING VOLTAMMETRY TO TRACE METAL ANALYSIS

Harry Bloom and Barry N. Noller

Chemistry Department, The University of Tasmania

Box 252C, Hobart, Tasmania, Australia 7001

INTRODUCTION

The spectacular developments which have taken place in quantitative analysis of trace metals over the past 10 to 15 years have resulted mainly from techniques which require a perturbation of an outer electron or electrons of atoms either in the gaseous state or dissolved in a mercury/solution interface.

Atomic absorption spectrometry (AAS) requires that a sample be converted to an atomic vapour and that the free atoms be able to absorb electromagnetic radiation at their resonance energies. This process is very selective and with modern instrumentation, enables trace quantities of the order of 0.01 ppm to be determined with good precision and large throughput of samples. The use of automatic aspiration of sample, multichannel analysers, automatic background correction for non-atomic light absorption etc. makes this method one which can be carried out by a person with little academic training. The rapid escalation of use of AAS bears testimony to the convenience and wide applicability of this technique, which was invented in Australia[1].

In contrast to the now commonplace use of AAS for trace element analysis, electrochemical methods which normally require that an element be deposited from solution on an electrode (usually mercury) and then anodically transferred back into solution with the simultaneous loss of an electron (or electrons), are not in general use. This is partly because the technology and methodology of electrochemical methods have not been developed to the stage which will permit them to be used routinely by semi-skilled operators, hence they do not find as much favour for routine analysis. Solution and electrode preparation, as well as careful control of deposition and stripping conditions, are necessary in electrochemical methods. Small changes in conditions often render the results meaningless and a suitably trained specialist chemist must be employed to ensure that the methods are properly applied.

The most sensitive electrochemical method used is Anodic Stripping Voltammetry (ASV) which can be applied using a hanging mercury drop electrode (HMDE) or a "thin film electrode" (TFE). (Stulíková[3] has shown that the so-called "thin film" on the glassy carbon electrode consists of small mercury drops which depend in size and number on deposition conditions). One of the main stimuli for these methods is T.M. Florence of the Australian Atomic Energy Commission who with G.E. Batley recently published notes[2] of a one day course held at the University of New South Wales. In the foreword to this excellent manual, the authors state (in part) "At a total instrumentation cost of around $5000, an ASV instrument is remarkably good value when compared with other trace analysis instruments." Whilst this is undoubtedly true, the instrument can be extremely misleading when used by a person who does not understand its many pitfalls. In this article we will attempt to present the positive and negative aspects of the ASV method and to compare it with AAS in applications performed at the Chemistry Department of the University of Tasmania and elsewhere.

ASV - THEORY AND INSTRUMENTATION

Trace metals are taken into solution from environmental samples and the solution is made up to a suitable concentration of supporting electrolyte e.g. 0.1 M KNO_3. The solution is de-aerated by bubbling in oxygen-free nitrogen. Stirring is necessary during deposition and stripping. The rate of stripping is one of the factors which determines the sensitivity. The micro-electrode used in this research was the TFE rotated at 1500 rpm using highly polished glassy (pyrolytic) carbon coated *in situ* by deposition of a thin mercury film as the working surface. Peak height or more accurately area under the stripping peaks gave anodic current which by Faraday's Law gives a direct measure of amount of metal electrodeposited onto the electrode. The TFE is more sensitive than the HMDE but it is better to use both electrodes to obtain complete information on any one system[2].

Identification of the metal deposited is achieved by measuring the potential between the working electrode and a reference (e.g. standard calomel electrode (SCE)). The counter (or auxiliary) electrode is usually a small coil of platinum wire. Considerations of the theory of deposition and stripping have been given by several authors including Ellis[4] and Copeland and Skogerboe[5].

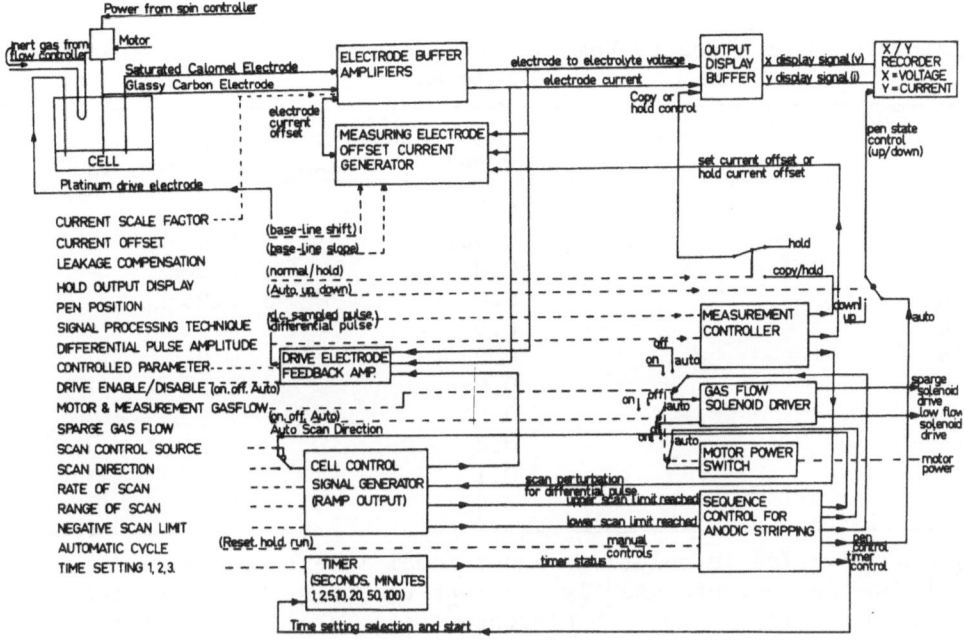

Figure 1.   Block Diagram of ASV Instrument.

Instrumentation varies according to the need. In this study the electronics were designed in this university[6] and the glassy carbon was manufactured by Tokai Electrode Manufacturing Co., Tokyo (Grade GC-20). The instrument can be used for differential pulse and other polarographic techniques. A block diagram of the instrumentation is shown in Figure 1.

COMPARISON OF SENSITIVITIES FOR VARIOUS ELEMENTS

ASV and Non-Flame AAS are two of the most sensitive analytical techniques known for a number of elements. Although the number of elements which are determinable by ASV is limited (not all elements dissolve in mercury), those which can are generally highly sensitive and compare favourably with atomic absorption (Table 1). While sensitivity is not always the most important parameter with a method of analysis, it can be an advantage to analyze directly for trace quantities of elements without performing concentration steps which may lead to contamination[2].

TABLE   1

Sensitivities of Elements Readily Determined by ASV Compared with
Non-Flame AAS

| Element | Sensitivity[a] ng/ml | |
	ASV[b]	Graphite Furnace AAS[c]
Ag	0.05	0.01
Bi	0.005	0.4
Cd	0.05	0.008
Cu	0.03	0.06
Hg	0.05	8
In	0.03	0.04
Pb	0.05	0.2
Sb	0.01	0.5
Sn	0.2	0.2
Tl	0.1	0.1
Zn	0.03	0.003

(a) Sensitivities in solutions under ideal interference-free
conditions and not necessarily the most sensitive known for respective element. (b) Taken from Florence[7]. (c) Taken from L'vov[8].
Sensitivities for 1% absorption with L'vov furnace, calculated in
terms of a 10 microlitre sample from absolute sensitivities.

THE USE OF ASV IN ENVIRONMENTAL ANALYSIS AND OTHER ADVANTAGES
OF ASV CF. AAS

ASV is capable of supplementing, and in some cases replacing, AAS
as a method for environmental analysis. Some elements, e.g. Cd, Sn,
present difficulties by AAS and background non-atomic absorption leads
to large errors when determining many metals in such media as blood
and saline waters. For such investigations ASV is potentially very
important. ASV can be used to particular advantage with solutions of
high salt content which normally give rise to spectral interferences
in AAS. In ASV the salt solution becomes the conductive matrix and is
normally required. ASV also enables chemical forms of different elements to be differentiated directly, e.g. Sb(III) and (V), Sn(II) and
(IV), Fe(II) and (III), Tl(II) and (III), As(III) and (V). AAS
normally gives the total concentration of an element and pre-separa-
tion of chemical forms for separate determinations is required.

The application of the glassy carbon electrode in ASV requires
deposition of mercury from solution. Mercuric ion is normally added to
solution to allow the formation of amalgams with elements present in
solution. If mercury is present in a sample solution, this will be deposited from solution, in the absence of added mercury. During the
stripping cycle mercury will return to solution quantitatively provi-

ded suitable conditions exist. As most other metals are stripped off
before Hg there are very few inter-element problems. Hence ASV is potentially a technique for the analysis of low levels of Hg in solution
and should be directly applicable to the determination of labile Hg in
waters. Biological materials require destruction of organic matter to
free Hg for analysis. Various people have studied the analytical capabilities of stripping Hg from a glassy carbon electrode[3,17,18].

PREPARATION FOR ANALYSIS

The speed of analysis depends in part on the preparation steps
required. With ASV up to 4-6 elements can be analyzed for at one time.
AAS requires a change of hollow cathode lamp for each element.

Preparation for Waters

Waters are normally collected in acid-cleaned containers, filtered through 0.45 μm Millipore filters and acidified with high purity
HCl to pH <2. This procedure separates particulates from labile and
organically bound soluble metal complexes. Acidification of filtered
water releases bound metals. Florence[2,7] recommends using acid-
cleaned polyethylene bottles for seawater and other waters but for
mercury, Bothner and Robertson[21] recommend using Pyrex instead of
polyethylene containers because of leaching of Hg from the surface of
the latter material at lower pH. A number of elements in waters have
been determined by ASV, either directly or following concentration[2].
However, mercury in waters has not been determined by ASV to any great
extent and has been restricted to non-flame AAS and other techniques.
Provided sufficient mercury is present in solution, direct determination by ASV using the glassy carbon electrode should be applicable.
Kendall[17] recommends a two-step procedure for waters: i) ASV with a
bare glassy carbon electrode to determine Hg concentrations and ii)
addition of Hg^{2+} and determine the other metals.

If soluble organic complexing agents are present in solution, as
in waste waters, prior destruction of these may be required. Simple
acid hydrolysis followed by mild heating is generally sufficient treatment to free complexed metals but may cause losses of mercury. A technique suitable for releasing mercury in waters has been outlined by
Alberts *et al.*[22] which involves oxidation of organic matter with per-
sulphate in an ampoule under nitrogen and heated to 120°C. Where insufficient Hg is present in solution for direct determination concentration will be required. Chelex 100 (Bio-Rad Laboratories) has been applied to the concentration of Hg^{23} but gave only 85% retention as added
labile Hg^{2+} ion. Concentration may be achieved by electrodeposition on
Pt, Ag or Au wire from a large sample volume followed by dissolution
of Hg in nitric acid into a smaller volume[26].

Preparation of Biological Materials

Preparation of biological materials for analysis by ASV requires the destruction of organic matter, normally by dry ashing or wet acid digestion. In dry ashing, care is necessary to avoid losses of volatile elements and retention on crucible surfaces. Silica crucibles are recommended rather than Pt dishes for ASV[2] because Pt interferes with Hg. In wet acid digestions, high purity acids are needed to maintain low blank levels. For some elements, such as mercury, wet acid digestions using closed reflux systems are needed to prevent losses[9,20]. Complete oxidation of all organic matter is necessary because many organic species are electro-active and interfere with metal ion determinations by ASV. An alternative procedure described recently[10] uses predigestion in nitric acid followed by heating with a mixture of sodium and potassium nitrates to complete sample decomposition. The resultant melt is then dissolved in dilute nitric acid, pH adjusted and ASV performed directly for the desired elements.

INTERFERENCE EFFECTS

Batley and Florence[24] list the most important interference effects in ASV as follows:
i)   Stripping peaks may have similar peak potentials such as Pb and Sn, Bi and Sb and Tl and Cd.
ii)   Organic and inorganic impurities may form films over the mercury surface which may inhibit the metal deposition/stripping process.
iii) Formation of inter-metallic compounds at the electrode surface. In the first case, use of alternative supporting electrolytes is a limited means of gaining separation through complexing action. Where insufficient potential shift is achieved, a chemical separation may be required, e.g. Sn from Pb as Sn(IV) bromide[2]. It is also possible in some cases to shift the deposition potential and prevent one metal from depositing. The second case arises when soluble organic species are present in solution or insoluble inorganic compounds form on the surface of the mercury film. The third case is probably the most serious interference with the glassy carbon electrode. Inter-metallic compounds form with metals having negative stripping peak potentials, such as Zn, Cd and In, and remain dissolved in the mercury film[2]. The Zn-Cu inter-metallic compound, for example, forms when each metal is present at high concentration. Copeland *et al.*[25] found that addition of Ga to solutions of Cu and Zn prevented the formation of the Cu-Zn inter-metallic because the formation constant for Cu-Ga is larger. Another combination which is troublesome is the Zn-Ni inter-metallic.

APPLICATIONS OF ASV

Analysis of Lead in Blood

The determination of lead in blood is one of the most common applications of ASV and numerous procedures are available[11],[12],[13] and usually depend on acid digestion of blood to remove organic matter; typically sample volumes are less than 0.5 ml. Acid digestions system include $HClO_4/H_2SO_4$, $HNO_3/HClO_4$, $HNO_3/HClO_4/H_2SO_4$ and $HClO_4$ alone.

The digestion method for the current work[12] used 0.1 ml aliquots of whole blood dispensed with an Oxford pipette (disposable plastic tips) which were added to 0.6 ml acid digest mixture (24:24:1 of HNO_3: $HClO_4$:H_2SO_4, high purity acids) in silica decomposition vessels[14] having a "tall form" structure (volume $\approx$ 10 ml). The digestions took approximately 2 hours to reach the H_2SO_4 residue and following cooling and dilution with 1 ml distilled water transferred directly to the ASV cell. Supporting electrolyte and mercuric ion were added and ASV performed using parameters given in Figure 2. The calibrations for standards over the range 0 to 30 ng Pb were found to be linear (Figure 3). Accuracy of the technique was established by analyzing five 0.25 g quantities of NBS SRM 1577 "Bovine Liver" following the same procedure as for blood which gave a concentration of 0.30 ± 0.04 µg/g Pb compared with the certified value of 0.34 ± 0.08 µg/g Pb.

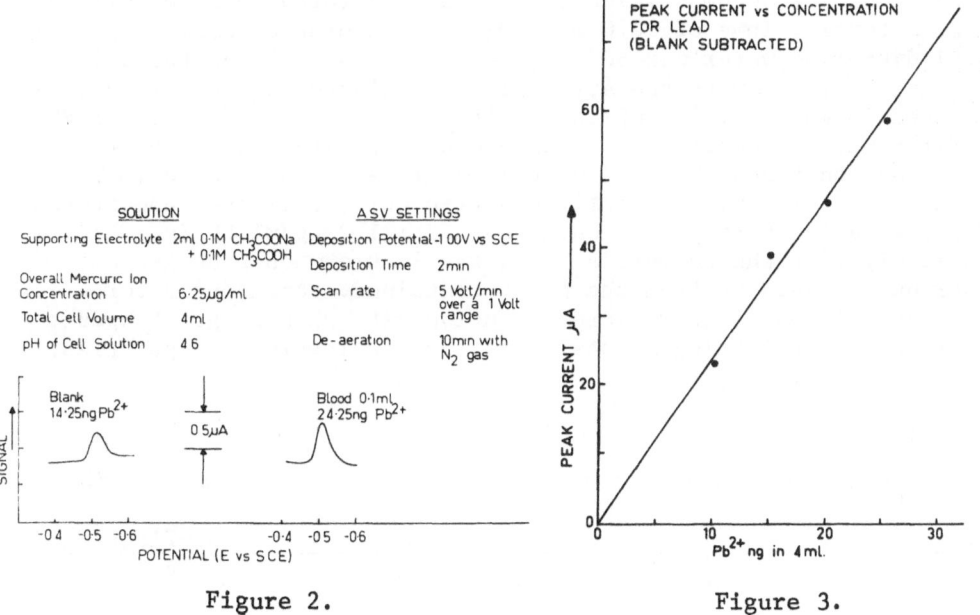

Figure 2.   Figure 3.

Current/Potential Curves for Lead and Calibration Line.

Analyses of various blood samples are shown in Table 2. Lead in
the blood samples was also determined by non-flame AAS using the Varian
Techtron Model 63 CRA with carbon tube. This technique uses 0.1 ml
whole blood in 0.1 ml 5% Triton X-100 solution and has been described[15]. Apart from extraneous results caused by incomplete digestions,
the chief disadvantage of the ASV technique, was the time to complete
an analysis (2-3 hours). Whilst the use of an aluminium block would
reduce the time of digestion to approx. 30 minutes, this is still a
long preparation time compared with the 5 minute preparation for the
non-flame AAS technique. The role of ASV in this laboratory has been
one of comparative technique. Alternative techniques such as solvent
extraction combined with flame AAS are not as accurate for analysis
of lead in normal human blood i.e. below 20 μg Pb/100 ml.

Analysis of Reagents, Lead in Nitric Acid

Normal reagent grade chemicals usually contain elements at levels
which are too high for blanks in ultra-trace analysis. What becomes important as outlined by Barnard and Dudley[16] is now not "how little is
present" but rather "how much is present". ASV is particularly useful
for detecting "how much" of elements outlined above are present in reagents because of its high sensitivity. This application of ASV is demonstrated with the determination of blank Pb levels in nitric acid.

5.0 ml aliquots of HNO_3 were added to 0.5 ml H_2SO_4 (BDH Aristar
Grade) in silica decomposition cells and evaporated down to fumes of
H_2SO_4. A total volume of 10.0 ml HNO_3 was evaporated. Following cooling,
9.5 ml distilled water was added to each silica cell and the solution
transferred directly to the ASV cell and analyzed for lead as described
for blood (Figure 2). 2 samples of nitric acid were analyzed for Pb.
The first was a commercially available high purity HNO_3 (BDH Aristar
Grade) and contained 12.4 ± 2.0 ng/ml Pb (n = 6) as received. The
second sample was double distilled nitric acid, collected and stored in
a Nalgene Teflon bottle and contained 3.7 ± 1.2 ng/ml Pb (n = 4). It
is unlikely that the commercial HNO_3 was impure when packaged but
rather had leached Pb from the glass container. Acids used for ultra-
trace analysis may require more attention with packaging. Reagents
will require more stringent standards for low levels of impurities.

TABLE 2

Application of ASV to Determination of Lead (Whole Blood Pb μg/100 ml)

Sample	ASV	Non-Flame AAS
1.	10.7	9.7
2.	11.7	8.8
3.	9.9	10.2
4.	19.9	20.0

Analysis of Mercury by ASV

Luong and Vydra[18] examined various electrolytes for Hg analyzed
by ASV but found potassium thiocyanate and sodium perchlorate to be
the most suitable electrolytes. An optimum value of pH 2 was required
because above this double peaks occurred and no peak occurred at pH
>6.0. In the presence of complexing electrolytes, such as thiocyanate,
the Hg stripping peak moved to a less positive potential.

Kendall[17] found 0.1 M $HClO_4$ to be the most suitable electrolyte
for Hg but silver (Ag^+) coincided with Hg and caused an interference.
50 ppm of chloride was also found to prevent the deposition of mercury.
Apart from mercuric ion, methylated mercury could also be detected
with about the same sensitivity as for free mercury. Kendall[17] also
found de-aerating of Hg sample solutions was unnecessary and was confirmed in this work[19]. Five minute de-aeration with N_2 gas had little
influence on the height of the Hg peak and only after 10 minutes was
there any significant increase. The analysis of Hg by ASV in this
work (and previously) was performed using 0.1 M $NaNO_3$ and 0.15 M HNO_3
in a total volume of 4 ml which gave a peak potential at +0.45 V vs
S.C.E. ASV instrumental settings are shown in Figure 4.

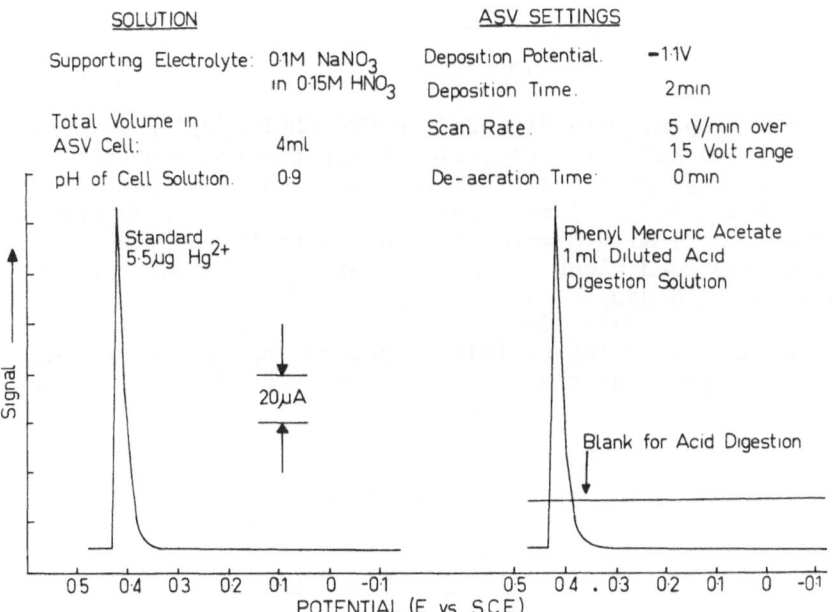

Figure 4.   Current/Potential Curves for Mercury.

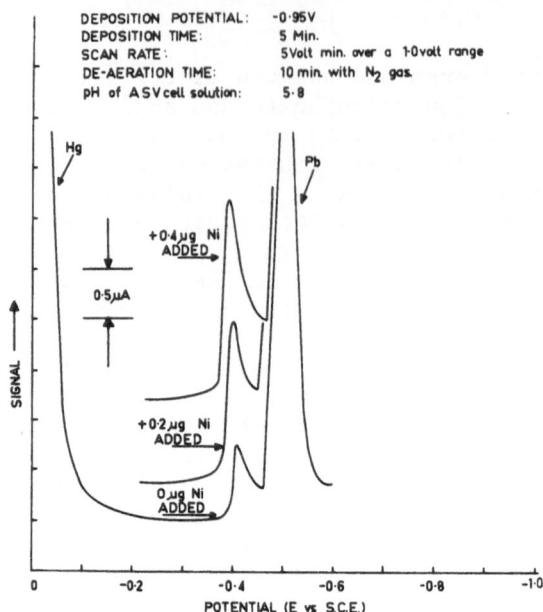

Figure 5.   Current/Potential Curves for Nickel.

Analysis of Mercury in Organic Materials

Various samples were digested in $HNO_3/HClO_4/H_2SO_4$ acid mixture using the closed refluxing digestion apparatus and procedure described by Gorsuch[9]. Digestion solutions were diluted to an overall concentration of 0.3 M HNO_3. 1 ml aliquots together with 2 ml supporting electrolyte plus 1 ml H_2O were transferred to the ASV cell and analyzed using the conditions shown in Figure 5. Standards were Hg^{2+} solutions in 0.3 M HNO_3.

Phenyl mercuric acetate (Alfa Products) and mercuric sulphide (BDH AR grade) gave quantitative recovery of Hg (Table 3). The same

TABLE 3

ASV Determinations of Mercury (% Recovery of Mercury)

	Phenyl Mercuric Acetate (59.6% Hg)	Mercuric Sulphide (86.2% Hg)
1.	105.5 ± 5.6 (n = 8)	100.9 ± 2.6 (n = 6)
2.	101.8 ± 5.8 (n = 6)	105.7 ± 3.7 (n = 4)
3.	102.7 ± 2.2 (n = 6)	100.3 ± 3.3 (n = 4)
4.	102.8 ± 3.9 (n = 6)	97.1 ± 5.4 (n = 4)

solutions were also analyzed by non-flame AAS and gave comparable re-
coveries of Hg. Samples of sediments from the Derwent River, Hobart,
with organic and inorganic Hg, were analyzed. Hg was determined in
two samples by ASV using standard additions (Hg by ASV was approximately 49 µg/g) and also by non-flame AAS (but the latter gave higher
concentrations). The current vs potential scan for Sample 1 showed a
single peak (0.175 V vs S.C.E.) but the scan for Sample 2 had 2 distinct peaks (0.175 and 0.195 V vs S.C.E.). Addition of Hg and Ag to
Sample 2 confirmed that the stripping peak at 0.195 V was due to Ag
but remained unresolved in Sample 1. The presence of silver was confirmed by flame AAS; Samples 1 and 2 contained 4.1 and 5.7 µg/g Ag
respectively.

Analysis of Nickel by ASV

Nickel is one of the elements less commonly determined by ASV.
Luong and Vydra[18] found that low concentration of Ni interfered with
the determination of Hg using a glassy carbon electrode. The interfering effect of Ni could be eliminated by shifting the deposition potential to -0.75 V vs S.C.E. at which potential Ni was not deposited but
the sensitivity of Hg was decreased. The application of this phenomena
was investigated with a cell feed solution for electrolytic zinc production[19]. The solutions (pH 5.4) contained a huge concentration of Zn
as $ZnSO_4$ (typically 117,000 ppm Zn as determined by flame AAS upon
dilution) in comparison with only a few ppm or less of Ni. To prevent
deposition of zinc, which caused the current-potential curve to remain
off-scale during the stripping cycle the deposition potential was set
to -0.95 V vs S.C.E. The addition of KCNS solution to the ASV cell
(0.1 M overall) produced good resolution of the Ni stripping peak
(Figure 5). Analysis for Ni was performed by the standard addition
technique. The ASV cell solution volume was 5 ml which contained 2 ml
cell feed solution, 0.5 ml KCNS stock solution and 20 µg Hg overall.
The addition of Ni produced a linear response and gave a stripping
peak for Ni at -0.415 V vs S.C.E. The initial concentration of Ni in
this solution was 0.26 ppm. Analysis by flame AAS was virtually impossible because the high concentration of $ZnSO_4$ caused severe burner
blockage and light scattering. The use of controlled potential deposition to keep Zn in solution prevented interference with the Ni determination and demonstrated this unique feature of ASV.

CONCLUSION

The role of ASV in trace analysis is twofold. Firstly ASV is
readily available as a comparative technique for AAS which is the
"work horse" of trace metal analysis. Secondly, ASV has particular
application where AAS is not well suited, such as solutions with high
salt concentration, multi-element analysis, and analyses where the
preparation step for AAS is longer.

The future of ASV will be determined by its ability to compete with AAS and other techniques in the realm of multi-element analyses and also in process control applications.

ACKNOWLEDGEMENTS

The authors wish to acknowledge the assistance of Mr. D. Richardson in performing the experimental work and financial assistance from Tasmanian Department of Environment.

REFERENCES

1.   Walsh, A., Spectrochim. Acta, 7 108 (1955).
2.   Florence, T.M., and Batley, G.E., *Anodic Stripping Voltammetry and Polarography*, Notes for a One-Day Course held at University of New South Wales, Kensington, 16th May, 1975, and papers therein.
3.   Štulíková, M., J. Electroanal. Chem., 48 33 (1973).
4.   Ellis, W.D., J. Chem. Educ., 50(3) A131 (1973).
5.   Copeland, T.R., and Skogerboe, R.K., Anal. Chem., 46(14) 1257A (1974).
6.   Grainger, A., Chemistry Department, University of Tasmania.
7.   Florence, T.M., *6th Federal Convention, Australian Water and Wastewater Association*, Technical Papers, Melbourne, May, 1974.
8.   L'vov, B.V., Spectrochim. Acta, 24B, 53 (1969).
9.   Gorsuch, T.T., Analyst, 84 135 (1959).
10. Holak, W., J. AOAC, 58 777 (1975).
11. Princeton Applied Research, Application Note AN-106.
12. Environmental Science Associates Inc., Application Note M-14.
13. Duic, L., Szechter, S., and Srinivasan, S., J. Electroanal. Chem., 41 89 (1973).
14. Šinko, I., and Kosta, L., Intern. J. Environ. Anal. Chem., 2 167 (1972).
15. Noller, B.N., and Bloom, H., Aust. J. Med. Technol., 6 100 (1975).
16. Barnard, A.J.,and Dudley, R.W., Industrial Research, 15 34 (1973).
17. Kendall, D.R., Anal. Lett., 5 867 (1972).
18. Luong, L., and Vydra, F., J. Electroanal. Chem., 50 379 (1974).
19. Low, H.V., Honours Thesis, University of Tasmania.
20. Feldman, C., Anal. Chem., 46 1606 (1974).
21. Bothner, M.H., and Robertson, D.E., Anal. Chem., 47 592 (1975).
22. Alberts, J.J., Schindler, J.E., Miller, R.W., and Carr, P.W., Anal. Chem., 46 434 (1974).
23. Riley, J.P., and Taylor, D., Anal. Chim. Acta, 40 479 (1968).
24. Batley, G.E., and Florence, T.M., J. Electroanal. Chem., 55 23 (1974).
25. Copeland, T.R., Osteryoung, R.A., and Skogerboe, R.K., Anal. Chem., 46 2093 (1974).
26. Reimers, R.S., Burrows, W.D., Krenkel, P.A., J. Water Pollut. Contr. Fed., 45 814 (1973).

IMPROVED INSTRUMENTATION AND PROCEDURES FOR INTERPRETING

GALVANOSTATIC POTENTIAL-TIME TRANSIENTS

P.H. Davies, E.J. Frazer, M. Skyllas and B.J. Welch

Department of Industrial Chemistry
School of Chemical Technology
The University of New South Wales
P.O. Box 1, Kensington, N.S.W. 2032

1.   INTRODUCTION

The resurgence in galvanostatic transient studies that arose from the work of Gierst and Juliard (1) has taken two major directions - as an analytical tool by employing the Sand equation, and in mechanistic studies of electrode reactions.   Analytical applications of chronopotentiometry have not developed widely because of the limited accuracies obtainable.   Problems can be encountered because of the need to ensure that mass transfer is only by diffusion;   electrode kinetic effects also introduce complications.   Compensating for electrode capacitance (2) and maintaining uniform characteristics of the electrode surface are other problems that can introduce errors.   Mechanistically the method has proven to be more useful when supplemented by other techniques or when using the more recently developed current-reversal procedure.

In molten salt media, surface study techniques have seldom been practical because of design restrictions that result from the heating element.   Consequently mechanistic studies have relied heavily on chronopotentiometry and similar transient techniques.   Researchers have rarely been able to rigorously apply all tests possible however, not only because of the limitations outlined above, but also because of ohmic contributions to the measured potentials.   The latter arise from experimental difficulties in positioning Luggin capillaries, and the higher current densities that are generally used in molten salt systems.

In this contribution we describe instrumentation that permits the recording of transients in digital form while switching the current to and from zero, with or without reversal and/or delay. Switching is done in such a manner that the switch points are always within the recording scale so that ohmic effects can be accurately measured. This procedure enables accurate determination of working-reference potentials so that diagnostic checks such as the variation of quarter wave potential with current or solute concentration can be made. Using a captive digital processing system we are able to analyse the curves under programme control introducing error compensation procedures as desired.

2. CONSTANT CURRENT PROGRAMMING SYSTEM

A constant current was generated by using a potentiostat in the standard galvanostatic mode (3), as is illustrated in the enclosed section of Figure 1. The potential drop between the selected resistor (R_S) that is positioned between the working electrode and ground corresponds to the potential applied to the positive input of the differential amplifier. When this potential is constant, the current is constant. The differential power amplifiers used (NS0061, Fairchild μ791) have similar operating characteristics to the μA741 and are capable of delivering output currents greater than 500 mA.

The circuit diagram of the potential programmer (P) constructed is given in Figure 2. The output of the summing amplifier (A4) is zero except during the time that immediately follows triggering one of the NE555 linear integrated circuit timers (T1 or T2). These timers give a constant output voltage for a period 1.1 $R_i C_i$ secs. once triggered by a negative pulse. The output rise and fall times are each approximately 100 nS and the fall of one timer can be used to trigger another giving a total switching time of

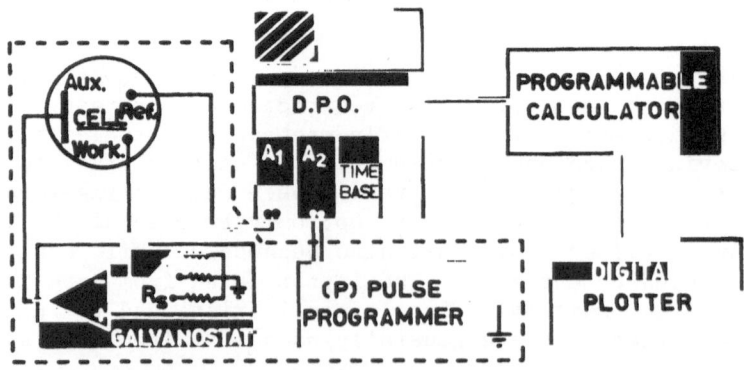

Fig. 1. System for Chronopotentiometry with Digital Recording and Processing.

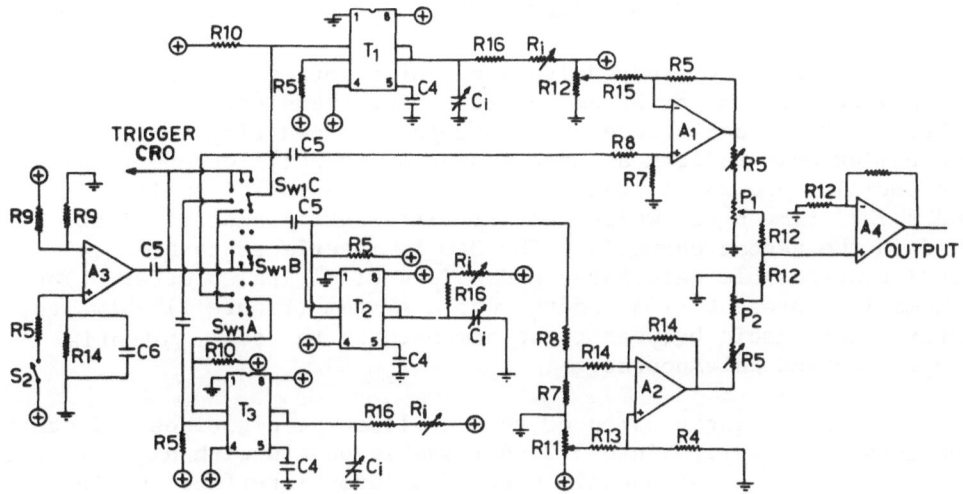

Fig. 2.   Circuit Diagram of the Potential Programmer.

less than 200 nS.   The divided constant output voltage of one timer
(T1) is connected to a non-inverting voltage follower (A1) and the
divided output of the second timer (T2) goes to an inverting
follower (A2).   Each of these amplifiers have zero offset adjustment and their outputs are adjusted to give 0 to +1 volt and 0 to
−1 volt through ten turn potentiometers P1 and P2 respectively.
These potentiometers, together with the variability available in
the time constant, enable a wide range of pulse heights and lengths
to be selected.

Initial triggering is from the changing output of amplifier A3
on opening switch S2.   The three-pole six position switch S1 ensures
that only one timer is triggered at any time. Variable delay between
the triggering of the timers T1, T2 can be accomplished with the
third timer T3 if so desired.   With the second switch (S1) we are
able to generate the following galvanostatic pulse programmes:
anodic only, cathodic only, anodic-cathodic and *vice versa* anodic-
delay-cathodic and *vice versa*.   This instrument has been used
successfully for transition times as short as 5 mS and as long as
15 seconds.   These are not instrumental limits however, but reflect
the range required in our work.   A typical current programme
together with the working reference electrode potential response
is illustrated in Figure 3.

3.   TRANSIENT RECORDING AND PROCESSING SYSTEM

The measurement and analysis of chronopotentiograms has been
streamlined to a major extent by recording transients on a
Tektronix 7704A dual trace oscilloscope combined with a P7001
processing unit to form a digital processing oscilloscope (DPO).
The oscilloscope was fitted with a 7A13 differential amplifier
and a 7B71 time base while another differential amplifier was
used in the second channel.   The DPO is interfaced to both a
Tek31 programmable calculator and a Tek4661 digital plotter.   The
schematic representations of the whole system (Figure 1) shows the
inter-relationship between these components, the cell, potential
programmer and galvanostat.

The processing section of the oscilloscope digitizes and stores
the acquired waveform into 512 addressable points with digital values
from 0 to 1023.   The oscilloscope-calculator interface enables
operations with stored waveforms to be performed under programme
control and up to four waveforms may be stored and displayed at
any time.   Operations such as curve smoothing, integration,
differentiation, summing and multiplication of waveforms may be
performed at the touch of a button.   The programmes are normally
stored on magnetic tape cartridges and fed into the calculator at

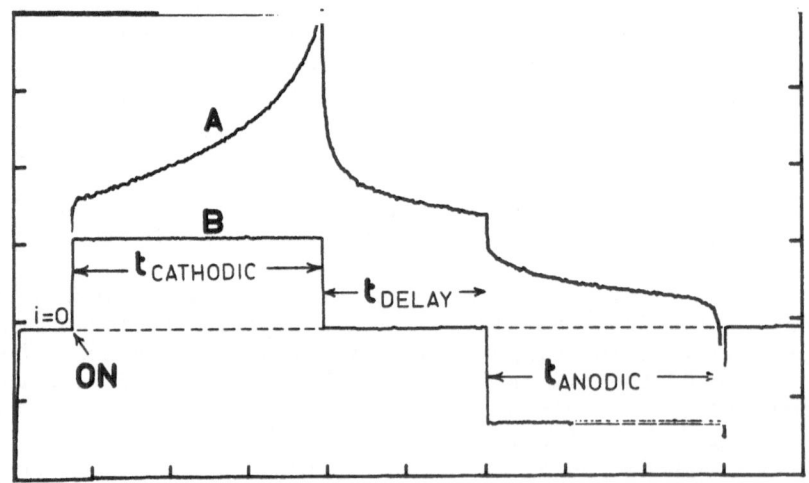

Fig. 3.   Typical Cell Current Programme and Electrode Response
          Curves.
          Curve A, Ref-W Potential, x=50mS/Div, y=50mV/Div;
          Curve B, Current Programme, x=50mS/Div, y=10mA/Div.

the commencement of an experiment.   The functions of the basic
programme routines which have been developed are summarised in
Table 1. Numerous other routines for more specialized calculations
have also been developed and these may be called from magnetic tape
when required. A typical example is the calculation of the shape
of the transient when allowing for electrode capacitance.

TABLE 1

Functions of the Basic Routines Used in the Digital Processing System

ROUTINE	DESCRIPTION
Data Input	Stores information used continuously such as solute concentration, electrode area, and temperature.
Set Zero	Reads the true position of 0 volts of the differential amplifier.
Differentiate	Differentiates the stored waveform; useful when estimating the transition time (τ).
Current	Automatically measures the imposed current i from its waveform acquired on second channel of DPO.
To Tape	Stores a waveform and scale factors on magnetic tape cartridge for future use.
From Tape	Transfers waveform from tape to DPO.
Copy	Copies curve displayed on DPO, including scale factors, using the digital plotter.
Soluble Reversible	Plots logarithm $(\tau^{\frac{1}{2}}-t^{\frac{1}{2}})/t^{\frac{1}{2}}$ versus E from 0.1τ to 0.9τ. Voltage scale magnified fourfold to assess linearity. Automatic linear regression calculates slope and displays RT/nF on screen.
Insoluble or Irreversible	Plots logarithm $(\tau^{\frac{1}{2}}-t^{\frac{1}{2}})$ versus E then as per Soluble.
Analysis	Computes and prints a full record of useful chrono information such as iR drop, $E_{\tau/4}$, τ, $i\tau^{\frac{1}{2}}$, Dn^2, $i\tau$.
Reveral Soluble	Plots appropriate logarithm function of τ reverse and t versus E, then as per Soluble.
Reversal Insoluble	As per Reversal Soluble using equation 1.

4. APPLICATIONS

The two major advantages of such a system are the speed with
which data can be processed and the improved accuracy that is
obtained by applying correction factors.   The pulse programmer
also increases the range of kinetic schemes that can be tested.
In a contribution such as this we can only exemplify a few of the
programmes.   Experimental results have been obtained for the
deposition of lead from $PbCl_2$ dissolved in molten LiCl-KCl eutectic,
and also for the oxidation of the sulphide ions (from PbS) to sulphur
in molten $PbCl_2$-NaCl eutectic.   In each case glassy carbon working
electrodes (surface area approx. 0.08 cm^2) were used and potentials
were measured versus Pb/Pb^{2+} reference electrodes.   The solvents
were purified and tested according to standard procedures before
addition of the solute.   All experimental measurements were made
at 710 ± 2 K.

Deposition of Lead

In the LiCl-KCl eutectic and on an inert electrode, lead ions
are reduced according to the reaction, Pb(II) + 2e = Pb, in a
reversible manner to form an insoluble product (4).   For such a
reaction in the absence of capacitance effects the Sand equation
should be obeyed and the ratio $i\tau^{\frac{1}{2}}/C$ should be constant.   Thus, a
plot of $\tau^{\frac{1}{2}}/C$ versus $1/i$ should be a straight line passing through
the origin.   Figure 4 showing the expected relationship was plotted
using data from the output of the *Analysis* programme.

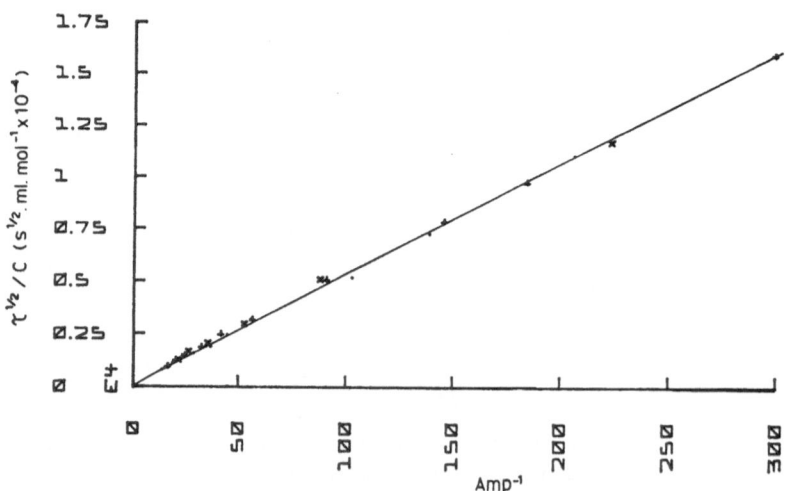

Fig. 4.   Sand Equation Test: $x = 5.75 \times 10^{-5}$ mol. Pb^{2+}/ml.,
+=8.60×10^{-5} mol.Pb^{2+}/ml.,.=1.15×10^{-4} mol.Pb^{2+}/ml.

TABLE   2

Variation of $E_{\tau/4}$ with ln i   (Conc = 1.15 x 10^{-4} moles/ml)

i,mA	26.95	34.93	41.01	48.19	52.92	59.00	72.01
ln i	3.29	3.55	3.71	3.88	3.97	4.08	4.28
$E_{\tau/4}$.V	0.048	0.056	0.056	0.060	0.053	0.048	0.055
$(\bar{E}_{\tau/4}-E_{\tau/4})$,mV	+6	-2	-2	-6	+1	+6	-1

For a reversible reaction the quarter wave potential should be independent of current.   Table 2 shows the variation of $E_{\tau/4}$ with current, again using data obtained from the *Analysis* programme. In this case the deviation from the mean $E_{\tau/4}$ is never more than ±6 mV.   Most importantly, the *Analysis* programme contains a sub-routine which automatically reads and corrects the apparent $E_{\tau/4}$ value for the ohmic contribution.   While manual reading and ohmic correction of $E_{\tau/4}$ values are quite difficult, this programme allows values to be consistently quoted to better than ±10 mV.

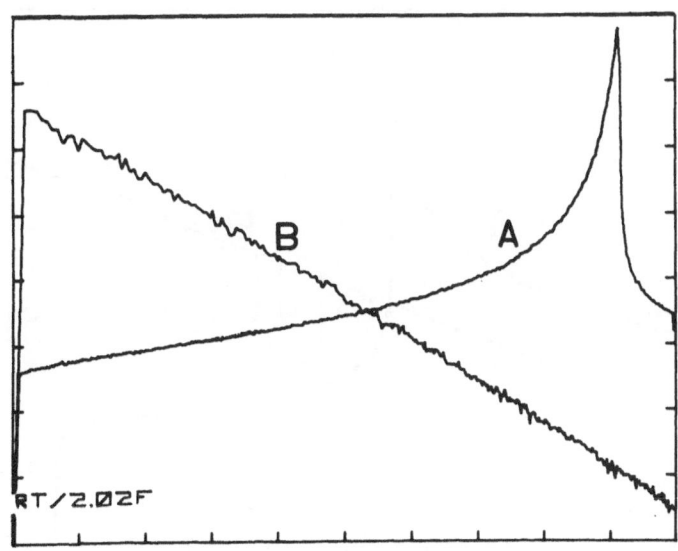

Fig. 5.   Cathodic Chronopotentiogram for 5.75x10^{-5} mol.Pb^{2+}/ml,
          i = 3.1 mA.
          Curve A, x=100mS/Div, y=50mV/Div;
          Curve B, E vs ln ($_{\tau}^{\frac{1}{2}}$-t$^{\frac{1}{2}}$) from 0.9 to 0.1 τ.

In any mechanistic investigation, the shape of the chronopotentiogram is invaluable in gaining knowledge about reaction
reversibility, product solubility and the number of electrons
involved in the reaction.   For the selected example one would
expect a plot of ΔE vs ln $(\tau^{\frac{1}{2}} - t^{\frac{1}{2}})$ to be linear with a slope of
RT/2.00F.   Such plots were performed using the *Insoluble* programme,
and Figure 5 shows a typical transient and the corresponding
logarithmic plot from 0.1 to 0.9 τ.   The slope of the line of best
fit calculated by the automatic linear regression was RT/2.02F which
is very close to the expected value.   In this example 334 points
were read and the correlation coefficient for the data was found
to be −0.9992.   The main advantages illustrated here are the speed
and ease of analysis and the accuracy obtainable using automatic
linear regression.

It is also possible to analyse the shape of the chronopotentiogram upon current reversal, provided reversal occurs at a time
approximating the forward transition time.   The potential time
relationship then is given by the following equation:

$$E = E' + RT/nF \ln i \{\tau - (\tau + t)^{\frac{1}{2}} - (1 + \theta) t_r^{\frac{1}{2}} \} \qquad [1]$$

where τ = forward transition time, t_r = time after reversal, i =

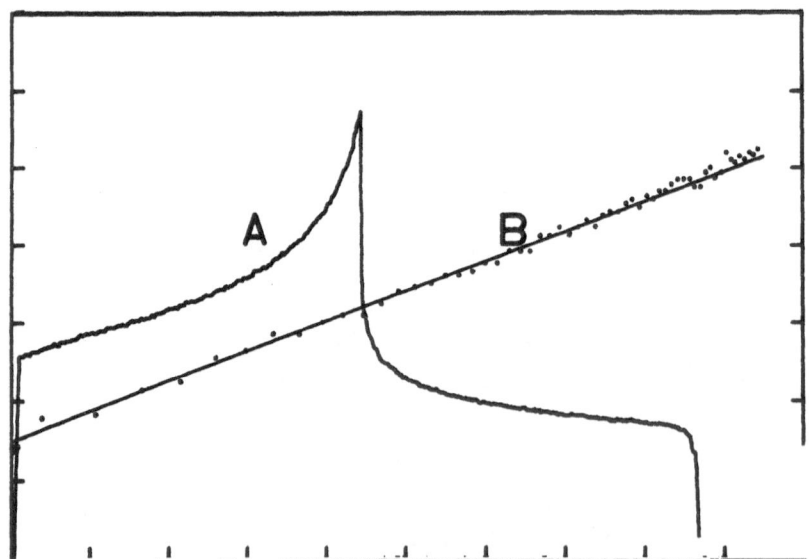

Fig. 6.   Chronopotentiometric Reduction and Oxidation of
          8.6×10^{-5} mol. Pb^{2+}/ml, i = 6.8mA, $\Theta = 1$
          Curve A, x=100mS/Div, y=50mV/Div;
          Curve B, E vs ln $\{\tau - (\tau + t)^{\frac{1}{2}} - (1+\theta) t_r^{\frac{1}{2}}\}$

forward current and θ = ratio of forward to reverse current. The
derivation of this equation will be described elsewhere(5). A
typical reversal curve has been analysed here using the *Reversal
Insoluble* programme (Figure 6).   The logarithmic relationship was
plotted directly on the digital plotter while a linear-regression
was simultaneously performed.   In this case the calculated slope
was RT/2.05F and the correlation coefficient -0.9965, again demonstrating the speed and accuracy of analysis obtainable.

Oxidation of Sulphide Ions

   It has recently been shown that sulphur anodically evolved
when electrolysing PbS dissolved in $PbCl_2$-NaCl melts has a limited
solubility in the solvent (6) while it also undergoes a chemical
reaction with the melt (7).   These two facets may be readily
established by utilizing our flexible system for programmed
current reversal.

   Beyerlein and Nicholson (8) have shown that the relationship
between the transition time on current reversal (τ_r) and the
forward electrolysis time (t_f) is given by the relationship

$$\tau_r/t_f = \theta^2/ \; \{ \; (1 + \theta)^2 - \theta^2 \; \}$$ [2]

if the product of the reversible reaction is soluble.   However if
the product of the reversible electron transfer is completely
insoluble then the relationship becomes

$$\tau_r/t_f = \theta$$ [3]

For a product having a limited solubility equation [2] should be
obeyed for low values of t_f.   For longer values of t_f (or t_f/τ_f)
local saturation would result and thereafter equation [3] would
tend to be obeyed.   Figure 7 presents the results obtained for
various forward times, currents, and current ratios.   To enable
comparison on a similar relative scale the ordinate used in each
case is the ratio t_f/τ_f.   The curves drawn through the experimental points intersect the ordinate at the theoretical values
(from equation [2]) in each instance.   The transition to an
insoluble product is evident from the steady increase in the ratio,
although it never attains the theoretical value.   This behaviour
can be partly attributed to the irreversible reaction

$$S_\nu \text{ (insol)} + \text{Melt} \rightarrow \text{ SOLUBLE PRODUCT}$$ [4]

That this process occurs, can be clearly established by introducing
various delay times between the forward and reverse electrolysis.
Experimental results are presented in Figure 8.   The decrease in
reverse transition time as the delay time is lengthened is directly

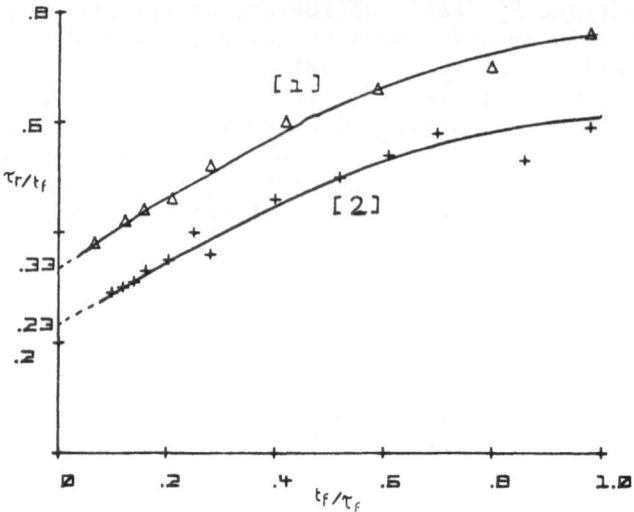

Fig. 7   Plot of τ_r/t_f versus t_f/τ_f
Curve 1, $\theta = 1$; Curve 2, $\theta = 0.75$

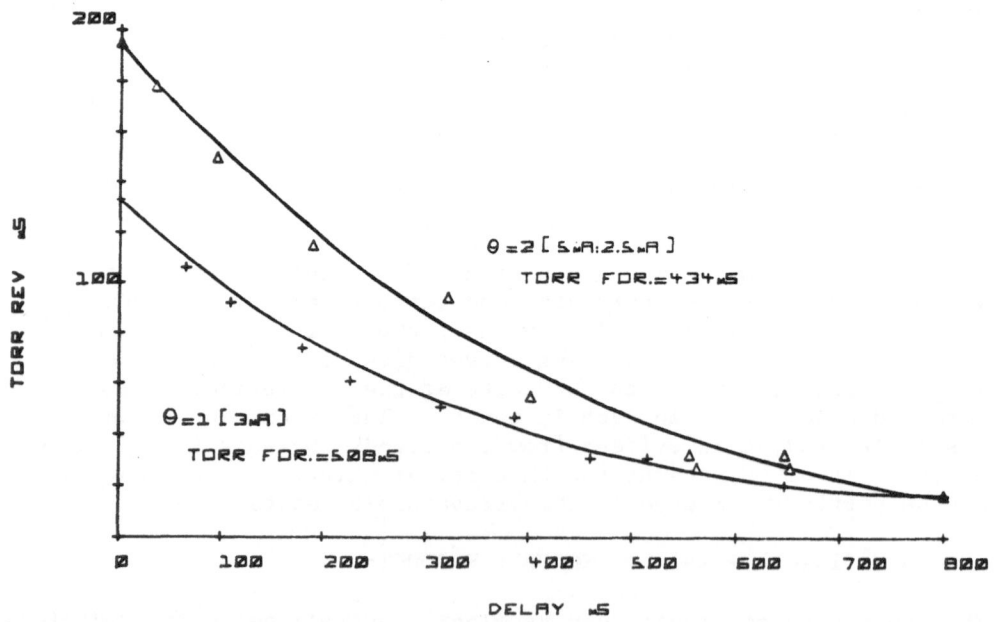

Fig. 8   Variation of τ_r with Delay Time at Constant τ_f.

related to the rate of removal of the sulphur from the electrode surface. Such a curve can be tested against theoretical curves for various kinetic schemes to evaluate rate constants and reaction mechanisms.

Because each switching of current is accompanied by an ohmic potential change, step current changes are easily identified on the stored transient. Therefore curves can be analysed and ratios calculated under programme control. The reproducibility can be affected by the prior history of the electrode, and the quality of results depends on the experimental procedure.

Capacitance Effects

Various graphical methods have been proposed to determine the diffusive transition time from experimental curves (2). From the description of the system developed it is apparent that appropriate corrections can be made under programme control minimising errors due to subjective judgements. Our interest has extended to the determination of the influence of capacitance on interpretations of the shape of galvanostatic potential/time curves - an aspect of chronopotentiometry often neglected previously. The method available for treatment depends on whether the capacitance-electrode potential relationship is known or not.

The capacitance current (i_c) at any point is given by the relationship

$$i_c = CdE/dt \qquad [5]$$

where C is the differential capacitance of the double layer. Normally the capacitance current tends to a constant value in the range 0.1 to 0.9 τ. If we were to assume it to be constant over the whole transition, although not necessarily having the same value as given above, then from the Sand equation the following relationship would hold

$$i_c/i = (\tau_m^{\frac{1}{2}} - \tau_s^{\frac{1}{2}})/\tau_m^{\frac{1}{2}} \qquad [6]$$

where τ_s is the transition time in absence of capacitance, τ_m is the transition time as determined by the maximum in the dE/dt plot, and i the imposed current. When the above conditions are obeyed, it can be shown that for an insoluble product or irreversible reaction plots of ln $(\tau_m^{\frac{1}{2}} - t^{\frac{1}{2}})$ versus ΔE tend to linearity with slopes similar to those predicted in the absence of capacitance. It should be noted that the method of determining the transition time is different to that normally employed. Figure 9 presents a series of simulated chronopotentiograms for the reversible deposition of lead from 1 x 10^{-5} mol. Pb^{2+} ml^{-1}, with a diffusive transition time τ_s of 0.9 millisecs and various values of electrode

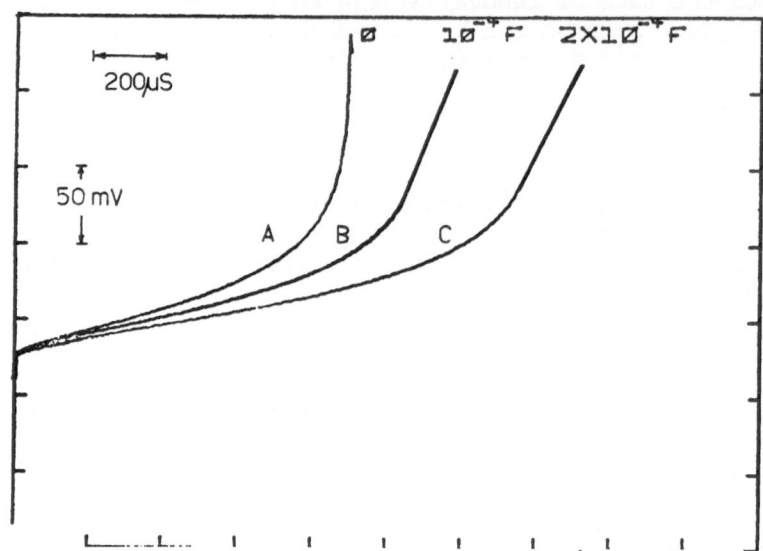

Fig. 9   Simulated Chronopotentiograms for Reduction of 10^{-5} mol.
Pb(II)/ml.τ_s = 0.9 mS:

Curve A, Cap = 0;   Curve B, Cap = 1×10^{-4} Farads;
Curve C, Cap = 2×10^{-4} Farads.

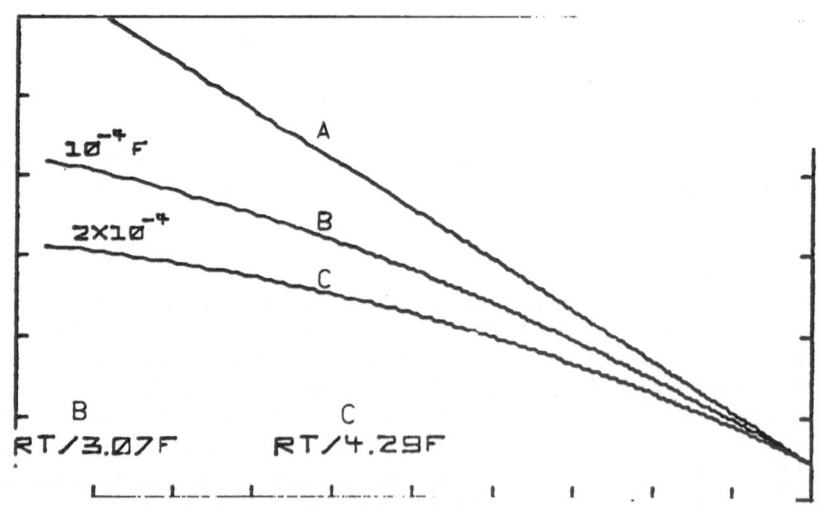

Fig. 10   Plot of Change in Potential versus Logarithmic Functions
for Curves A, B, & C of fig. 9.

capacitance.   Potential time analyses of these curves are shown
in Figure 10.   Analysis of each curve using the *Insoluble* routine
and the appropriate value of τ_m produced curve A, with the theoretical
slope of RT/2F.   If curves B and C of Figure 9 are analysed using
the transition time obtained by the extrapolation technique (i.e. τ_s)
then curves B and C of Figure 10 result.   The slopes then
(RT/3.07F, RT/4.29F) are considerably different to the theoretical
value.

Although using τ_m instead of τ_s passively compensates for the
influence of capacitance when interpreting the shape of chronopotentiograms, this does not apply to other diagnostic tests.   For
example, the quarter wave potential should be independent of current
for the reaction scheme being considered, but a trend is observed
using either transition time if no compensation is made for
capacitance.   More rigorous application of all diagnostic tests
can however be made if an approximate value of electrode capacitance
is known.

De Vries (9) has developed a distortion parameter, K, given
by the expression

$$K = \frac{2\,i\,C\,RT}{n^3 F^3 \pi\, DM^2} \tag{7}$$

for the influence of capacitance on the shape of galvanostatic
potential-time curves.   the above equation D is the diffusion
coefficient of the ion ($cm^2\ s^{-1}$) and M the solute concentration
(mol. ml^{-1}).   Equation [8] can also be expressed in terms of τ_s
using the Sand equation.   By employing the latter modification
to equation 7 , equation [6], and the appropriate boundary
conditions, the galvanostatic potential-time equation becomes (10),

$$E = Const + RT/nF\ ln\quad \{\ \tau_s^{\frac{1}{2}}/(1-8.6\ RTC/nF\tau_s\ i\ ) - t^{\frac{1}{2}})\ \} \tag{8}$$

This equation has been tested for each of the chronopotentiograms
in Figure 9.   The resulting logarithmic plots each coincided with
curve A of Figure 10, giving the theoretical slope (RT/2F).
Equation [8], being a complete potential-time relationship, can
also be used for other diagnostic tests and consequently is a
useful method for compensating for electrode capacitance.   When
electrode capacitance is known, this equation can be incorporated
into our DPO software to overcome one of the basic obstacles to
precise chronopotentiometry.

5.   CONCLUSIONS

Two major problems in the application of chronopotentiometry
have been emphasised, namely, electrode capacitance and the ohmic
contribution to the measured potential.   Both problems have been

alleviated to some extent by the recording and analysis of transients using a digital processing system incorporating specially developed software.   The range of kinetic schemes which can be investigated has also been extended by the introduction of a potential programmer allowing both reversal and delay programmes.   The general features of the total system have been exemplified by experimental studies on Pb (II) in LiCl-KCl, sulphide ions in $NaCl-PbCl_2$, and a theoretical study of the influence of capacitance on transient shape and analysis.   While there is obviously room for further software sophistication, the system both simplifies routine chrono-potentiometry and also enables fast and accurate mechanistic inter-pretations of potential-time transients, with the added ability to introduce error compensation procedures when required.

6.   ACKNOWLEDGEMENTS

Purchase of equipment used in this project was assisted by a grant from Comalco Limited.   We are also indebted to Mr. G.J. Houston and Mr. O.P. Farrell for experimental and general assistance.

7.   REFERENCES

1.   L. Gierst and A. Juliard, *J. Phys. Chem.*, 57 (1953) 701.

2.   P. Bos and E. Van Dalen, *J. Electroanal. Chem.*, 45 (1973) 165.

3.   S. Grimnes, *Chem. Instrumentation*, 5, (1973) 141.

4.   D. Inman and J. O'M. Bockris, *J. Electroanal. Chem.*, 3 (1962) 126.

5.   P.H. Davies, M. Skyllas and B.J. Welch, to be published

6.   B.J. Welch, A.J. Cox and P.L. King in J.H.E. Jeffes and R.J. R.J. Tait (Eds.), *Physical Chemistry of Process Metallurgy: The Richardson Conference*, Inst. of M & M, 1974, p.33.

7.   G.J. Houston, M. Skyllas and B.J. Welch, *Abstract No. 78 of Fourth Australian Conference on Electrochemistry*, Adelaide Feb. 1976.

8.   F.H. Beyerlein and R.S. Nicholson, *Anal. Chem.*, 40 (1968) 286.

9.   W.T. De Vries, *J. Electroanal. and Interfacial Chem.*, 17 (1968) 31.

10. P.H. Davies, *Thesis Univ. of N.S.W.* (1975).

THE ELECTROCHEMISTRY OF SULPHIDE MINERALS

E. Peters

Department of Metallurgy, The University of B.C.

Vancouver, B.C.   V6T 1W5

INTRODUCTION

A sulphide mineral can be used as an electrode because it is an electronic conductor, similar to a metal. However, in order to understand its electrochemical behaviour, it is necessary to take into account at least two characteristics not found with normal pure metal electrodes: (1) the presence of several components (sulphur and at least one metal), and (2) the specialized electronic properties, usually those of a semi-conductor (although the conductivity of sulphides is sometimes as high as that of some metals).

THERMODYNAMICS

The electrode potential of a metal sulphide can be written in the same way as that of a pure metal, but an extra Nernst term is needed to take into account the chemical potential of the metal in the form of a sulphide, viz

$$M^{n+}_{(aq)} + ne \xrightleftharpoons{\text{MS}} M^{o}_{(MS)}$$

$$E = E^{o}_{1} - \frac{2.3RT}{nF} \log a_{M^o} + \frac{2.3RT}{nF} \log a_{M^{n+}} \qquad [1]$$

where a_x is the activity of x with respect to the standard state (the standard state of a metal is a pure metal).

According to the Gibbs phase rule, a simple sulphide mineral, MS, will have one degree of freedom at constant temperature and pressure, which can be identified with the chemical potential of

the metal $\mu_{M(MS)} = -RT \ln a_M$.   As long as a second phase containing only M and S does not form, this chemical potential is an independent variable, in terms of Equation [1].   The sulphide mineral
will have a stoichiometry linked to the chemical potential, and
there will be an upper and lower limit to the chemical potential
linked to the limits of composition as described in a two component
phase diagram*.   Thus the single electrode potential of a sulphide
mineral is an indeterminate value within a range with defined upper
and lower limits.   Where the phase diagram contains only one sulphide phase, such as lead sulphide, the lower limit corresponds to
the potential of the pure metal electrode, and the upper limit corresponds to the potential of the elemental sulphur/sulphide couple.
If the solution contains metal ions, the sulphide activity can be
obtained from the metal ion activity and the solubility product of
the metal sulphide.

The electrode potential can also be written in terms of an
appropriate electrochemical reaction for sulphur, e.g.

$$S^o_{(MS)} + 2H^+ + 2e \rightleftharpoons H_2S(g)$$

$$E = E^o_2 - \frac{2.3RT}{2F} \log f_{H_2S} + \frac{2.3RT}{2F} \log a_{S^o} - \frac{2.3RT}{F} pH \qquad [2]$$

where f_{H_2S} is the fugacity of H_2S gas and a_{S^o} is the activity of
elemental sulphur in the solid sulphide phase.   In fact, the potential of equation [2] is exactly the same as that of equation [1]
because of the thermodynamic link between a_{S^o} and a_{M^o} in the solid
and between f_{H_2S} and $a_{M^{n+}}$ in the aqueous phase.

Ternary mineral systems, such as Cu-Fe-S or Ni-Fe-S have one
more component, and therefore one more degree of freedom.   Equations can still be written in the form of Equation [1] or [2], but
the link between a_{M^o} and a_{S^o} in the solid is tied to an additional
variable: the activity of the second metal.   The thermodynamic
treatment involves a three-component Gibbs-Duhem relationship.

At constant solution composition, the two component sulphide
system becomes invariant in electrochemical potential when there
are two non-aqueous phases, (i.e. metal + sulphide, sulphur +
sulphide, or two different sulphides), in equilibrium with each
other.

---

*   It is useful to consider the composition range of a sulphide
mineral as a solid solution of a pure component in a stoichiometric
sulphide compound, and it is also useful to write the formula in
terms of variable metal, which best represents the structural adjustments, i.e. pyrrhotite is $Fe_{1-\delta}S$ where the stable range at room
temperature is $0 < \delta < 0.14$.   It may also be possible to have super-
saturated solid solutions, i.e. $\delta < 0$ or $\delta > 0.14$.

For a three component sulphide, the equivalent requirement is for three phases. Since the equilibration of three solid phases with each other is a far slower process than that of only two solid phases, it is not useful to discuss the detailed thermodynamics for the three-component sulphide system; rather, it is more useful, in most cases, to consider a pseudo-2 component process. For example, for the mineral chalcopyrite, the upper limit of thermodynamic stability in acid solutions is expressed by the equation

$$CuS + S^o + Fe^{++} + 2e \rightleftharpoons CuFeS_2$$

$$E = E_3^o + \frac{2.3RT}{2F} \log a_{Fe^{++}} \tag{3}$$

which involves three solid phases: covellite (CuS), sulphur (S°), and chalcopyrite. In fact, covellite is never observed as a decomposition product of chalcopyrite in oxidation or anodic experiments[*], the observed reaction being

$$2S^o + Cu^{++} + Fe^{++} + 4e \rightleftharpoons CuFeS_2$$

$$E = E_4^o + \frac{2.3RT}{4F} \log a_{Fe^{++}} a_{Cu^{++}} \tag{4}$$

which is a relatively reversible reaction involving only two solid phases. Thus, the evidence is that sulphur and metal will separate in anodic or oxidizing experiments relatively rapidly at low overvoltages, but that separation of two metals from each other in a three component sulphide does not occur during elemental sulphur formation.

On the cathodic side, separation of iron and copper (or nickel) does occur from the ternary minerals chalcopyrite ($CuFeS_2$) and bornite (Cu_5FeS_4), but not from pentlandite ($NiFeS_{1.8}$), i.e.

$$CuFeS_2 + \frac{12}{5} H^+ + \frac{4}{5} e \rightleftharpoons \frac{1}{5} Cu_5FeS_4 + \frac{6}{5} H_2S + \frac{4}{5} Fe^{++}$$

$$E = E_5^o - \frac{2.3RT}{F} \log f_{H_2S}^{3/2} a_{Fe^{++}} - 3\frac{2.3RT}{F} pH \tag{5}$$

However, in this case, as in equation [4], there are only two solid phases (and possibly a gas phase, H_2S).

---

[*]   The reaction $Cu^{++} + CuFeS_2 \rightarrow 2CuS + Fe^{++}$ occurs slowly in acid solutions at temperatures above 180°C, but covellite is oxidized rapidly by oxidants more powerful than Cu^{++}.

Irreversible Oxidation of Sulphur to Sulphate

It is common to describe the electrochemical thermodynamics of sulphide mineral-water systems in terms of E_h–pH diagrams, such as Fig. 1. However, during electrochemical experiments sulphur is not oxidized to sulphate except at large overpotentials. Therefore, it is not very useful to use those lines on these diagrams that represent electrochemical production of $SO_4^=$ or HSO_4^-, at least in acid solution. However, the production of sulphate from sulphide minerals during anodic polarization at substantial overvoltages is an important electrochemical study.

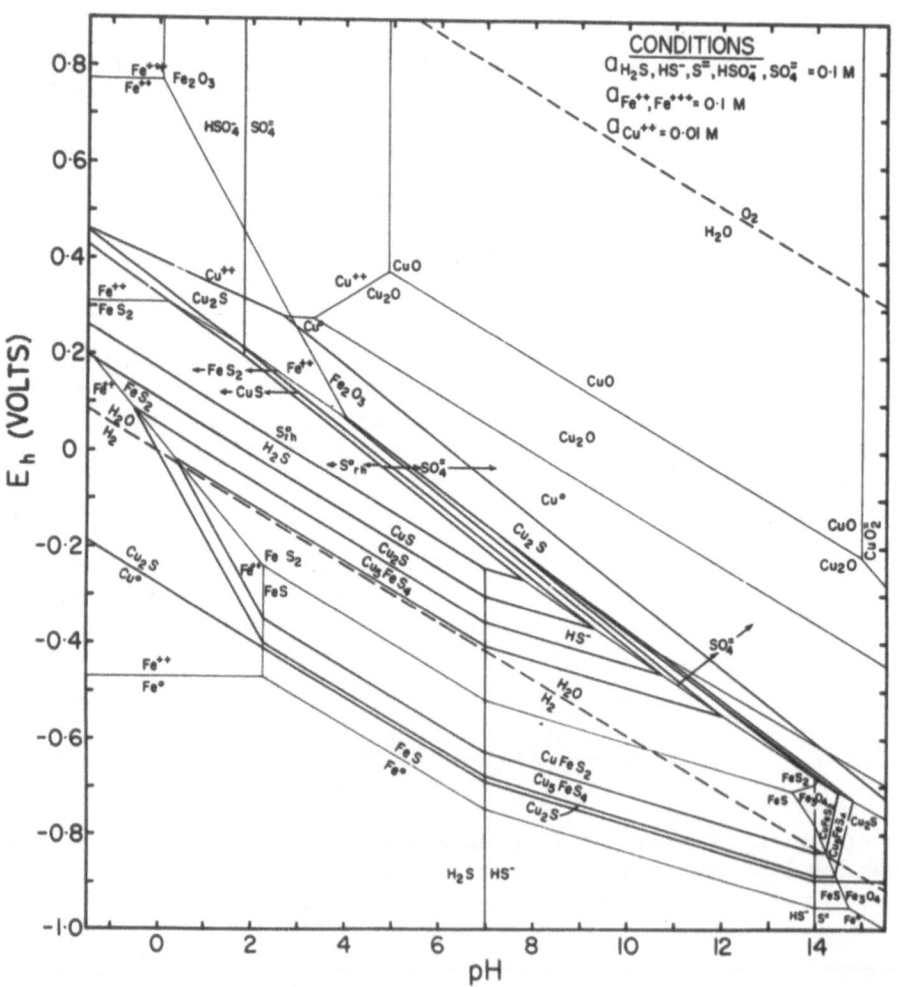

Fig. 1.   The E_h–pH Diagram for the Cu–Fe–S–H_2O System.

EXPERIMENTAL PROCEDURES WITH SULPHIDE MINERALS

The measurement of rest potentials or mixed potentials of sulphide minerals is straightforward when using high impedance voltmeters. However, when doing polarization studies, it is necessary to take two sulphide properties into account:

(a) Sulphide minerals may be of low electrical conductivity, so that IR values through an experimental specimen may be appreciable.

(b) Sulphide minerals are semi-conductors and may have non-ohmic junctions with either the current supplying electrical lead or the electrolyte-mineral interface. It is useful to compare a typical galvanostatic or potentiostatic set up with the equivalent electrical circuit (Fig. 2).

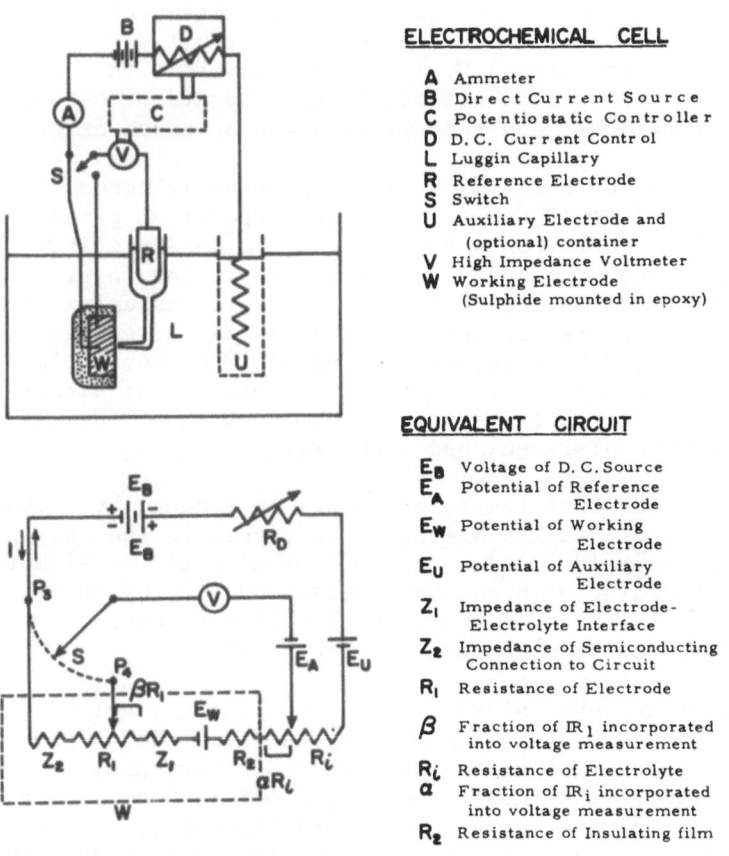

ELECTROCHEMICAL CELL

A Ammeter
B Direct Current Source
C Potentiostatic Controller
D D.C. Current Control
L Luggin Capillary
R Reference Electrode
S Switch
U Auxiliary Electrode and
   (optional) container
V High Impedance Voltmeter
W Working Electrode
   (Sulphide mounted in epoxy)

EQUIVALENT CIRCUIT

E_B Voltage of D.C. Source
E_A Potential of Reference
       Electrode
E_W Potential of Working
       Electrode
E_U Potential of Auxiliary
       Electrode
Z_1 Impedance of Electrode-
       Electrolyte Interface
Z_2 Impedance of Semiconducting
       Connection to Circuit
R_1 Resistance of Electrode
β Fraction of IR_1 incorporated
       into voltage measurement
R_i Resistance of Electrolyte
α Fraction of IR_i incorporated
       into voltage measurement
R_2 Resistance of Insulating film

Fig. 2. Experimental Apparatus and Equivalent Circuit for Polarization Studies on Sulphide Minerals.

With the switch S at P_3 in the equivalent circuit, (a three wire system common to most potentiostats), the high impedence volt-meter V sees IR potential drops from the current circuit across Z_2, R_1, Z_1, R_2, and αR_i. The value of αR_i is very small if the Luggin capillary is very close to the working electrode, but the remaining impedances are incorporated into the three-electrode cir-cuit. Usually, the purpose of making measurements is to determine the current-voltage properties of the E_w-R_2-Z_1 combination, and the value of Z_2 and R_1 must be known as a function of current before the polarization curve can be interpreted. The non-ohmic character of Z_2 presents the largest potential difficulty here. In some cases it is negligible and the three-wire system works fairly well if R_1 is not too large.

A four-wire system (switch S at P_4) overcomes the problem pre-sented by Z_2, and we have conducted many measurements in this way. Any contact impedance above the junction to R_1 is of negligible significance when the voltmeter V makes its measurements at currents of less than 10^{-9} amperes in the voltage circuit. The Wenking pot-entiostats used in our U.B.C. laboratories permit this four wire circuit to be set up when needed, because positions P_3 and P_4 are external connections, and the switch S can be added externally.

In general, sulphide minerals have specific conductivities of the order of 10^3 to 10^{-3} Ω^{-1} cm^{-1}. For specimens of about $1cm^3$, the value of R_1 corresponds to the specific conductivity. At cur-rents of $I=1ma/cm^2$ the contribution of βR_1 to the voltmeter could therefore be in the range of 10^{-6} to 0.5 volt, assuming $\beta \simeq 0.5$. At the top of this range, the potentiostat may give unstable control. However, specimens with resistances below 10 Ω provide $I\beta R_1$ values low enough to permit stable control. Operation at higher currents is not always practical because the ohmic surface resistance R_2 cannot always be distinguished from βR_1.

The resistance of the specimen can be reduced by thinning, and if higher current studies are necessary, this is the only practical method of keeping $I\beta R_1$ under control for minerals of higher resis-tivity. The value of β can be reduced modestly by judicious loca-tion of P_4 on the mineral specimen, but it can never be reduced to negligible values as can α in the electrolyte.

For high current polarizations, the limiting value of E_B fre-quently provides an upper limit of current. This is especially so when Z_2, R_2, or R_i are high even when R_1 is very small. Recently we found a pyrite specimen exhibiting about 1000 ohms of resistance between platinum wires pressed into drilled holes. This fell to about 3 ohms between faces when the specimen was shadowed with gold on opposite faces by sputtering (and polishing off any gold depositing around the circumference). Clearly, the large value was z_2, and it was lowered by increasing the area of contact. The potentiostat had a maximum value of 90 volts for E_b, so the maximum current was less than 9 ma, until the gold contact was introduced. There can also be a problem with R_i becoming too large for the voltage source E_b if a salt bridge is used to isolate the electrolyte around the auxiliary electrode. This is often done to avoid auxiliary electrode discharges from diffusing to the working electrode and providing an extraneous electrochemical process interfering with that being studied. In general, a 90 volt potentiostat can handle this better than a 20 volt instrument, but at 1 ampere current with a highly conductive electrode, the electrolyte resistance including the salt bridge must be kept below 90 Ω even with a 90 volt machine.

Preparation of Working Electrodes from Sulphide Minerals

Natural minerals must be treated with care to avoid altering their composition and structure during electrode preparation, and traditional methods of electrode preparation must be treated with caution. In general, sulphides are vulnerable to (a) heat (b) oxidation, and (c) introduction of mobile ions such as Ag^+, Hg^+, and Cu^+ from electrical contacts. It is advisable not to use conducting cements based on silver, nor to use mercury or soldered connections. Our method is to use either platinum wires, press-fitted into carefully drilled holes, or graphite based conducting cement on a surface shadowed with a thin layer of gold, for electrical connections.

Specimens are easily overheated and/or oxidized during polishing of flat surfaces. For example, we have observed that polished faces of chalcopyrite are nearly passive to an anodic potential, or to ferric ion oxidation, while rough, saw cut specimens are reactive. The mineral pyrrhotite oxidizes almost instantly to a superficial extent in air, and becomes unreactive as a consequence. In both cases there is strong evidence that an invisibly thin layer of magnetite (Fe_3O_4) accounts for the irreversible anodic process. In one study of pyrrhotite, it was observed that any chemical treatment capable of removing magnetite or avoiding its appearance (strong HCl; cathodic treatment; cutting, grinding, or polishing in an inert atmosphere) resulted in a very reactive mineral, both chemically and electrochemically. The simplest of these treatments to apply to a polished specimen is a cathodic treatment ($\sim10^{-2}$ coulombs/cm^2). This is known to activate pyrite in oxygen free acid as well as pyrrhotite in solutions sufficiently acid to permit H_2S evolution. It might also work for chalcopyrite, for which it has not been tested.

SELECTED MINERAL SYSTEMS

Galena, PbS

Brodie[1] studied some electrochemical reactions of the galena-dilute perchloric acid system in 1966-67. The E_h-pH diagram of this system is given (after Majima[2]) in Fig. 3. The system was a three-electrode set-up, and the electrical connection to the galena specimen was a piece of platinum foil cemented to the galena with a silver-conductive cement, the whole being encased in an epoxy-type resin, except for the polished surface. An Electroscan 30 (Beckmann Instruments) with potentiostat was the instrument used in this work.

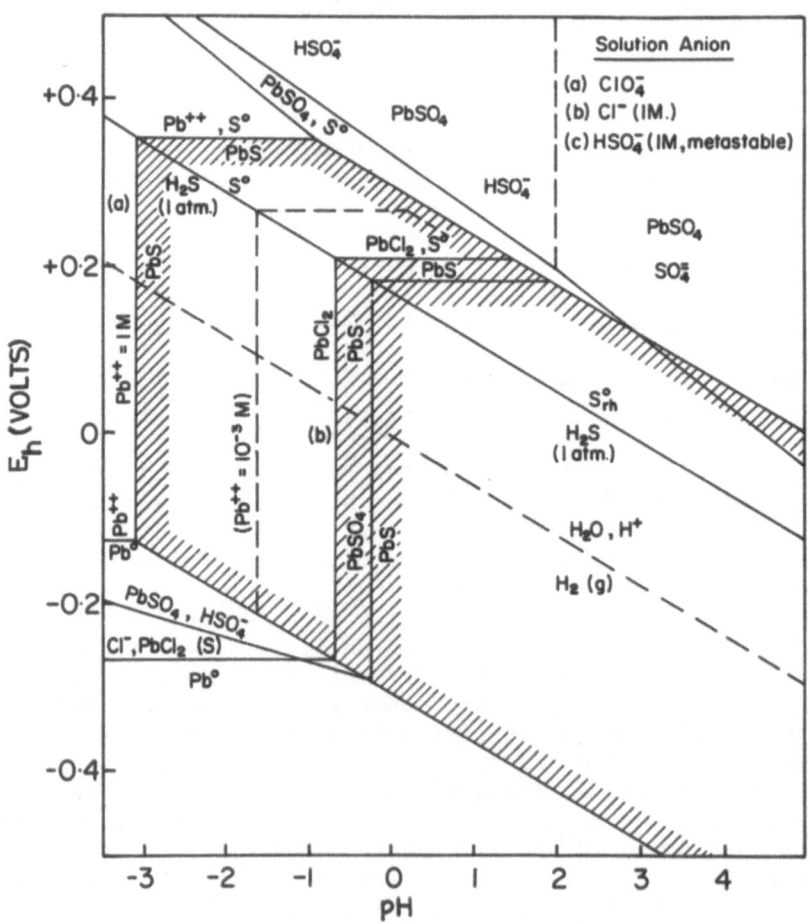

Fig. 3.   The E_h-pH diagram for the Pb-S-H$_2$O system.

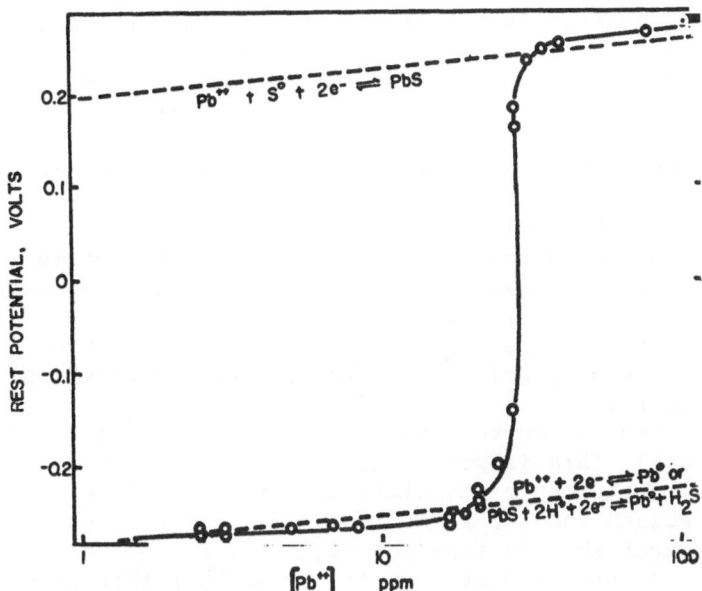

Fig. 4.   Rest potentials of PbS in HClO₄ solutions after cathodiz-
ing, then anodizing for successive intervals.

Galena exhibits essentially reversible behaviour in this sytem,
in spite of significant variation in its observed "rest" potential*.
Fig. 4 is a kind of electrochemical "titration" curve on an electrode that was previously cathodized.   Potential measurements were
made and solution samples taken when the current was interrupted.
It is clear that, within experimental precision**, the galena
specimen normally exerts either the two-phase potential Pb/PbS or
the two-phase potential PbS/S°, according to equations [1] and [2].
The single phase region extends for about half a volt and within
this region the potential can be altered by very small currents.

This indicates that under the conditions of polarization in
this experiment the single-phase PbS range is very narrow, either
because of a very small stoichiometry range for PbS at room temperature, or because the nucleation and growth of a second phase (sulphur

---

*   The rest potential is the observed potential on open circuit
without reference to the distinction between "reversible" and
"mixed" potentials.
**   In 1M.HClO₄ solution, the calomel electrode potential was
corrected downwards by about .030 volts due to the effect of H^+ on
the liquid junction potential and the effect of $KClO_4$ precipitation
at the junction.   This correction may be in error by up to ± 0.015
volt.

or lead) on the surface is kinetically faster than the solid-state
diffusion of lead or sulfur in PbS, as needed to alter the composition within the stoichiometry range to a significant depth from the
surface.

When the current is not interrupted on the galena electrode
polarization curves are obtained as shown in Figs. 5(A) and 5(B).
These curves have the following characteristics:
    (1) The anodic curves are fairly similar, but current
densities above about 1 ma/cm^2 are difficult to sustain because of
build-up of sulphur, and so steady-state values tend to level off.
Irregularities in the sweep curves can be accounted for by either
impurities or morphological changes (shrinkage and cracking) on
the specimen surface.
    (2) The cathodic curves show considerable spread for different
scanning speeds.   This is most likely due to differences in nucleation rate and morphology of metallic lead formed by the cathodic
reduction of galena and the formation of H_2S gas bubbles.   There
may also be a contribution from the electrode resistance at lower
sweep rates or at steady state, as the normally p-type galena
switches to n-type on the lead-saturated surface and develops a
zone of minimum electrical conductivity in the region of the n-p
junction.   At high sweep rates, this zone, which must develop by
solid-state diffusion, may fail to grow to a significant thickness.

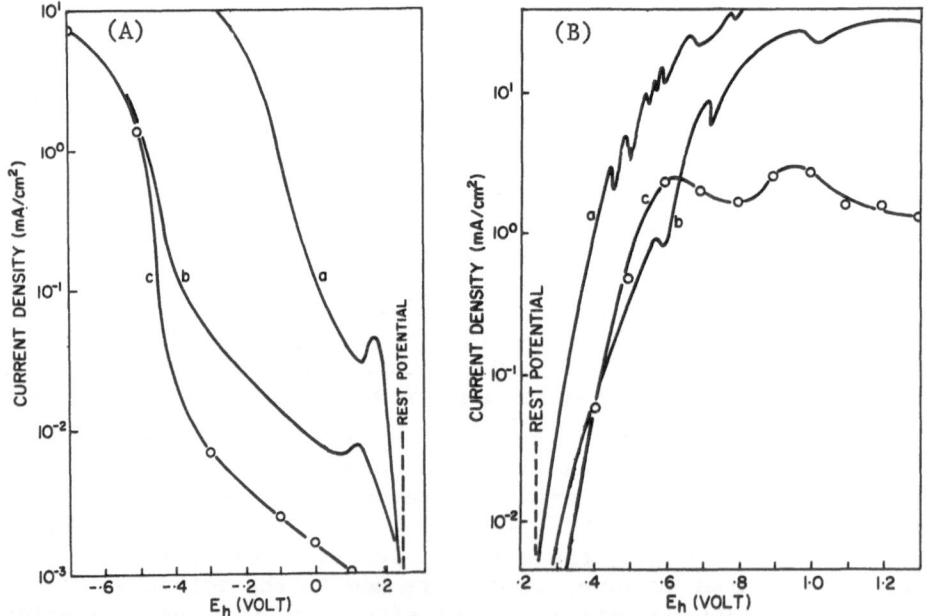

Fig. 5.   Cathodic (A) and anodic (B) polarization curves for PbS
           in 1M $HClO_4$.   Curves (a) 4mV/sec, (b) 2mV/sec, (c) steady
           state.

Fast cathodic sweeps lead to a peak in the polarization curve centered at +0.15 volts. This peak can be regarded as a transient, representing the disappearance of a specific amount of reducible impurity on the electrode surface. The area under this peak (when plotted on a current instead of a log-current base) divided by the scanning rate, provides a measure of the number of coulombs per cm^2 involved, in this case about 2.5×10^{-4} coulombs/cm^2 (curve a) or 5×10^{-5} coulombs/cm^2 (curve b). A monatomic layer of elemental sulphur would account for 3.2×10^{-4} coulombs/cm^2 (neglecting surface roughness) and similar films of unidentifiable reducible adsorbed substances (i.e. oxygen, thiosulphate, sulphite) would require similar quantities of current. As long as the experiment is concerned with bulk decomposition of a sulphide phase, such peaks falling below 10^{-3} coulombs/cm^2 can be safely ignored as an artifact of surface active species of unknown origin*.

In the study of sulphide electrodes, it is important to determine over a fairly substantial degree of mineral decomposition, the extent to which current is consumed by reactions other than those represented by the reversible mineral decomposition chemistry. For this purpose, the "current efficiency" is measured, i.e. H_2S evolution vs. coulombs in the cathodic reaction, and Pb^{++} dissolved in the anodic reactions. For galena, these experiments were done potentiostatically, for current consumptions in the range of 10 to 100 coulombs.

The results are shown in Fig. 6. The cathodic reaction exhibits only a small excess of current, which may be due to hydrogen discharge. The high hydrogen overvoltage on metallic lead suppresses this, once the galena becomes coated with lead. The anodic reaction also usually consumes a small excess of current, the amount increasing with potential (above 0.8 volts) and is assumed due to sulphate formation by the reaction

$$PbS + 4H_2O \rightarrow PbSO_4 + 8H^+ + 8e \qquad [6]$$

At 1.2 volts, the anodic ratio is 2.33, and this can be accounted for if 4% of the galena decomposed via reaction [6] instead of by elemental sulphur formation.

The galena electrode may be regarded as one of the most reversible among sulphide minerals. Both its anodic and cathodic reactions are capable of being sustained under galvanostatic conditions in 1M perchloric acid at current densities of more than 1 mA/cm^2 without more than 200 mv of polarization. Such polarization as occurs is undoubtedly due to diffusive processes in the pores filled with electrolyte, enhanced by morphological relationships between the reacting and product solid phases.

---

* Of course, they are of consuming interest to investigators studying the properties of surfactants.

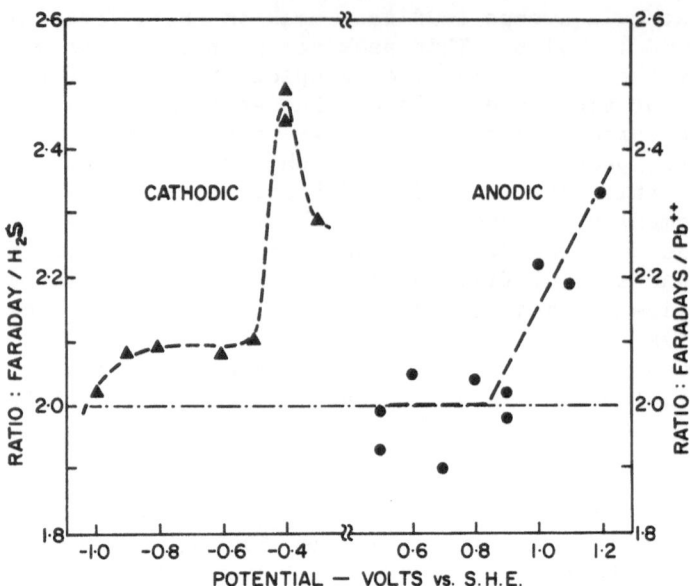

Fig. 6.   Current efficiencies for anodic and cathodic processes
          potentiostatically controlled in 1M HClO$_4$.

Pyrite, FeS$_2$

The pyrite electrode stands out in sharp contrast with galena,
because of the following characteristics, described in the literature by Majima[3]:

(1) The maximum thermodynamic stability range on the anodic
side is described by the equation

$$FeS_2 \rightarrow Fe^{++} + 2S^\circ + 2e \qquad\qquad [7]$$

and the standard potential of this reaction is about 0.340 volts
vs. S.H.E.. Yet the mineral exhibits a rest potential of about
0.62 volts and does not decompose at an appreciable rate on standing in acid solutions.

(2) Coulometric studies at 1 ma/cm^2 of anodic current at a
potential of about 1.05 volt indicate that the only electrode
reaction of consequence is

$$FeS_2 + 8H_2O \rightarrow Fe^{+++} + 2SO_4 = + 16H^+ + 15e \qquad\qquad [8]$$

(3) Coulometric studies on the cathodic site indicates that
between 0 and 0.62 volts the only reaction capable of maintaining
a significant current is reduction of oxygen, or other dissolved
oxidants, viz:

$$O_2 + 4H^+ + 4e \rightarrow 2H_2O \qquad\qquad [9]$$

(4) At negative potentials, pyrite decomposes cathodically according to the equation

$$FeS_2 + 4H^+ + 2e \rightarrow Fe^{++} + 2H_2S(g) \qquad\qquad [10]$$

However, this process takes place at low current efficiencies, and hydrogen is simultaneously discharged at potentials in the range -0.2 to -0.5 volts, accounting for 65 to 80% of the cathodic current. At pH = 0 the reversible standard potential for equation [10] is +0.002 volts, very close to the hydrogen electrode. At 1 mA/cm^2 of hydrogen discharge current, the overpotential is about 0.26 volts, about the same value as tungsten, and below that of silver, nickel, bismuth, and iron. Pyrite can therefore be regarded as a mineral of low hydrogen overvoltage.

The E_h-pH diagram and potentiostatic polarization curves for pyrite are shown in Figs. 7 and 8. The difference between the two specimens in Fig. 8 is due to the use of a three electrode system, with fairly large values for either Z_2, or R_1, or both in one of the specimens.

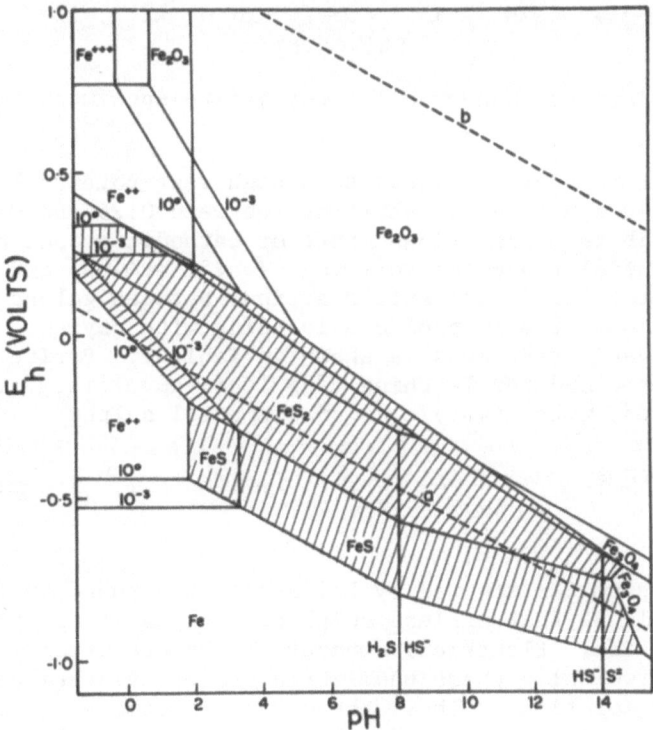

Fig. 7. E_h-pH diagram for the Fe-S-H$_2$O system at 25°C, 0.1M Fe^{++}.

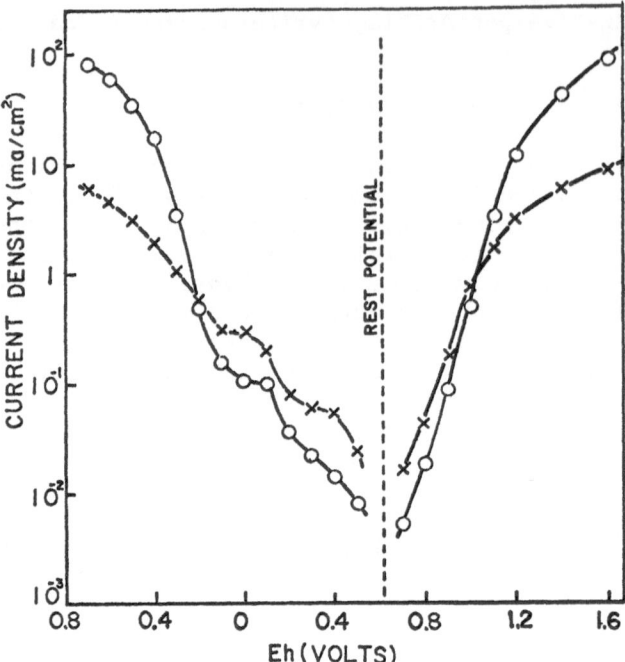

Fig. 8.   Polarization curves for two pyrite specimens in 1M HClO$_4$.

The fact that pyrite exhibits a high rest potential and decomposes at an insignificantly slow rate between 0.25 and 0.62 volts, but also exhibits a significant rate of cathodic reduction of oxygen in this potential range (up to 1 ma/cm^2 at 1 atm. O$_2$ and temperatures above 45°C) results in galvanic reactions between galena and pyrite under conditions of acid pressure leaching with oxygen. The evidence for such galvanic reactions is shown in Table I. Pyrite is galvanically protected and the leaching of galena, covellite, or sphalerite is accelerated, with high yields of elemental sulphur.

Copper Sulphides: Covellite (CuS), Digenite (Cu$_{1.8}$S), and Chalcocite (Cu$_2$S)

This system was studied by Etienne[4,5] who found that it is neither as simple as that of lead sulphide, nor is it as irreversible as iron sulphide. Electrical conductivities are usually high, and most of the two-solid phase equilibria are electrochemically reversible in acid solutions. Thus, the anodic reactions had much the same properties as galena, indicating relatively reversible processes for the following reactions:

$$Cu_2S \rightarrow Cu_{1.8}S + 0.2Cu^{++} + 0.4e \qquad\qquad [11]$$

Table I

Galvanic Effect with Pyrite
During Oxygen Pressure Leaching

$1M. HCLO_4$; 480 psi O_2(initial); 4 hrs.
$-270 + 325$ mesh minerals.

Minerals g/l.	Temp. °C	Assays. $Fe_{tot.}$	Fe^{++}	(g/l). M^{++}	S° (residue)
110 FeS_2	100	9.1	8.1	—	3.4
170 PbS	100	—	—.	28.6 Pb	4.1
110 FeS_2; 170 PbS	100	3.8	3.8	54.3 "	11.0
110 FeS_2	80	4.2	2.5	—	2.6
170 PbS	80	—	—	18.3 "	2.8
110 FeS_2; 170 PbS	80	1.7	1.7	89.5 "	17.3
110 FeS_2; 170 PbS	60	1.0	1.0	64.0 "	11.9
105 CuS	100	0.2	—	21.9 Cu	5.1
110 FeS_2; 105 CuS	100	—	1.7	39.5 "	11.5
95 ZnS	100	0.4	0.4	3.6 Zn	1.9
95 ZnS; 110 FeS_2	100	6.5	5.9	14.7 "	12.4

$$Cu_{1.8}S \rightarrow CuS + 0.8Cu^{++} + 1.6e \qquad [12]$$

$$CuS \rightarrow S° + Cu^{++} + 2e \qquad [13]$$

In all cases, currents of over 1 ma/cm^2 could be sustained. The cathodic reactions were not studied, but the reverse reactions of expected cathodic processes were studied by anodizing copper in H_2S saturated solutions, viz:

$$yCu + H_2S \rightarrow Cu_yS + 2H^+ + 2e \qquad [14]$$

This study not only exhibited the essentially unpolarized nature of this reaction (at currents of 0.4 to 1.2 mA/cm^2 and temperatures of 40 to 70°C), but also permitted measurements of a chrono-potentiometric type to be used in evaluating the solid state ionic transport of copper ions through the copper sulphide scale that was formed, as follows.

In an experimental configuration such as shown in the three wire setup (Fig. 2) the anodic galvanostatic reaction [14] was conducted with the potential V recorded as a function of time. A schematic representation of the potential time curve was obtained as shown in Fig. 9. Two steady-state processes are evident on the curve during which dE/dt is constant, and these conform to the development of a uniform scale of Cu_2S or $Cu_{1.8}S$ on the electrolyte phase boundary side of the scale. The potential gradient through the scale maintains the necessary ionic flux. As long as the overvoltages

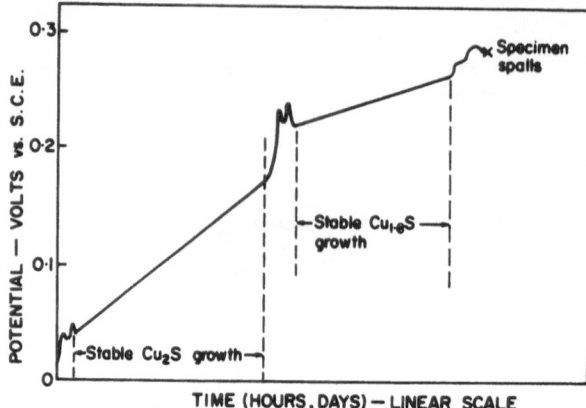

Fig. 9   Potential-Time Curve for Anodic Growth of Copper Sulphide
        Scale at Constant Current in H_2S Saturated Acid Solution
        (Schematic).

could be regarded as constant, the potential (for equation [14]
could be regarded as equal to

$$E = E^{\circ}_{14} - \frac{RT}{2F} \ln \left\{ \frac{a^y_{Cu} \cdot a_{H_2S}}{a_{Cu_yS} \cdot a^2_{H^+}} \right\} + \eta \qquad [15]$$

and $$\frac{dE}{dt} = - \frac{RT}{2F} \cdot y \cdot \frac{d(\ln a_{Cu})}{dt} \qquad [16]$$

That is, the potential-time curve is interpreted as wholly due to a
linear decline in the chemical potential of copper on the outside
surface of the sulphide scale.   It can be further shown that the
flux of Cu^+ ions through the scale is related to this gradient, as
follows:

$$I^2 M/2Fd = \sigma_{Cu^+} dE/dt \qquad [17]$$

where I is the ionic current due to Cu^+, M is the molecular weight
of Cu_yS, d is the density of Cu_yS, and σ_{Cu^+} is the specific ionic
conductivity of the surface (i.e. newly forming) scale.   All of the
change in potential with time is assigned to the last-formed scale,
because under galvanostatic conditions the underlying scale should
conform to steady-state conditions.* (Footnote next page)

TABLE II

Ionic Transport Properties of Chalcocite and Digenite from Anodic Scale Growth Measurements

Conditions					
Temperature($^\circ$C)	30	40	51	62	70
Current Density(ma/cm^2)	0.20	0.40	0.40	1.20	1.60
Properties: Chalcocite Cu$_2$S					
dE/dt (mv/min)	0.041	.084	.053	0.222	0.282
σ_{Cu^+} (ohm^{-1} cm^{-1})X10^5	0.8	1.6	2.5	5.2	7.4
D$_{Cu^+}$ (cm^2 sec^{-1})*X10^{11}	2.9	6.0	9.8	21.5	31.3
Properties: Digenite(Cu$_{1.8}$S)					
dE/dt (mv/min)	0.014	0.028	0.017	0.083	0.083
σ_{Cu^+} (ohm^{-1} cm^{-1})X10^5	1.8	3.5	5.7	10.6	18.8
D$_{Cu^+}$ (cm^2 sec^{-1})*X10^{10}	0.68	1.4	2.4	4.5	8.2

* $D_i = \dfrac{\sigma_i\ V_m\ RT}{n^2F^2}$ V_m = molar volume.

Equation [17] is true when the sulphide has an ionic conductivity that is much smaller than the electronic conductivity, applicable to all binary copper sulphides. The results are summarized in Table II.

Anodic chronopotentiometric experiments were also conducted on chalcocite, digenite, and covellite anodes in acid solution. The potential-time plots obtained are shown in Fig. 10. The anodic processes correspond to those of equations [11], [12], and [13]. Potential variations in Fig. 10 are close to actual.

Since sulphur is not oxidized (i.e., dissolved) it is useful to consider the molar volume changes in going through these equations: Cu$_2$S (27.5 cm^3) Cu$_{1.8}$S (25.4 cm^3), CuS (20.4 cm^3), and S$^\circ_{rh}$ (15.5 cm^3). Thus the volume shrinkage for equation [11] is 7.6%; that for equation [12] is 19.7%, and that for equation [13] is 24.0%. A model for the growth of the reaction product would therefore be consistent with Fig. 11. This morphological model permits consideration of copper transport from the reactant-product mineral interface to the bulk aqueous solution either as Cu$^+$ ions by solid state diffusion

---

*   This would not, of course, be true under conditions where y is changing, because the flux of Cu$^+$ ions, not the total current, must be kept constant to justify stability, and the (nominal) Cu$^+$ flux is proportional to 2(y-1).

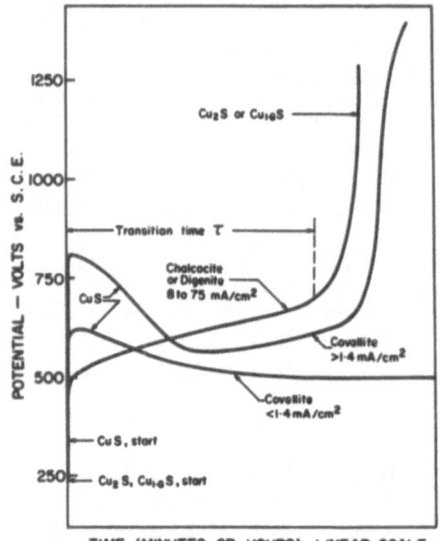

Fig. 10 Potential-Time Plots for
Chalcocite (or Digenite)
and Covellite During
Anodic Decomposition at
Constant Current (Sche-
matic).

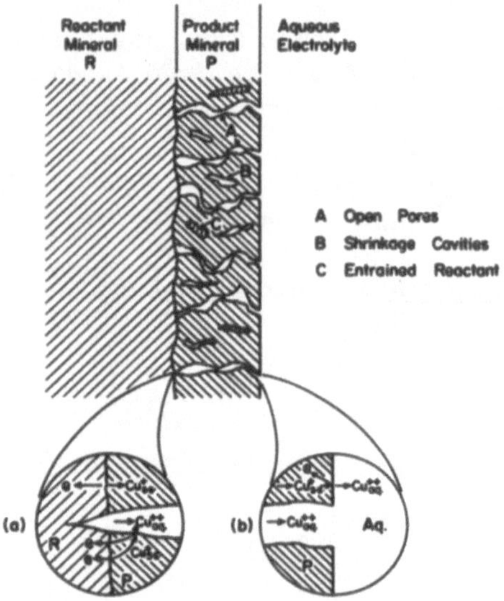

Fig. 11 Model of Anodically
Formed Scale on Copper
Sulphides, Showing Mor-
phology of the New Phase
and Pore Formation.

through product sulphide, or by aqueous pore diffusion through the
shrinkage volume of the product (or even by a hybrid process in which
short-range transport is by solid state diffusion and long-range
transport by pore diffusion in the aqueous electrolyte).

Where the reactant phase is digenite (or chalcocite) and the
product phase is covellite, Etienne[4] found that a gradual rise
in potential, averaging 187 mv. was observed before a transition
time τ, and this was considered the potential gradient across the
product (covellite) phase. The onset of the sharp potential rise
was interpreted as the point at which the covellite product was no
longer able to support the (constant) copper ion flux. At 55°C and
current densities between 7.5 and 75 ma/cm^2, Etienne found the relationship:

$$I^2\tau = 2.5 \text{ A}^2 \text{ cm}^{-4} \text{ sec} \qquad [18]$$

This is equivalent to the product thickness (proportional to $I\tau$)
being inversely proportional to the current density at transition.

A completely solid-state process, with Cu$^+$ ions diffusing
through a covellite layer, would require a conductivity for Cu$^+$ ions

(through a covellite solid of 20% porosity) of $\sigma_1 = 8 \times 10^{-4}$ Ω^{-1} cm^{-1}, or about 14 times the value shown in Table II for digenite at 51°C. The ionic conductivity of covellite is not known, but it is probable, on structural grounds, that it is much less than that of digenite or chalcocite. Thus, solid state diffusion of Cu^+ through the covellite coating is excluded as a part of the mechanism, except for short-range processes such as shown in the bottom part of Fig. 11(a).

The remaining possibility is that pore diffusion of aqueous species accounts for the main mechanism, with the transition occurring when the bottom of the pore crystallizes out $CuSO_4 \cdot 5H_2O$ due to saturation. This corresponds to a concentration of 2.2M at the bottom of the pores at 55°C. If copper sulphate were a completely dissociated salt with an average equivalent conductance of about 100 Ω^{-1} cm^2 eq^{-1}, and an average concentration in the pores of 1M, the polarization across the covellite surface would amount to only about 15 mv. However, $CuSO_4$ is known to be only weakly dissociated, and Etienne calculated a polarization of 187 mv if the dissociation constant has a value of about 0.01 - a realistic value in this case.

Pore diffusion of Cu^{++} and $CuSO_4$ species can thus account for the polarization of chalcocite and digenite during conversion to covellite by galvanostatic anodization in acid solution, and the abrupt termination of the steady-state polarization at time τ is most likely due to $CuSO_4 \cdot 5H_2O$ crystallizing in the bottom of the pores.

The equivalent covellite process leading to elemental sulphur formation is an entirely different problem, in spite of the similarities of the polarization curves (Fig. 10). Experimentally, covellite forms a sulphur film, and there is an initially severe overvoltage, even at very low current densities (0.75 ma/cm^2) that decays with time by more than 100 mv.

No clear-cut model for covellite decomposition emerges from these experiments. Possibly, covellite becomes depleted in copper at the electrolyte interface (i.e. $CuS \rightarrow Cu_{1-\delta}S + \delta Cu^{++} + 2\delta e$) and then sulphur nucleates, causing shrinkage (i.e. $Cu_{1-\delta}S \rightarrow (1-\delta)CuS + \delta S_{rh}^\circ$). The polarization may then be largely due to a kind of concentration polarization on the mineral side. This could be quite substantial if the nucleation of sulphur is difficult and the reduced activity of copper associated with supersaturation of sulphur in the $Cu_{1-\delta}S$ covellite phase was substantial. The sulphur would normally nucleate and grow heterogeneously to provide porosity and access of the solution to the covellite surface, but at excessive current densities the sulphur might coat the covellite protectively causing an abrupt rise in potential as sulphur is forced to oxidize to sulphate by the current flow.

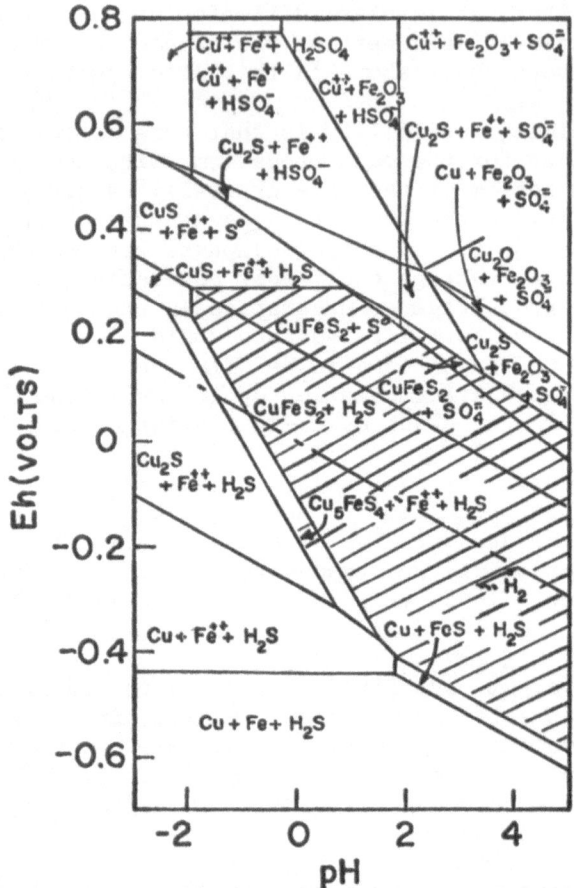

Fig. 12 E_h-pH Diagram for Chalcopyright Decomposition Without
H_2S as a reactant, and Without FeS_2 as a Product (After Majima[4]).

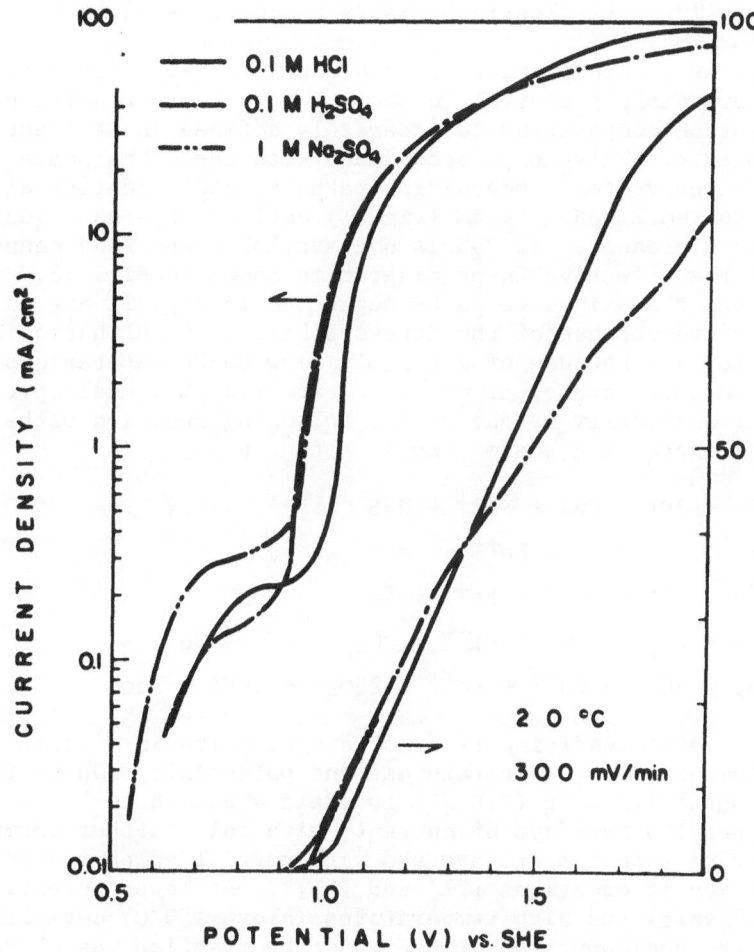

Fig. 13   Anodic Polarization Curves for Chalcopyrite.   Logarithmic
and Linear Scales.

Copper - Iron Sulphides: Chalcopyrite, $CuFeS_2$

Chalcopyrite is the most important sulphide mineral of copper, and also the most difficult of the copper sulphides to treat hydro-metallurgically. Its electrochemistry is complicated by the following factors:

1. It is a ternary mineral, and would normally not exert a singular reversible potential in the absence of other solid phases unless solution composition is accurately defined in at least three solute contents. This is in accordance with the Gibbs phase rule.

2. Chalcopyrite decomposition paths in the oxidative direction according to the E_h-pH diagram (Fig. 1) call for H_2S as a reactant and FeS_2 as a product. If H_2S is not available and FeS_2 cannot be nucleated, the effective E_h-pH diagram is shown in Fig. 12.

3. Even the oxidative paths suggested in Fig. 12 are not precisely observed because of the irreversibility of sulphur oxidation (to sulphate) and because of generally slow Cu-Fe separations. At pH=0 at room temperature, it might be expected that chalcopyrite would react anodically by one of the following reaction paths (given in order of increasing thermodynamic potential).

$$CuFeS_2 + 2H^+ \rightarrow CuS + Fe^{++} + H_2S \qquad\qquad [19]$$

$$CuFeS_2 \rightarrow CuS + S° + Fe^{++} + 2e \qquad\qquad [20]$$

$$CuFeS_2 \rightarrow 2S° + Cu^{++} + Fe^{++} + 4e \qquad\qquad [21]$$

$$CuFeS_2 + 4H_2O \rightarrow CuS + Fe^{++} + SO_4^= + 8H^+ + 8e \qquad\qquad [22]$$

$$CuFeS_2 + 8H_2O \rightarrow Cu^{++} + Fe^{++} + 2SO_4^= + 16H^+ + 16e \qquad\qquad [23]$$

In a careful anodizing study of chalcopyrite in a variety of solution compositions, temperatures, and potentials, Jones[6] found the reaction at 1.5 volt (S.H.E.) to yield about 28 to 34 equivalents of copper per 100 Faradays of current, with only sulphur forming as a new phase at room temperature and with partial sulphur oxidation (a combination of equations [21] and [23]). At lower potentials (0.6 to 0.7 volt) and high temperatures (above 150°C) covellite was formed on the chalcopyrite surface, and the reaction was close to equation [20]. It was also found that chalcopyrite electrochemistry varied significantly with the nature of the mineral specimen and its method of preparation. Polycrystalline material suffered from heavy grain boundary penetration in sulphate (but not so much in chloride) solutions. Carefully polished specimens were found to be unreactive, as compared to a fresh, saw-cut surface. It was difficult to obtain reproducible polarization curves with chalcopyrite.

Polarization curves for chalcopyrite always exhibit a form such as shown in Fig. 13. A limiting current, or at least a small current/potential slope is found between about 0.7 and 1 volt (vs. S.H.E.) and then a sharp rise in current occurs, that is often linear with potential.

The dependence of the limiting current on scan rate indicates that this process is probably a superficial phenomenon. At 20°C in 0.1 M HCl only .01 to 1 coulombs/cm^2 are consumed in this region, and there is considerable evidence for a dissolution of much more iron (and possibly sulphur) than copper. It is probable that this early reaction is represented by equations [20] or [22] and the current is limited by the rate at which CuS and S° can nucleate and grow. The total coulombs consumed represent only about .02 to 2 microns of chalcopyrite decomposition over a uniform surface. At the lower end of the range, the process could represent depletion of iron from the chalcopyrite surface layers, which might have been altered by the preparation procedure.

At higher potentials the current increases to substantial levels. The overvoltages are very high, and there is usually a linear relation between current and potential (see Fig. 13).

Jones used a four-electrode system (Fig. 2, S at P4) and obtained typical values of the equivalent resistance of 2 to 20 ohms for each square centimeter of surface, well within the range for the mineral resistance (βR_1), plus surface resistance (R_2) and solution resistance (αR_i). This would account for a moderate dependence on solution composition and a lack of dependence on scan rates. In view of the fact that in sulphate media especially, chalcopyrite seemed to preferentially corrode deeply in the grain boundaries while the surface appeared passive, the existence of a much larger value for αR_i than expected from positioning of the Luggin capillary in the cell is possible. The inertness of the polished surface is attributed to thin oxide (copper ferrites or magnetite) yielding a high value of R_2.

CONCLUSIONS

This paper presents some of our experimental procedures and findings on the electrochemistry of sulphide minerals. It is clear that these minerals have many of the properties of metals and semiconductors, but differ mainly by having many more options for decomposition paths when anodized (or, for that matter, cathodized). Some binary sulphides, like galena, chalcocite, and digenite, react along reversible paths anticipated from thermodynamic calculations. Others, like pyrite, seem to have passive, or at least fairly irreversible properties, and low overvoltages for hydrogen discharge or oxygen reduction. The properties of covellite and chalcopyrite have been discussed in similar contexts, but the conclusions are not as clear. We may find that most sulphides can be characterized according to their reversibility along thermodynamically predictable reaction paths that yield elemental sulphur, or a mineral of higher sulphur activity.

ACKNOWLEDGEMENT

Work described in this paper was supported mainly by the National Research Council of Canada, with contributions from Cominco, Ltd., and from the International Copper Research Association.

REFERENCES

1. J.B. Brodie, *The Electrochemical Dissolution of Galena*, M.A.Sc. Thesis, Department of Metallurgy, The University of British Columbia, Vancouver, B.C., 1967.
2. H. Majima and E. Peters, *Electrochemical Mechanisms for the Decomposition of Metal Sulphides*, Unpublished manuscript, 1969, 24 pages.
3. E. Peters and H. Majima, Can. Met. Quarterly, 7(3) (1968) 111.
4. A. Etienne, *Electrochemical Aspects of the Aqueous Oxidation of Copper Sulphides*, Ph.D. Thesis, Department of Metallurgy, The University of British Columbia, November 1967.
5. A. Etienne, J. Electrochem. Soc., 117 (1970) 870.
6. D.L. Jones, *The Leaching of Chalcopyrite*, Ph.D. Thesis, Department of Metallurgy, The University of British Columbia, 1974.

OXYGEN REDUCTION ON SULPHIDE MINERALS

T. Biegler, D.A.J. Rand and R. Woods

CSIRO Division of Mineral Chemistry, P.O. Box 124
Port Melbourne, Vic. 3207, Australia

Sulphide ores constitute the major economic source of nonferrous base metals. The reasons for interest in the electrochemistry of the sulphide minerals in such ores have been dealt with by Peters (1) elsewhere in this volume. Briefly, electrochemical mechanisms underlie many processes of practical importance in metal recovery from sulphides, such as the weathering (2-4) and leaching (5) of the ores and the production of sulphide mineral concentrates by froth flotation (6,7). These involve corrosion-like processes consisting of coupled anodic-cathodic reactions. Reduction of dissolved oxygen is the major cathodic component in certain of these systems, *e.g.* oxygen pressure leaching, the weathering of ore bodies as a consequence of differential aeration, and flotation with thiol collectors. It is the aim of the work presented here to provide a basis for understanding the chemistry of this reaction.

We have studied the reduction of oxygen on a range of electrically conducting mineral sulphides and, in more detail, the kinetics and mechanism on the iron sulphide mineral, pyrite (FeS_2). The general approach has been to establish the voltammetric behaviour of the minerals using rotated electrodes prepared from specimens which were usually of natural origin. Care must of course be taken in the selection of suitable specimens for such studies. Many minerals are difficult to find in a pure, massive condition, but it is usually possible to obtain a natural sample out of which an electrode can be fashioned with a surface which is, for all practical purposes, a single homogeneous sulphide.

Various methods for fabricating electrodes from such material have been employed (1, 8-13). We favour the use of silver-epoxy conducting cement for attachment of a contact wire followed by

291

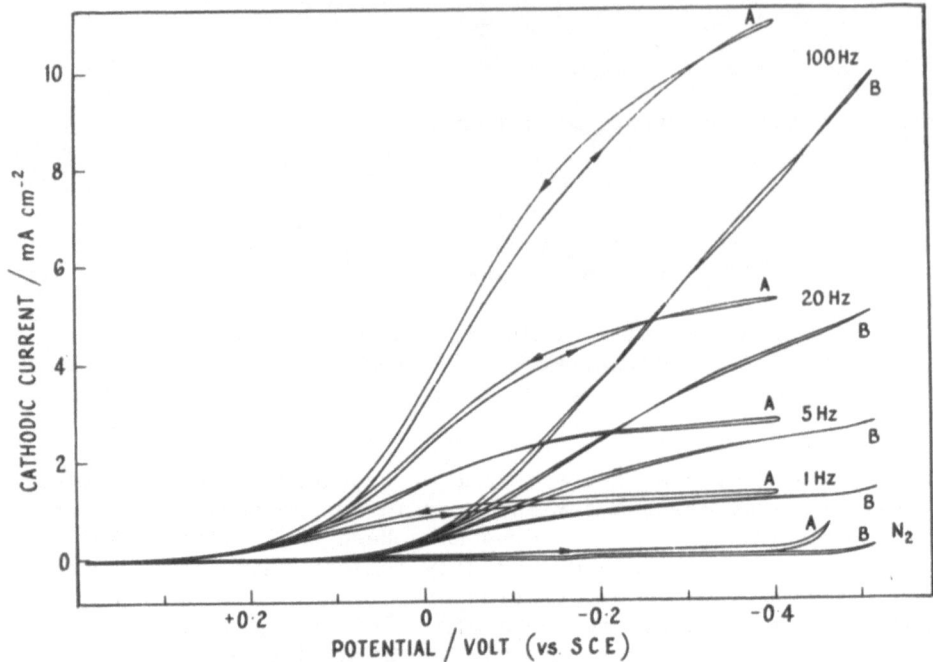

Fig. 1   Cyclic voltammograms for (A) rough and (B) polished pyrite
         in O_2-saturated 1M $HClO_4$. Sweep rate 20 mV s^{-1}, rotation
         speeds (Hz) as shown (Ref. (12)).

encapsulation of the specimen in an epoxy resin (*e.g.* Araldite D).
The specimen is located so that its centre is close to the axis of
rotation, and the cast is machined to suit the particular rotator
assembly being employed. The electrode surface is prepared by
grinding on silicon carbide paper and polishing with diamond and/or
alumina polishing media. The working surface is generally rectangular or irregular in shape, with an area less than 1 cm^2; electrodes
of such shape give diffusion currents in accordance with rotating
disc electrode theory.

VOLTAMMETRIC BEHAVIOUR IN ACID SOLUTIONS

Certain features are common to most of the sulphides studied
and are exemplified by the cyclic voltammograms for pyrite shown
in Fig. 1. At very low rotation speeds (W), the oxygen reduction
reaction shows a limiting current. This current is proportional
both to oxygen concentration and to $W^{\frac{1}{2}}$ (Fig. 2) and corresponds
to the diffusion current calculated from the Levich equation for a

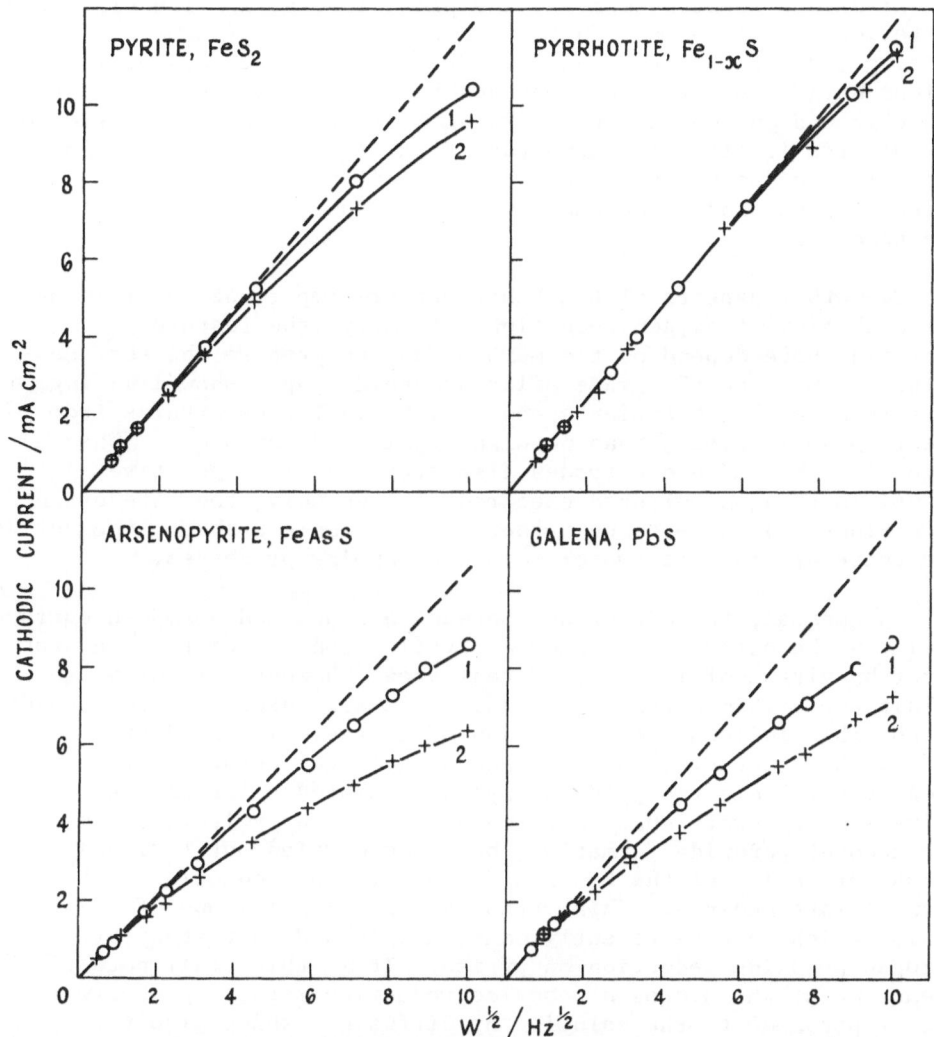

Fig. 2 Dependence of maximum oxygen reduction current on rotation
speed for (1) rough and (2) polished surfaces of four
minerals. Solution is O_2-saturated 0.1M $HClO_4$ except for
pyrite (1M $HClO_4$).

4-electron reduction process. In other words, under conditions
where the limiting current is observed, all oxygen diffusing to
the surface is reduced to water:

$$O_2 + 4H^+ + 4e \rightarrow 2H_2O \tag{1}$$

At higher rotation speeds, the plots in Fig. 2 all deviate from the Levich relationship. In fact, under these conditions a limiting current is not observed. The current plotted, i_M, is the maximum oxygen reduction current reached just before the sharp rise in background current due to hydrogen evolution and/or reduction of the sulphide itself. The departure from linearity in these plots is simply a consequence of the inaccessibility of the limiting current region, and it depends on both the sulphide and the electrolyte used.

Two other aspects of the behaviour are important in arriving at a mechanism of oxygen reduction. Firstly, the features described above depend on the method used to prepare the surface. Surfaces ground on 600 grade silicon carbide paper show limiting currents at higher rotation speeds, with smaller departures from linear i_M vs $W^{\frac{1}{2}}$ plots, than do highly polished surfaces.* The oxygen wave on rough electrodes also starts at more positive potentials than on polished electrodes. Secondly, the wave often shows kinks and inflections rather than a simple S-shape, suggesting that there are in fact two or more overlapping processes.

In general, the difference between a rough and a smooth surface is simply the outcome of higher activation-controlled rates at the former by virtue of its greater real area. However, in order to explain the voltammograms observed, it is also necessary to consider the possible influence of formation of a stable intermediate. In fact, we have detected the presence of hydrogen peroxide after potentiostatic electrolysis in oxygen-saturated solution. An analysis of the wave shape for rough and smooth pyrite, taking into account peroxide formation, has been carried out (12) and allows derivation of the kinetic parameters for reduction of the peroxide intermediate. This method gives a Tafel slope of -245 mV, a value which we have recently confirmed in a direct study of hydrogen peroxide reduction on pyrite. It is this small potential dependence of the hydrogen peroxide reduction rate, together with loss of peroxide to the solution by diffusion, which give the wave its drawn-out shape and prevent the current reaching the 4-electron limiting value as the rotation speed is increased. Increase in surface area by roughening accelerates the rate of peroxide reduction (proportional to real area) relative to diffusion into the bulk (proportional to projected area) and therefore favours the

---

* Similar behaviour was recently reported by Morcos (14) for oxygen reduction on rough and smooth palladium surfaces and attributed to a higher specific activity of electrodeposited palladium. However, the data as presented tell nothing about the palladium activities, since comparison of specific electrocatalytic activities must be made on the basis of *activation*-controlled currents. A more likely explanation of his results is the one we propose here for sulphides.

approach to a 4-electron current. According to these arguments, the activities of sulphides in general as peroxide reduction catalysts should contribute to the detailed voltammetric behaviour of the oxygen reduction process on these minerals.

KINETIC PARAMETERS IN ACID SOLUTION

For all the sulphides studied, there is a potential region at the foot of the oxygen wave where the current does not depend on rotation speed and is therefore under activation control. We have used measurements, both sweep and steady-state, in this region to derive kinetic data for the reaction. For example, on pyrite in 1M $HClO_4$ or 1M H_2SO_4, Tafel plots are reasonably linear over 2-3 orders of magnitude of current, the Tafel slope is close to -130 mV, and the exchange current is around 10^{-11} A cm^{-2}. On other sulphides (15) the kinetic parameters are often not as readily or reproducibly determined as on pyrite, e.g. the range of linear Tafel plots may be shorter, there may be more hysteresis on sweeps, or there may be continuous activity changes with time.

Despite these difficulties, it is still possible to obtain kinetic measurements which reliably reflect the catalytic activity of sulphide minerals for oxygen reduction. Activation-controlled currents are summarized in Fig. 3 for the range of minerals studied. Within the current range of Fig. 3(A), the minerals galena (PbS) and pentlandite ((Fe,Ni)$_9$S$_8$) have Tafel slopes similar to that for pyrite, with exchange currents in the range 10^{-11} A cm^{-2} (pyrite, pentlandite) to 10^{-14} A cm^{-2} (galena). Higher slopes on arseno-pyrite (FeAsS), bornite (Cu$_5$FeS$_4$), chalcopyrite (CuFeS$_2$), covellite (CuS), and pyrrhotite (Fe$_{1-x}$S), a lower slope on chalcocite (Cu$_2$S), and changes in apparent Tafel slope with potential for some of these sulphides suggest that different rate-determining steps and/ or mechanisms may be operating. In view of the obvious differences in the way in which the current for different sulphides depends on potential, it would generally be appropriate (if somewhat arbitrary) to make activity comparisons within the range of experimentally accessible currents rather than to compare, for example, exchange currents obtained by extrapolation.

EFFECTS OF pH

The influence of changes in supporting electrolyte pH have been studied in detail for pyrite (12). The shape of the oxygen wave changes somewhat as the pH is raised, but there is little change in the position of the foot of the wave with respect to the S.C.E. The Tafel slope decreases from around -130 mV in molar acid solutions to -67 mV at pH 12. There is also a change in the ranking of activities of the various sulphides as the pH is

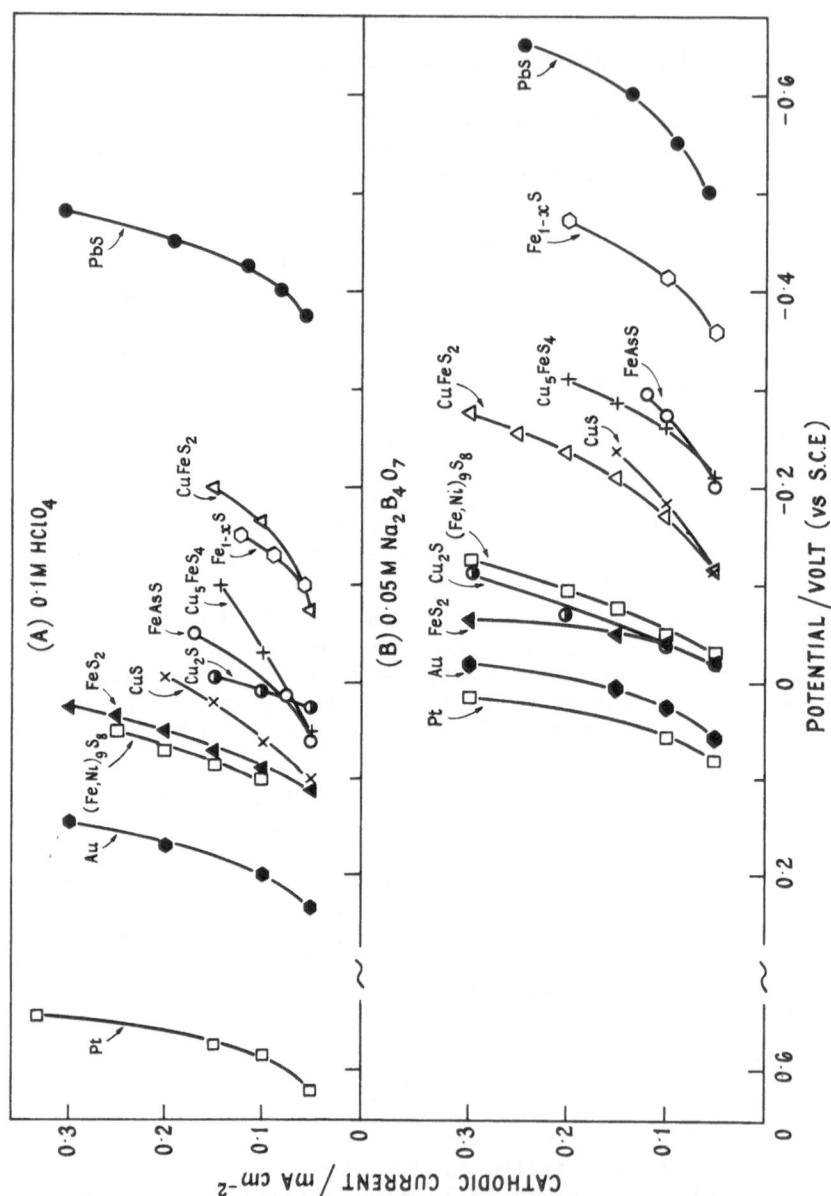

Fig. 3   Activation-controlled currents at the foot of the O_2 reduction wave on various electrodes in O_2-saturated acid (A) and alkaline (B) solutions (Ref. (15)).

increased (*cf*. Fig. 3, (A) and (B)).

RATE-DETERMINING STEP

In acid solutions, the Tafel slope of -130 mV for pyrite and the lack of pH dependence of the current suggest that the rate-determining step is

$$O_2 + e \rightarrow O_2^-$$ (2)

At pH values above $\sim$ 4, the reversibility of this reaction increases and the rate of the next stage

$$O_2^- + 2H^+ + e \rightarrow H_2O_2$$ (3)

enters into the kinetics (12). On other sulphides there is a tendency for increased Tafel slope with pH (15), but there are not sufficient data available for mechanistic analysis.

SURFACE REACTIONS OF THE SUBSTRATE

Although essentially a complicating factor in the study of oxygen reduction, the reactivity of the sulphide itself can produce some interesting and revealing effects on voltammograms. At the monolayer level and below, very little is known about surface reactions of sulphides, but we have noted, for example, that with pyrite in acid solution small current peaks appear on sweeps in the absence of oxygen (Fig. 2, ref. 12). It is suggested that these represent charge transfer processes involved in the formation of surface layers of various iron-sulphur stoichiometries; many metal-sulphur systems show numerous stable phases, and it seems likely that an even greater range of compositions would be possible at the surface monolayer level. In neutral and alkaline solution, involvement of oxide layers needs also to be considered. Such changes in surface composition may well contribute to the complexity of detail in the voltammetric behaviour of the oxygen reduction reaction.

The bulk oxidation and reduction processes which sulphide minerals may undergo are better understood. Thermodynamic data for these are usually presented in the form of Eh-pH diagrams (*e.g.* ref. 5), and surface phase changes on a sulphide electrode are expected to occur in accordance with such data. For example, the oxygen reduction reaction on galena exhibits phenomena consistent with the decomposition reactions expected for this mineral. When cyclic voltammograms for galena in acid solution are measured only near the foot of the oxygen wave, there is little hysteresis and the curves are reproducible (Fig. 4(A)).

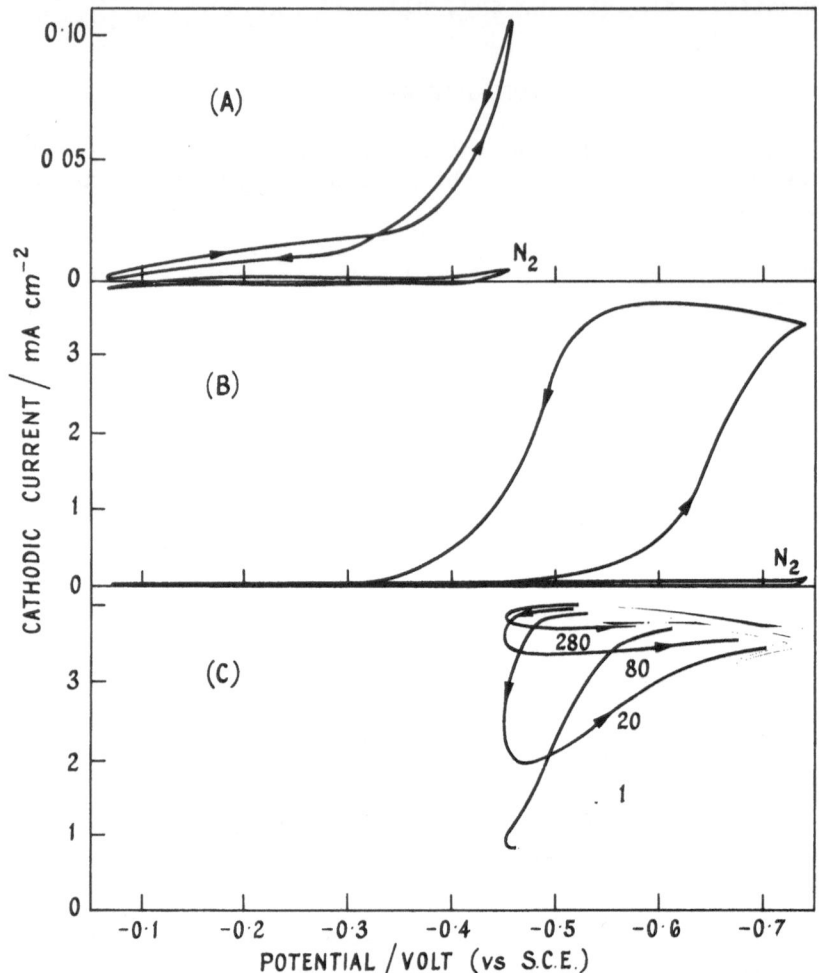

Fig. 4   Cyclic voltammograms over various potential ranges for
         polished galena rotated at 20 Hz in O_2-saturated 0.1M
         $HClO_4$. Base curves in N_2 shown for (A) and (B). Curves
         in (C) taken after number of cycles indicated.

However, if the cathodic limit is extended in order to cover the
complete wave (Fig. 4(B)) the electrode becomes much more active
on the return (positive-going) sweep. These curves are also
reproduced with cycling, but if the potential scan is restricted
to a range at the cathodic end of the wave (Fig. 4(C)) the
electrode shows a steadily increasing activity with cycling until
the current in the whole range becomes diffusion-controlled.

These results are interpreted in terms of reduction of the galena surface to elemental lead,

$$PbS + 2H^+ + 2e \rightarrow Pb^o + H_2S \qquad (E_o = -0.339 \text{ V}) \qquad (4)$$

and of dissolution of lead,

$$PbS \rightarrow Pb^{2+} + 2e \qquad (E_o = -0.126 \text{ V}) \qquad (5)$$

which occurs at much more cathodic potentials than anodic decomposition of galena,

$$PbS \rightarrow Pb^{2+} + S^o + 2e \qquad (E_o = +0.354 \text{ V}) \qquad (6)$$

For the voltammograms in Fig. 4, no lead is formed on curve (A) and the sweep is representative of a galena surface. In (B) lead forms at the cathodic end of the negative-going scan according to eq. (4) and, as we have shown in separate experiments, is a more active oxygen reduction catalyst than galena. The lead redissolves during the return scan, and the electrode surface thus regains its original composition. In (C) lead forms without redissolving and therefore builds up to give a roughened surface with the properties of lead.

Under conditions where the metal formed reacts anodically to yield an oxide (*e.g.* in alkaline solution) the sulphide surface cannot return to its initial composition during cycling. The formation and growth of an oxide layer accounts for the continuous changes in activity found with cycling in such solutions and can be minimized by restricting the cathodic potential limit to avoid reduction to the metal.

ACTIVITY AND SEMICONDUCTING PROPERTIES

The possible role of semiconducting properties in determining various aspects of mineral sulphide reactivity has often been noted (9,16). In the work discussed above none of the effects generally associated with semiconductor electrode processes (*i.e.* photosensitive limiting currents) was observed, and we attribute this to the high charge carrier concentrations in the sulphides used. Nevertheless, the oxygen reduction reaction would appear to provide a favourable model system for the detection of a semiconductor effect, since it is an electrode reaction which is electrocatalytic in nature and which, therefore, might be sensitive to the properties of the substrate.

The approach adopted (17) was to examine the influence of conduction type on activity for oxygen reduction with pyrite electrodes made from six natural specimens and two synthetic

crystals. Pyrite occurs naturally as p-type or as metallic n-type
(high charge carrier concentration, small thermoelectric voltage,
low resistivity). The synthetic material had n-type semi-
conductivity, and hence three different electrical types were
covered. Activities, as given by oxygen reduction currents at a
fixed potential, and Tafel slopes are shown in Table 1.

There is no systematic dependence of either kinetic parameter
on semiconducting type, although there are certainly differences
in activity between some of the specimens. Of the various factors
which might have an influence (*e.g.* stoichiometry, purity, grain
size, orientation, crystal defects), we consider that impurities
in the pyrite are the most likely cause of the differences in
activity. In fact, the specimen with highest activity was one
which had been doped with nickel, replacing 2.6% of the iron.
Moreover, as other sulphides are generally less active than pyrite,
minor impurities must be present as substituents in the pyrite
lattice rather than as sulphide inclusions if they are to produce
a measurable effect on activity. Of course, impurities also
generally determine the electrical type of pyrite, but the results
clearly show that this property is not the primary basis for the
observed differences in activity.

CONCLUDING REMARKS

The kinetics and mechanism of the oxygen reduction reaction
on more conventional (metal) electrodes has been the subject of
long controversy (18), so that it would be unduly optimistic to
expect initial investigations on novel electrode materials to
yield conclusive results. Nevertheless, the reaction on sulphide
minerals can be studied by standard electrochemical methods which
provide information relevant to certain reactions of practical
importance involving sulphides.

The high oxygen reduction activity of pyrite explains, though
by no means quantitatively at this stage, the acceleration of
oxygen pressure leaching of minerals such as galena and covellite
observed (5) on addition of pyrite. The flotation of certain
iron-containing sulphides with xanthates requires oxygen in an
amount which changes in a manner consistent with the order of the
mineral activity for oxygen reduction; *viz.* pyrite requires less
than chalcopyrite which in turn needs less than arsenopyrite or
pyrrhotite (19). The formation of acid in mine drainage waters,
a significant environmental problem (20), results mainly from
pyrite corrosion for which oxygen reduction is one of the cathodic
processes. Our results suggest that the intrinsic activity of
pyrite for this process will not vary much with locality, but as
the effects on anodic activity are still unknown it is not yet
possible to predict whether a particular mine will, for chemical

TABLE 1

Kinetic Data for Oxygen Reduction on Pyrite Electrodes

Electrical type	In 0.1 M HClO$_4$		In 1 M H$_2$SO$_4$	
	Current at 0.1 V vs S.C.E. /μA cm^{-2}	$-(\partial E/\partial \log i)$ /mV	Current at 0.1 V vs S.C.E. /μA cm^{-2}	$-(\partial E/\partial \log i)$ /mV
* n-type semi.	50	121	41	130
* n-type semi.	186	148	338	117
n-type metallic	46	125	56	128
n-type metallic	46	135	71	129
n-type metallic	29	141	35	131
p-type semi.	51	154	78	123
p-type semi.	40	145	73	158
p-type semi.	142	159	123	118

* Synthetic pyrite crystals.

reasons, be more susceptible to this problem.

In general, further progress in the chemistry of these mixed-potential processes involving sulphides will depend on the development of a thorough understanding of both the anodic and the cathodic reactions involved.

REFERENCES

1. E. Peters, *This Volume*, chap. V.
2. M. Sato and H.M. Mooney, Geophysics, 25 (1960) 226.
3. M. Sato, Econ. Geol., 55 (1960) 1202.
4. E.H. Nickel, J.R. Ross and M.R. Thornber, Econ. Geol., 69 (1974) 93.
5. E. Peters, in D.J.I. Evans and R.S. Shoemaker (Eds.), *International Symposium on Hydrometallurgy*, AIME, New York, 1973, p. 205.
6. A. Granville, N.P. Finkelstein and S.A. Allison, Inst. Mining Met. Trans., Sect. C, 81 (1972) 1.
7. R. Woods in M.C. Fuerstenau (Ed.), *Proceedings of the A.M. Gaudin Flotation Symposium*, AIME, New York, 1976.
8. R. Tolun and J.A. Kitchener, Inst. Mining Met. Trans., 73 (1964) 313.
9. G. Springer, Inst. Mining Met. Trans., Sect. C, 79 (1970) 11.
10. S. Chander and D.W. Fuerstenau, J. Electroanal. Chem., 56 (1974) 217.
11. M.J. Nicol, Inst. Mining Met. Trans., Sect. C, 84 (1975) 206.
12. T. Biegler, D.A.J. Rand and R. Woods, J. Electroanal. Chem., 60 (1975) 151.
13. H. Tributsch, Ber. Bunsenges. Phys. Chem., 79 (1975) 570.
14. I. Morcos, J. Electrochem. Soc., 122 (1975) 1492.
15. D.A.J. Rand, *Oxygen reduction on sulphide minerals. Part III. Comparison of activities of various copper, iron, lead, and nickel mineral electrodes*, to be published.
16. A.P. Prosser in M.J. Jones (Ed.), *Mineral Processing and Extractive Metallurgy*, Institution of Mining and Metallurgy, London, 1970, p. 59.
17. T. Biegler, *Oxygen reduction on sulphide minerals. Part II. Relation between activity and semiconducting properties of pyrite electrodes*, J. Electroanal. Chem.,in press.
18. J.P. Hoare, *The Electrochemistry of Oxygen*, Interscience, New York, 1968.
19. I.N. Plaksin and S.V. Bessonov in *Proceedings of the Second International Congress on Surface Activity*, Vol. 3, Butterworth, London, 1957, p. 361.
20. N.V. Blesing, J.A. Lackey and A.H. Spry in M.J. Jones (Ed.), *Minerals and the Environment*, Institution of Mining and Metallurgy, London, 1975, p. 341.

THE KINETICS AND MECHANISMS OF THE NON-OXIDATIVE DISSOLUTION OF

METAL SULPHIDES

P.D. Scott and M.H. Nicol

National Institute for Metallurgy

Johannesburg, South Africa

INTRODUCTION

The base metals form a series of sulphides of widely varying composition, many of which occur abundantly in nature. Conventional processing of these minerals has, until rather recently, involved pyrometallurgical operations which convert the sulphides directly to metal in some cases (e.g. copper and lead) or to oxides in others (e.g. nickel and zinc). Alternative hydrometallurgical routes have received much attention in recent years as a result of many factors, the principal of which is the necessity of treating low-grade ores in a relatively pollution-free manner. The first step in such a process is the leaching of the sulphide mineral by a process which has, in most cases, involved the oxidation of the sulphide to the corresponding metal ion and elemental sulphur or, in some instances, to sulphate. Little attention has been given to an alternative dissolution reaction which one can write in the general form

$$MS + 2H^+ = M^{2+} + H_2S$$

and which we have termed "non-oxidative" dissolution in that no change occurs in the formal oxidation states of the reactants. In the case of predominantly ionic sulphides such as those of the alkaline and alkaline-earth elements, the above reaction simply involves the transfer of ions across the double-layer from the lattice to the solution. However, the base metal sulphides have varying degrees of covalent character from the partially ionic ZnS to the almost metallic iron and nickel sulphides. In the case of these solids, ionic transfer would necessarily have to be preceded by other step(s) involving varying degrees of electron transfer.

In both cases, however, the potential of the solid relative to that
of the solution is expected to play a significant role in determining the rate of dissolution and it is predominantly this aspect
which will be explored in this paper.

Thermodynamic data relevant to the non-oxidative reaction for
the sulphides used in this work is given in Table I.

It is apparent that, with the exception of PbS, there are no
thermodynamic restrictions to dissolution. In the case of galena,
this restriction can be minimized by the use of high concentrations
of acid and confining the measurements to the initial states of the
reaction.

EXPERIMENTAL

All experiments were carried out using rotating discs of the
appropriate mineral sealed into PTFE holders. As indicated in
Table I, a selected natural sample of galena was used while the
nickel and iron sulphides were synthesized and characterized
using established procedures[1]. The rates of dissolution were
conventionally followed by atomic absorbtion analysis of the
solution as a function of time. In the case of galena, an

TABLE 1

Thermodynamic Data for the Reaction $MS + 2H^+ = M^{2+} + H_2S(aq)$

MS	ΔG^o_{298-1} (kJ mol^{-1})	ΔG^*_{298-1} (kJ mol^{-1})
PbS (galena, natural)	44.93	− 0.96
NiS (millerite, synthetic)	15.84	−30.05
FeS (troilite, synthetic)	−11.50	−57.39
Fe_9S_{10} (hexagonal pyrrhotite, synthetic)	−48	−107[**]

[*]Calculated for unit activity H_3O^+ and 10^{-4}M M^{2+} and 10^{-4}M H_2S

[**]Approximate values calculated using data for Fe_7S_8

amalgamated silver ring concentric with the disc was used to
monitor the lead ions produced. Electrochemical experiments were
carried out in conventional cells using potentiostats and signal
generators assembled from integrated-circuit operational
amplifiers. All solutions were continually purged with high-
purity nitrogen. All potentials were measured and are quoted versus
the saturated calomel reference electrode. For convenience in a
direct comparison with the electrochemical measurements, the
dissolution rates have been expressed as equivalent current densities. Unless otherwise stated, all experiments were carried out at
25°C.

RESULTS

Troilite, FeS

The variation of the dissolution rate of a rotating FeS disc
electrode with potential in 0.1M HCl solution is shown in Fig.1
together with a typical cyclic voltammogram obtained under the
same conditions. Steady-state faradaic currents (also shown in
Fig. 1) measured during dissolution were small (approximately 2%
of the equivalent dissolution current). At potentials anodic to
-0.1V, the rates were found to be extremely slow and no measurable
dissolution occurred at potentials above 0V. It can be seen that
the rate of dissolution increases rapidly with increasing cathodic
potential up to -0.3V and thereafter decreases slowly at more
cathodic potentials. Also apparent is the fact that the steady-
state currents measured during dissolution follow very similar
trends with potential as does the cyclic voltammogram which has a
cathodic peak at a potential very close to that corresponding to
the maximum rate of dissolution. The magnitude of the anodic peak
at about -0.05V is dependent on the limit of the cathodic excursion
and, as is apparent from Fig. 1, reflects a charge considerably
less than that constituting the cathodic peak (500 to 1000 μC cm^{-2}).

The rest potential of FeS in 0.1M HCl solution was always
greater than 0.1V and, under these conditions, no dissolution
occurred. However, a short (as little as five to ten seconds) cathodic pulse to potentials less than -0.1V initiated dissolution and
thereafter the reaction continued at open-circuit with the potential
remaining between -0.05 and -0.15V. Similar behaviour was found in
experiments with more concentrated solutions of HCl (or HClO$_4$) but,
in addition, it was noted that dissolution in acids at concentrations
above about 3M was spontaneous in that the cathodic pulse was
unnecessary to initiate reaction although an induction period (which
could be as long as an hour) was observed during which no dissolution
occurred and the potential drifted slowly in a cathodic direction.
The end of this induction period could be observed as a fast potential excursion to about -0.05V and subsequent rapid dissolution.

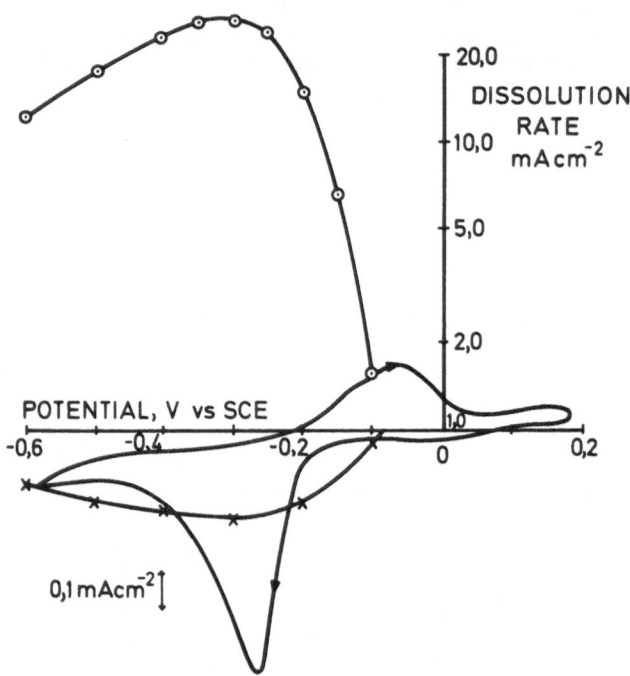

Figure 1.   Comparison of the cyclic voltammogram (50mV s^{-1}) for a
rotating (1000 rev min^{-1}) FeS disc electrode in 0.1M HCl solution
with the dissolution rate (Θ) (measured chemically) as a function
of potential.   Also shown are the steady-state faradaic currents
(X) measured during dissolution.

Hexagonal Pyrrhotite, Fe_9S_{10}

Preliminary experiments showed that free dissolution of pyrrhotite is only possible in concentrated (greater than about 3M) solutions of HCl. Thus, contrary to the observations made with troilite, a cathodic pulse in 0.1M HCl did not initiate dissolution, the potential returning immediately to values in excess of 0.1V. However, measurable dissolution did occur if the potential was held cathodic to about −0.1V. In 1M HCl solution, a cathodic pulse was effective in initiating slow dissolution, the rest potential remaining around −0.05V. Addition of sulphide (in the form of Na_2S) to the 1M HCl solution also resulted in a cathodic shift of potential and thereby initiated dissolution.

The currents measured during experiments in which the dissolution rate was determined at various potentials were considerably higher than those observed during the dissolution of troilite. The results of some of these experiments are shown in Table II for dissolution in 1M HCl.

More extensive data obtained in 0.1M HCl solution is shown in Fig. 2 which also gives typical cyclic voltammograms for pyrrhotite and rest potentials (measured five minutes after a cathodic pulse to −0.2V) in 0.1M and 1M HCl solution.

Several points emerge from a consideration of these results and a comparison with those obtained for troilite

TABLE II

The Rate of Dissolution of Fe_9S_{10} in 1M HCl at Various Potentials

Potential V vs SCE	Dissolution Rate (i_D) mA cm^{-2}	Faradaic Current(i) mA cm^{-2}	i_D/i
−0.05	4.0	0.46	8.7
−0.10	14.2	1.55	9.2
−0.20	40.4	4.10	9.8
−0.30	43.9	4.72	9.3
−0.30[*]	7.9	0.89	8.9

[*] 0.1M HCl solution

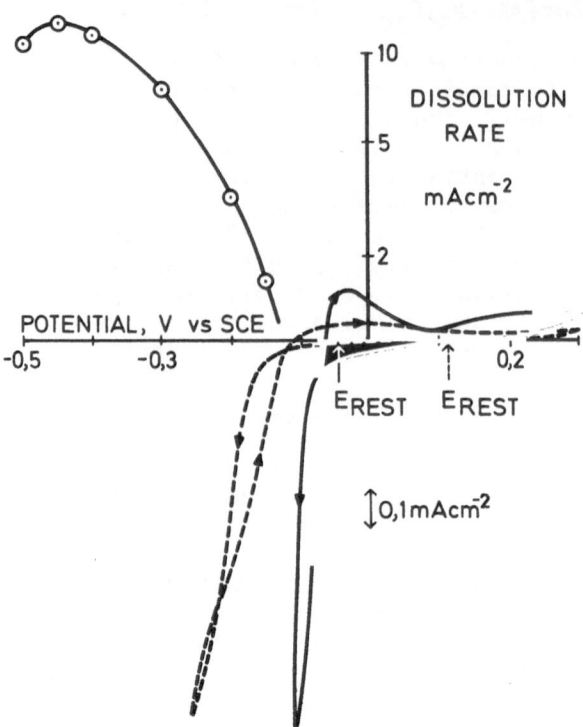

Figure 2, Cyclic voltammograms (10mV s^{-1}) for a rotating (500 rev min^{-1}) Fe$_9$S$_{10}$ disc electrode in (----) 0,1M HCl and (——) 1M HCl solution. Also shown are the rest potentials in the two solutions and the dissolution rate (⊙) (measured chemically) in 0.1M HCl as a function of potential.

(i)   The rate of dissolution is, as in the case of troilite, strongly dependent on potential with a maximum rate at −0.45V in 0.1M HCl (Fig. 2). This is about 0.1V more cathodic than the corresponding figure for troilite under the same conditions (Fig. 1).

(ii)   The ratio of the dissolution rate to the faradaic current necessary to sustain it is, as shown in the last column of Table II, very close to 9. This figure is consistent with reduction of pyrrhotite to troilite by the reaction

$$Fe_9S_{10} + 2H^+ + 2e = 9FeS + H_2S$$

followed by the dissolution of the troilite.

(iii)   The maximum rate of dissolution of pyrrhotite (12.4mA cm^{-2}) is reasonably close to that for troilite (27mA cm^{-2}) if allowance is made for differences in surface roughness.

(iv)   The rate of dissolution, at potentials anodic to the maximum, is approximately proportional to the concentration of HCl. Similar trends (not shown in this paper) were observed with troilite.

(v)   The cyclic voltammogram for pyrrhotite is similar in some respects to that for troilite although reduction occurs at potentials less cathodic (-0.15V) in the case of pyrrhotite. However, the peak at -0.25V found for troilite is not observed with pyrrhotite which exhibits a broad maximum (not shown) at about -0.4V with currents in excess of 1.5mA cm^{-2} at the peak. For both solids, this cathodic process is strongly dependent on the acid concentration. Both also give rise to an anodic peak close to 0V. In the case of pyrrhotite, this peak only appears in concentrated acid solutions and other experiments, not described in this paper, have shown that it corresponds to the oxidation of sulphide (presumably as H_2S) to sulphur. Similar observations were made with troilite in concentrated acid solutions.

Millerite, NiS

The rate of dissolution of a synthetic disc of NiS was found to be slow even in rather concentrated acid solutions at room temperature. Experiments were therefore conducted at 70°C and some typical results are shown in Fig. 3 from which it can be seen that the rate increases slowly with decreasing potential to about -0.2V in 3M HCl and -0.4V in 1M HCl after which the rate increases rapidly. The point at which this transition occurs can be correlated with an increase in cathodic current as shown in the cyclic voltammogram. The steady-state currents measured during dissolution were generally small but increased rapidly at potentials cathodic to the transition point. This increase can be attributed to the evolution of hydrogen at the more cathodic potentials. As a result of this, the ratio of the equivalent dissolution current to the measured faradaic current was variable in that it decreased from a value of 7 to 9 at low cathodic potentials to below 0.1 at potentials cathodic to the transition point. Additional experiments will be required to resolve the total cathodic current into its various components.

A more extensive study of the cyclic voltammetry of NiS in this potential region has revealed the presence of at least three peaks on an anodic sweep after holding at -0.3V for two minutes. This rather complex behaviour is not unexpected since nickel is known to form several stable metal-rich sulphides. Synthesis of some of these materials and electrochemical characterization of them along these lines is at present under investigation. As a result of the complexity of the system, the above results must be considered as being preliminary in nature.

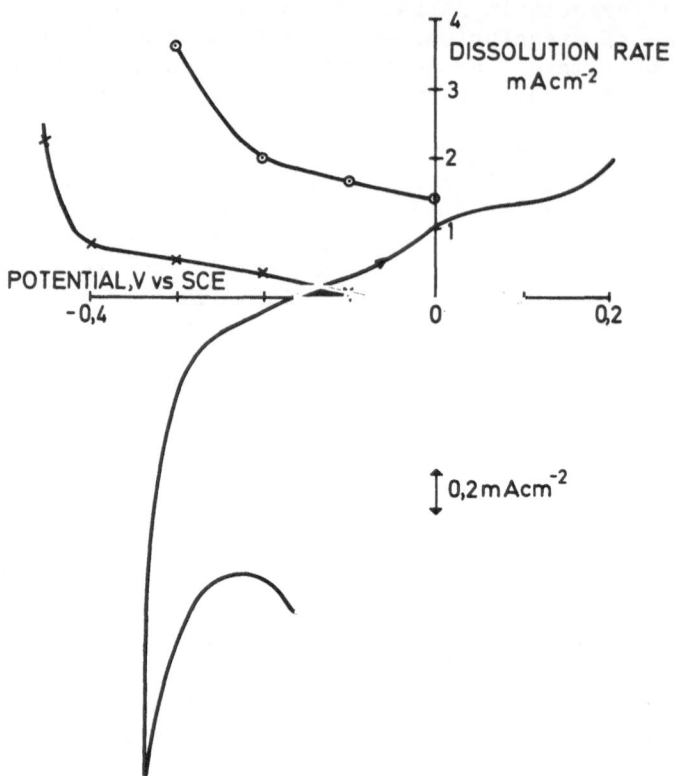

Figure 3.   Comparison of the cyclic voltammogram ($40mV\ s^{-1}$) for a
rotating (500 rev min^{-1}) NiS disc electrode in 3M HCl solution at
70°C with the rate of dissolution (measured chemically) under the
same conditions ($\ominus$) and in 1M HCl at 70°C (X).

Galena, PbS

          As in the case of NiS, the rate of dissolution of galena in
HCl solutions is considerably slower than that of the iron sulphides.
At high concentrations of acid and chloride (greater than about
2M HCl), previous work in this laboratory[2] has shown that the rate
of dissolution is governed by the mass-transport of the product ions
away from the surface of the solid.   Thus, a kinetic model based on
the assumption that the solubility product is satisfied at the
surface and that the rate of convective-diffusion from the surface
of a rotating disc is governed by the Levich equation predicts the
rate of dissolution over a range of some three orders in magnitude
to within a factor of two.   At low acid concentrations (less than
1M HCl), the rate was found to be independent of the disc rotation
speed and is therefore controlled by the rate of the chemical
reaction at the surface.

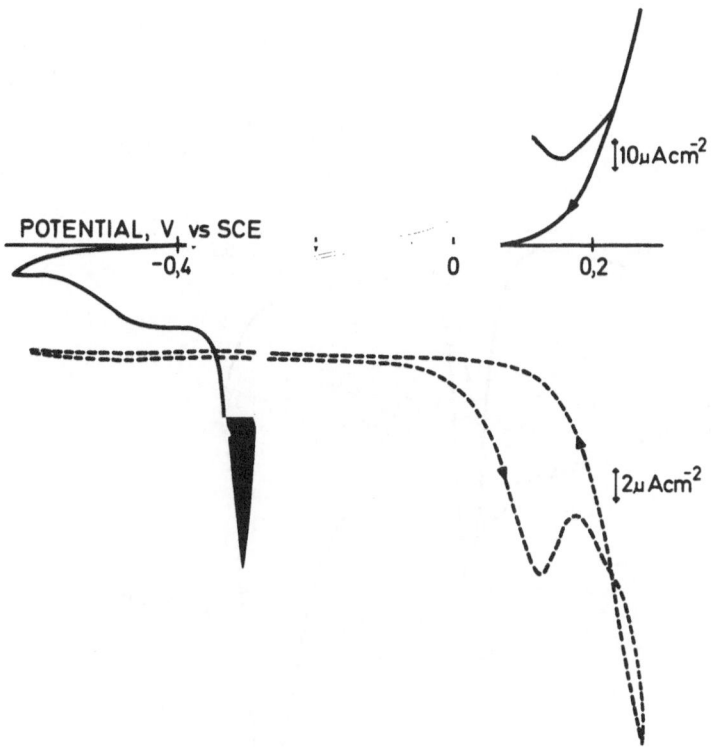

Figure 4. Cyclic voltammetry (10mV s^{-1}) for a rotating (1000 rev min^{-1}) PbS disc in a solution containing 0.1M NaClO$_4$ and 0.01M HCl and the corresponding currents (----) at an amalgamated silver ring held at −0.8V.

    Under these conditions, the rate of dissolution is so slow that measurements by normal chemical means was found to be tedious and often unreliable. The use of a rotating natural galena disc/ amalgamated silver ring electrode to monitor the lead ions produced by dissolution was found to be invaluable in following the potential-dependence of the dissolution reaction. The variation of ring current (due to the reduction to metal of lead ions produced at the disc) during a cyclic scan of the disc potential in 0.01M HCl solutions is shown in Fig. 4. Under these conditions of low HCl concentration, little non-oxidative dissolution of the PbS occurs and the ring response is that expected for (i) anodic dissolution of the PbS at potentials above 0V to produce Pb^{2+} and S^{o} and (ii) reduction of S^{o} at −0.3V on the cathodic half-cycle. The detailed interpretation of these details is beyond the scope of this paper.

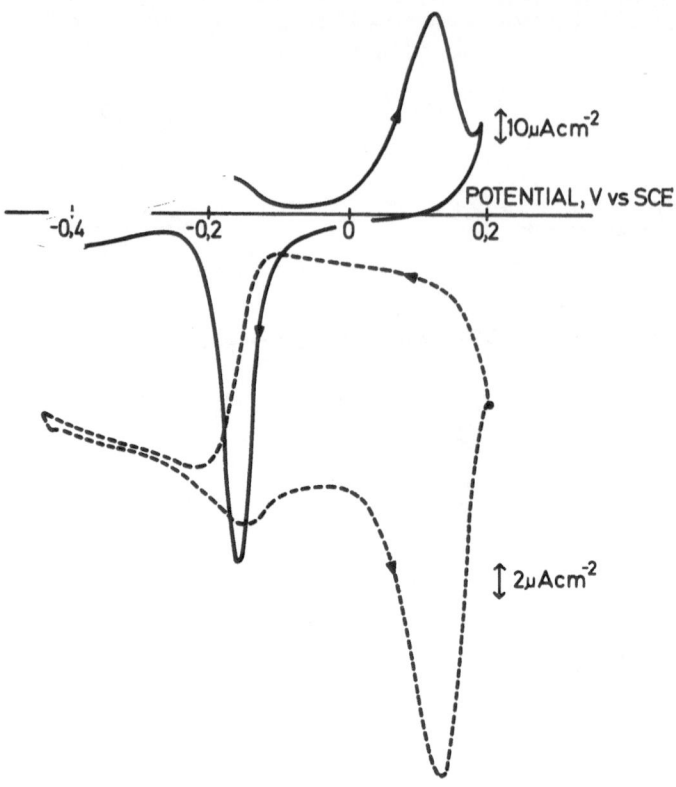

Figure 5.   Cyclic voltammograms (10mV s^{-1}) for a rotating (850 rev min^{-1}) PbS disc in 0.5M HCl solution and the corresponding currents (----) at an amalgamated silver ring held at -0.8V.

Similar experiments carried out in 0.5M HCl solution yielded the curves shown in Fig. 5. Although the cyclic voltammogram for the disc shows essentially the same features, the ring current exhibits a marked hysteresis between -0.2 and +0.2V. This behaviour is only explicable in terms of a parallel non-oxidative production of Pb^{2+} which is strongly dependent on the faradaic processes occurring at the disc. Thus, on the anodic sweep from -0.4V, the ring current indicates a slowly (in terms of potential) increasing rate of non-oxidative dissolution which drops to a low rate at a point between 0 and 0.2V. This slow dissolution is maintained during the cathodic sweep from 0.2V to -0.1V. Reduction of sulphur at -0.15V on the disc results in a simultaneous restoration of the original non-oxidative dissolution rate. Steady-state

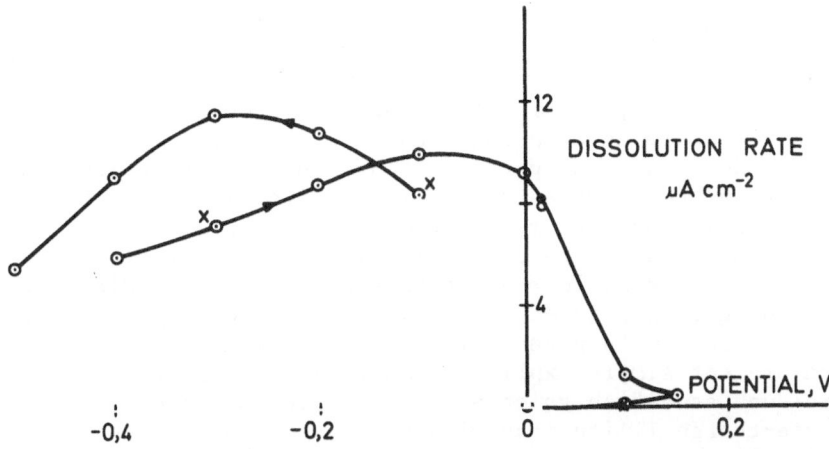

Figure 6. The rate of dissolution of galena in 0.5M HCl solution as measured by the ring current (Θ) and chemical analysis (X) as a function of potential.

ring current measurements of the non-oxidative dissolution rate under the same conditions are shown in Fig. 6. Also shown for comparison are two results obtained by chemical analysis which serve to confirm the accuracy of the ring electrode data. Comparison of this figure with that for the iron sulphides points to similarities in the potential-dependence of the rate of dissolution in that it increases rapidly from negligibly small values at potentials close to 0V and passes through a maximum rate before decreasing slowly again at more cathodic potentials. Steady-state faradaic currents at the disc were found to be negligibly small in the region +0.15V to -0.4V.

DISCUSSION

The interpretation of the above experimental findings, in particular, the potential dependence of the rates of dissolution of all the sulphides studied in terms of mechanisms can be discussed with the following possibilities in mind.

The Ionic Model

Engell[3] in 1956 and Vermilyea[4], some ten years later, proposed theories which quantitatively described the dependence of the dissolution rate of ionic crystals (predominantly oxide) on the potential. In terms of these theories, which are similar in most respects, the slow step is the transfer of ions across the

interface. The rates of both cation and anion transfer are
assumed to depend exponentially on the potential difference between
the solid (oxide) surface and the Helmholtz layer in the solution.
Quantitatively, they predict that the rate should be a maximum at
the freely-dissolving potential and that it should decrease with a
Tafel-like dependence on potential on anodic or cathodic polarization (relative to the freely-dissolving potential). At high anodic
overpotentials, the rate is limited by transfer of the anions and
by cations at cathodic overpotentials. These theories have been
reasonably successful in the interpretation of oxide dissolution
data[5] although, as pointed out by Vermilyea, the large changes in
dissolution rate with potential observed by Engell and other
workers are not simply explicable in terms of the theory. Other
factors such as cation valence changes, adsorption of other ions
and space-charge limitations within the solid have been invoked
to rationalize the orders of magnitude changes with potential that
have been observed[5].

Several factors make it unlikely that the results of the
present investigation can be simply treated in terms of an ionic
model. The most important of these are (i) the fact that the
sulphides involved can hardly be considered ionic solids, (ii) the
interesting observation that the potential region in which
appreciable dissolution occurs (cathodic to about -0.2V) is similar
for all the sulphides examined (possible reasons for this will be
enlarged upon below), and (iii) that appreciable dissolution at
open circuit only occurs in very strongly acidic solutions. Even
under these conditions, the maximum rate is only attained at
potentials cathodic to the freely-dissolving potential.

The Metallic Model

The highly covalent (almost metallic) nature of the bonding
in the sulphides suggests the possibility that dissolution in
acidic solutions may be similar to the corrosion of metals, i.e.
an electrochemical process involving simultaneous anodic dissolution
of the solid and cathodic reduction of protons. In its simplest
(and most naive) form, the analogy is complete if the sulphide is
viewed as an alloy and that anodic dissolution of the metallic
component occurs while reaction of adsorbed hydrogen atoms with the
sulphur replaces the normal recombination reaction on pure metals.
Direct reduction of the sulphur component is, of course, also
feasible. In general terms, the sulphide is therefore both oxidized and reduced.

A necessary condition for the feasibility of such a process
is that the two half-reactions have a common potential region in
which they are both electroactive. This simple approach cannot
be employed to explain (i) the maximum rates of dissolution found

at potentials several hundred millivolts *cathodic* to those
expected, on thermodynamic grounds, for the reactions

$$MS \rightarrow M^{2+} + S^o + 2e$$

for which the following standard reduction potentials (versus SCE)
can be calculated

FeS: $-0.16V$ PbS: $0.13V$ and   NiS: $-0.02V$

(ii) The absence, on the cyclic voltammograms, of processes having
sufficient activity to enable such a mixed-potential model to be
relevant.

On the other hand, the observations that (i) all the sulphides
have a common, broad region over which they dissolve which happens
to be cathodic to the potential required for reduction of S^o ($-0.10V$
versus SCE) and (ii) that continuous cathodic faradaic currents,
which constitute a small (but relatively constant) proportion of
the equivalent dissolution current, accompany dissolution in all
cases (this may not necessarily apply to PbS for which these
currents were too small to measure accurately), suggests that
there is a possible link between a faradaic process related to the
reduction of the sulphide and the rate of non-oxidative dissolution.

A mechanism which would satisfy these requirements involves the
formation, by reduction, of a non-stoichiometric surface intermediate which could then dissolve in the manner similar to that
described above. In the case of troilite and pyrrhotite, the only
known iron-rich sulphide is mackinawite (FeS_{1-x} , where $0.07 > x >$
0.04). There are two known metal-rich nickel sulphides Ni_3S_2 and
Ni_7S_6 while no lead sulphide phases of different stoichiometry have
been reported. Although no reliable thermodynamic data is available
which can be used to accurately predict the potential required for
the reactions

$$MS + 2xH^+ + 2xe \rightarrow MS_{1-x} + xH_2S,$$

a first approximation assuming that the free energies of formation
of MS and MS_{1-x} are similar (most of the available data indicates
this), results in a predicted standard potential equivalent to that
for the reduction of sulphur. This fact coupled with the observation that the faradaic currents are about 2%, 11% and about 12%
respectively of the dissolution currents for FeS, Fe_9S_{10} and NiS
suggests that reduction to the above (or similar metastable) phases
may be the requirement for dissolution. In the case of pyrrhotite,
the data is of sufficient accuracy to conclude that reduction to
troilite almost certainly precedes dissolution. In this regard, it
should be mentioned that open-circuit dissolution of both troilite
and pyrrhotite results in the production of a black residue (about

1% of starting material) which is not soluble even in concentrated HCl but which electron-microprobe analysis has shown to contain both iron and sulphur. Cyclic voltammograms run after a period of open-circuit dissolution have shown considerably increased anodic and cathodic processes which are possibly related to the presence of this insoluble residue.

Under open-circuit conditions, reduction of the sulphide is also possible if sufficient H_2S is available at the surface. This is apparent from the observations that the open-circuit potential during dissolution lies, in all cases, between -0.15V and 0V, i.e. in the region where measurements on pyrrhotite and galena have shown that H_2S is oxidized to sulphur. Also, the observation that addition of sulphide initiated dissolution of pyrrhotite is consistent with this view. This requirement can also be invoked to explain the necessity for a cathodic pulse to initiate dissolution in some cases.

It will be apparent from these considerations that much still remains to be resolved in terms of the mechanisms of these apparently simple reactions. Despite this, however, the practical implications of the dependence of the rates of these processes on the potential should have widespread application. For example, the observations of both Van Weert[6] and Ingraham[7] who found that the dissolution of both natural and synthetic pyrrhotites was subject to induction periods, is readily explicable on the basis of our findings.

ACKNOWLEDGEMENT

This paper is published by permission of the Director-General, National Institute for Metallurgy, Johannesburg.

REFERENCES

1.   P.H. Ribbe, (Ed.) *Sulfide Mineralogy*, Mineralogical Society of America, Vol. 1, 1974.
2.   P.D. Scott and M.J. Nicol, Trans. Inst. Min. Met., in Press.
3.   H.J. Engell, Z. Physik. Chem. N.F., 7 (1956) 158.
4.   D.A. Vermilyea, J. Electrochem. Soc., 113 (1966) 1067.
5.   J.W. Diggle, *Oxides and Oxide Films, Volume 2*, Dekker, New York, 1973, p.281.
6.   G. Van Weert, K. Mah and N.L. Piret, CIM Bulletin, 77 (1974) 97.
7.   T.R. Ingraham, H.W. Parsons and L.J. Cabri, Can. Met. Quarterly, 11 (1972) 407.

COMBINED ELECTROCHEMICAL AND PHOTOVOLTAIC STUDIES OF THE GALENA-
ELECTROLYTE INTERFACE

Paul E. Richardson and Edwin E. Maust, Jr.

College Park Metallurgy Research Center
Bureau of Mines, U.S. Department of the Interior
College Park, Maryland 20740

ABSTRACT

The electrophysical changes of a galena electrode undergoing
forced and open-circuit reactions have been studied using electrochemical measurements with simultaneous measurements of the surface
photovoltage (SPV). These measurements have provided some insight
into the relationship between surface reactions, the accompanying
changes in the stoichiometry of the electrodes, and the surface
potential. Measurements are reported for both polished galena surfaces and for fresh surfaces produced by in situ cleavage under
0.1 $\underline{M}$ sodium tetraborate. The SPV results indicate that at the
rest potential, the surface potential of degenerate n-type and
slightly p-type galena is large and negative, corresponding to an
electron-deficient or hole-rich space charge. During reduction,
the apparent flat-band potential occurs at -0.42 V (SCE) for the
p-type and at -0.65 V for the n-type electrode. The SPV measurements on cleaved surfaces indicate the surface potential is zero
at cleavage. They also show the progressive development of a
negative surface potential with subsequent open-circuit oxidation.
The results are interpreted using a model based on associating lead
vacancies with acceptor-like surface states and sulfur vacancies
with donor-like surface states, and assuming densities dependent
on past electrochemical treatment.

INTRODUCTION

During the 1950's, Plaksin et al.[1-5], proposed that the semiconducting properties of sulfides may have an influence on their

reaction with xanthate collectors. They demonstrated that oxygen was an essential reagent in the xanthate flotation of galena, and attributed its activating effect, as well as the effect of neutron irradiation and illumination on flotation, to changes in the type and concentration of charge carriers at the galena surface. Eadington and Prosser[6] have also studied the effects of stoichiometry on the oxidation rate of lead sulfide and observed an induction period for oxidation which decreased with decreasing lead content. They suggested that oxygen was rapidly chemisorbed during the induction period with the development of a positive space charge, and that a slower process, involving the reaction of lead sulfide with adsorbed oxygen, then produced a sulfur-rich surface which converted a layer near the surface from n- to p-type. There have been other experimental studies of the relationship between the semiconducting properties, surface reactions, and flotation characteristics of the galena-xanthate oxygen system. Most of this work, especially as it applies to the role of oxygen in the xanthate flotation of galena, has been reviewed in recent publications[7,8]. It is clear from these reviews that the role of oxygen is complex, with the theories falling into three categories which have been outlined by Woods[9]. These are: (1) the chemical model, wherein xanthate is considered to undergo metathetical reaction with oxidation products on the surface; (2) the mixed potential model wherein the cathodic process is considered to be the reduction of oxygen and the anodic process is the oxidation of xanthate; and (3) the semiconductor model wherein oxygen and other oxidizing agents are considered to convert the surface from n- to p-type, facilitating the reaction of xanthate with cationic sites.

From the standpoint of the electrochemical and semiconducting models, the single most important parameter characterizing the solid side of the galena-electrolyte interface is the potential difference across the space-charge layer, usually referred to as the surface potential. The importance of this parameter lies in the fact that it uniquely determines the surface concentration of electrons and holes and is a contributing factor in free energy changes associated with charge-transfer processes such as reduction, oxidation, and chemisorption occurring at the interface. Experimentally, little has been determined about either the sign or the magnitude of the surface potential for galena in aqueous solutions. Woods[9] found a transfer coefficient of 0.5 for the chemisorption of xanthate on galena implying that the solid has no space charge. Minden[10] has calculated a space charge width of less than 10^{-7} cm for small particles, whereas Poling[11] has calculated a more reasonable width of 10^{-4} cm.

Recent work in our laboratory[12] using photovoltage measurements has shown that significant potentials are developed across the space-charge layer of galena. The sign and magnitude of the surface potential are dependent on the density of free carriers in the bulk as well as on the past electrochemical history. In this paper, the

photovoltage experiments are discussed and analyzed in terms of the
small signal photovoltage in the presence of trapping, as developed
by Frankl and Ulmer[13]. The results are interpreted in terms of surface states associated with changes in the stoichiometry of galena
as a result of electrochemical reactions.

EXPERIMENTAL

The electrochemical cell and associated gas train have been
described elsewhere[12]. All tubing joints and cell ports were of
the spherical type and sealed with Viton "O" rings to minimize atmospheric leakage. The electrolyte was 0.1 $\underline{M}$ sodium tetraborate
which provided a buffered pH of 9.2. The cell solution was purged
continuously with ultra-high purity nitrogen passed over hot copper.
Experiments carried out in this system on the steady-state polarization behavior of platinum[14] have shown that there are diffusion
currents of the order of 1 $\mu A/cm^2$. These currents are believed to
arise from the residual oxygen resulting from cell leakage or as
an impurity in the purging gas.

The polished electrodes were rectangular ($\simeq$ 0.5 x 0.5 x 0.1
cm), natural crystals of galena cleaved along equivalent (100)
planes. An ohmic contact was obtained by spot welding a platinum
wire to one face. The samples were encapsulated in Araldite with
one complete (100) face ($\simeq$ 0.25 cm^2) exposed. This surface was
polished with 0.3 μ alumina and rinsed in high-purity water immediately prior to insertion in the cell. Conductivity and Hall
effect measurements on adjacent portions of the samples have shown
the n-type material had a typical carrier concentration of
$\simeq$ 2 x 10^{18} cm^{-3} and a mobility of $\simeq$ 590 cm^2/volt sec. The p-type
samples had a carrier concentration of $\simeq$ 6 x 10^{16} cm^{-3} and a
mobility of $\simeq$ 325 cm^2/volt sec.

The samples for the in situ cleavage experiments were rectangular ($\simeq$ 0.2 x 0.2 x 0.8 cm) natural crystals which were mounted
with about 0.4 cm projecting out of an epoxy encapsulation with the
edge of the epoxy outlining a (100) cleavage face. The sample,
mounted on the end of a hollow glass tube, was inserted in the electrochemical cell. A sharp blow on a solid glass rod inserted
through the hollow tube and resting on the projecting portion of
the sample usually resulted in (100) cleavage with the surface
nearly flush with the epoxy. About 95% of the total exposed electrode area after cleavage consisted of the fresh surface. Hall
effect measurements were not carried out on these samples. However, the behavior of the SPV under cathodic polarization showed
that all of the approximately 20 samples cleaved within the cell
were strongly n-type.

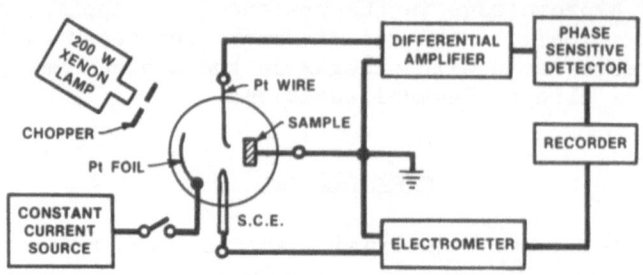

FIG. 1. - Electrodes and circuitry.

The electrodes and circuitry are shown schematically in Fig. 1. The galena electrode was reduced and oxidized by passing a constant current between it and a counter electrode, while simultaneously recording the potential between the sample and a calomel electrode. All electrode potentials reported herein are relative to a saturated calomel reference.

The photovoltage was generated with illumination from a 200 watt xenon-arc lamp filtered with a combination of Corning CS7-69 and CS2-64 filters which passed a 2,000 A band centered at 8,000 A. The illumination was chopped at 86 Hz. The photovoltage was detected as a potential difference between the sample and the platinum wire probe, amplified with a Keithley 604 differential amplifier with an input impedance of 10^{14} ohms, and measured using an oscilloscope, a transient digitizer or phase detector depending on the particular experiment.

RESULTS

Transient Response of Galena to Light

Before discussing the dependence of the SPV on the electrode potential and electrochemical history, it is advantageous to consider the transient response of the electrode to a single light pulse. This is done to establish the sign conventions used in this paper and to briefly discuss the origin of the photoeffects. Typical photoresponse curves for the p-type electrode are shown in Fig. 2 at two electrode potentials. Curve 2a was obtained at the rest potential and curve 2b was obtained after reducing the sample to -0.5 V. Both curves are characterized by a fast-initial response to illumination (both the rise and fall follow the incident light intensity) and a slower decay during the on period (constant

light intensity) and during the off period (dark). The slow decay
has not been studied in this work. Slow effects have been observed
on germanium[15] and have been discussed by Dewald[16] and Gerischer[17]
in terms of either a hole or electron-current across the interface
in the direction required to establish an equilibrium corresponding
to the changed illumination condition.

The fast photovoltage (SPV) can be attributed[18] to the change
in the surface potential arising from the sudden increase in the
steady-state minority-carrier concentration just beneath the space-
charge layer. The concentrations of electrons and holes in the
space-charge layer also change to keep the total charge in the
solid constant. If surface states exist, the altered carrier concentrations at the surface also cause a change in the surface-state
charge. In the absence of surface states, illumination always tends
to decrease the magnitude of the surface potential. The positive
SPV of Fig. 2a corresponds to the electrode becoming more negative
than the reference electrode during illumination. Relative to the
bulk of the solid, positive SPV's correspond to positive changes
in the surface potential. A possible band diagram for curve 2a is
illustrated in Fig. 2c. Cathodic bias tends to produce a positive
surface potential (Fig. 2d) and a negative SPV (Fig. 2b)

Dependence of Photovoltage on Electrode Potential

Open-circuit Measurements. At the rest potential a positive
photovoltage similar to that of Fig. 2a has been observed for all
of the galena electrodes we have studied. Usually, a steady open-
circuit potential in the range -0.1 to -0.2 V, and a steady, positive SPV are observed within seconds after inserting a freshly
polished electrode into the deoxygenated borate solution. The
range of steady-state potentials was probably due to differences
in the residual oxygen concentration in the cell. The positive
SPV indicates the surface potential is negative, that is, the space-
charge layer is electron deficient and hole rich relative to the
bulk. It will be shown below that fresh surfaces produced by cleavage under the electrolyte also have a positive SPV after the electrodes reach the steady-state potential, indicating the positive
SPV is not an artifact of polishing.

Typically, the magnitude of the open-circuit SPV on polished
electrodes ranged from 1 to 5 mV with our light conditions. However,
the magnitude depends on, among other factors, both the absolute
intensity of the light and the bulk lifetime of the photo-generated
carriers. These quantities were not determined, and for this reason
our primary interest will be with the relative magnitude of the SPV
with a constant light intensity.

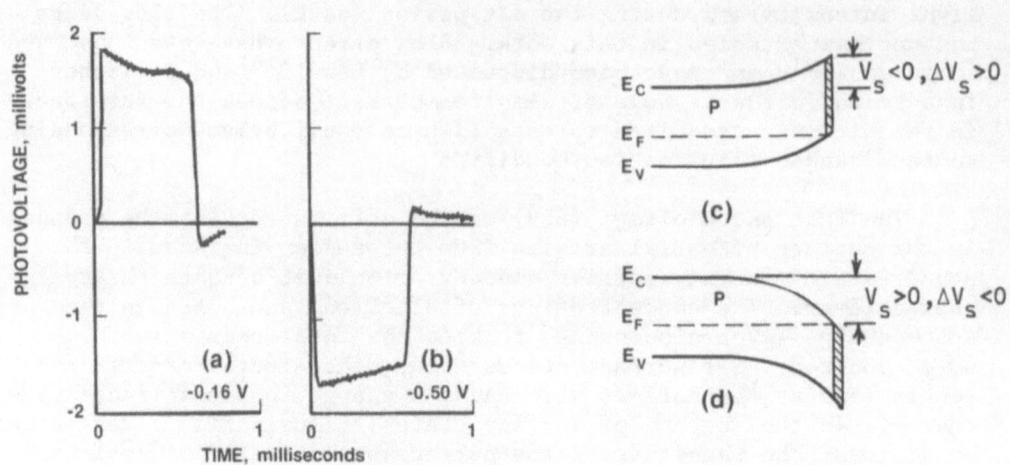

FIG. 2.- Transient response of p-type galena to illumination and
         possible band diagrams for the surface potential:
         (a) open-circuit, (b) following reduction. In the ab-
         sence of surface states a positive SPV (Fig. 2a) is ex-
         pected when the surface potential is negative (Fig. 2c).

SPV During Reduction and Oxidation. Fig. 3 shows continuous
recordings of the fast photovoltage as a function of electrode potential for the n- and p-type samples. The curve for the n-type
electrode was obtained by reducing the sample at a constant current
of 2.6 $\mu A/cm^2$ starting from the steady-state rest potential
(-0.105 V) and then reversing the current at -0.680 V to obtain the
oxidation segment. The initial transient at the start of reduction
(shown dashed) was due to the rapidly changing displacement current
at the instant of applying the current and to the time constant
(1 second) of the phase sensitive detector. The dotted curve shows
the behavior observed when the current was increased slowly from
zero at the start of reduction or oxidation.

The curve for the p-type electrode was obtained by slowly increasing the cell current at the beginning of reduction and oxidation to eliminate the current transients. The constant polarization current of 2.6 $\mu A/cm^2$ was reached at -0.2 V on the reduction
cycle and at -0.45 V on the oxidation cycle.

The following features of these curves are of interest:
(i) For the n-type electrode, the sign of the photovoltage was
always positive. A negative SPV is not expected for a degenerate
n-type sample because under a strong cathodic bias, which tends to
produce a positive surface potential, the electrode should be similar to a metal electrode with no significant space charge and with
the Fermi level lying at the conduction band edge. (ii) For the

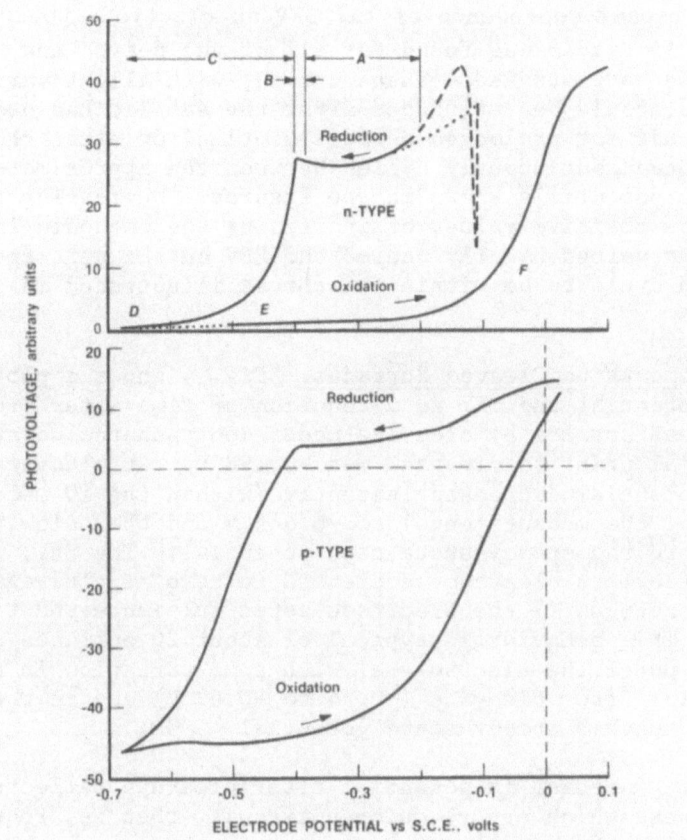

FIG. 3. - Dependence of the photovoltage on electrode potential
      for the n- and p-type electrodes.

p-type sample, both a positive and negative SPV were observed with
the sign consistent with that expected for the polarization. The
zero crossings of the SPV give the apparent flat-band potential
(FBP). For the p-type sample this occurs at -0.42 V during reduction and at -0.040 V during oxidation. The FBP is similar to
the point-of-zero-charge. At this point none of the interfacial
charge is compensated by the sample. (iii) For segment A during
reduction and for segment E during oxidation, the SPV was essentially
insensitive to polarization, that is, the space-charge layer was
shielded from the externally applied electric field. (iv) The
magnitude of the SPV differs from the two directions of polarization.
Measurements have also been made at several current densities between
0.5 and 5 $\mu A/cm^2$, and this difference cannot be attributed to the
magnitude of the current. Rather, the SPV was found to depend predominantly on past history (reducing or oxidizing treatments) and
on electrode potential and to have a linear dependence on light
intensity (not shown).

The functional dependence of the SPV on electrode potential
illustrated in Fig. 3 was found for all of the n-type and p-type
electrodes we have studied. These curves, with slight variations
in magnitude, could be reproduced after the samples had been left
on open-circuit for prolonged periods of time, or after the electrodes had been continuously cycled between the approximate anodic
and cathodic potentials shown in the figures. Decreasing the anodic
limit to less positive values or increasing the cathodic limit to
less negative values usually caused the SPV on the next reduction
or oxidation cycle to be within the curves illustrated in the figures.

Measurements on Cleaved Surfaces. Fig. 4 shows a plot of the
electrode potential and SPV as a function of time after producing
a fresh galena surface by cleavage under deoxygenated borate. The
rest potential prior to cleavage was -0.198 V. At cleavage, the
electrode potential decreased instantly (within the 10 msec time
resolution of the measurements) to -0.545 V and then slowly increased toward the steady-state rest potential. The SPV, which
was positive before cleavage, decreased to zero at cleavage and
remained at zero until the electrode potential increased to
$\gtrsim$ -0.45 V. This behavior is typical of about 20 surfaces produced
by cleavage under the electrolyte, with some variation in the initial potential after cleavage (-0.54 to -0.69 V) and in the time
required to reach a steady-state potential.

The slow increase in potential after cleavage reflects an oxidation process which occurs on open-circuit, that is, requires no

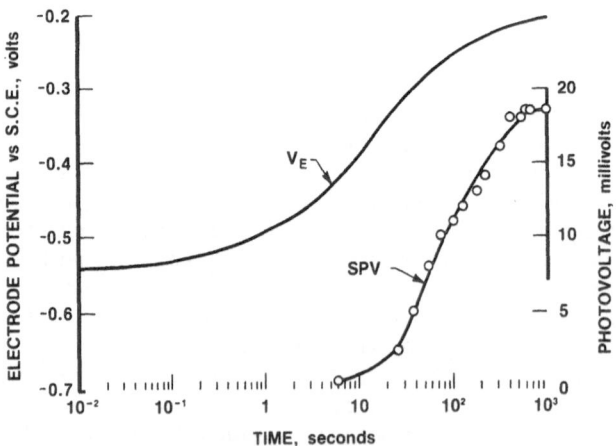

FIG. 4. - Electrode potential (V_E) and photovoltage (SPV) as a
          function of time after cleavage.

external current flow. The balancing cathodic process is believed
to be the cathodic reduction of oxygen. The variation among samples
with regard to the time required to reach the steady-state rest potential after cleavage may be due to either differences in the dissolved oxygen concentration from one cleavage experiment to the
next, or to differences in the samples. The process occurs spon-
taneously on a fresh surface, hence, it is not dependent on oxidation products or contaminants arising from prior treatment.

The time dependence of the SPV indicates the surface potential
was zero for a short time after cleavage and then became negative
as the electrode potential increased above -0.45 V. Subsequent
measurements under cathodic polarization showed that the photovoltage vanished or became extremely small at -0.68 V. The absence
of a negative SPV indicates that the cleaved samples were all similar to the polished n-type specimen, that is, degenerate n-type.

DISCUSSION

These experiments have shown that the surface potential is
strongly dependent on the electrochemical history of the electrodes
as well as on the subsequent instantaneous electrode potential.
This is not surprising, especially for a diatomic electrode which
is a nonstoichiometric semiconductor. First, because of the low
solubility of lead and sulfur in galena, about 0.1%[19], the passage
of a small charge may result in the situation where one or the
other constituent exceeds its solubility in the lattice. Once the
solubility is exceeded, distinct phases of elemental lead or sulfur
may exist. Before the solubility limits are reached, the stoichiometry may be expected to change continuously with polarization.
In fact, Brodie[20] has determined that galena in acid solutions
undergoes reactions under moderate anodic and cathodic potentials
which change the stoichiometry of the surface. Second, it is well
known that galena is a nonstoichiometric semiconductor[19] with the
carrier densities dependent on anion and cation vacancies. Lead-
rich crystals are n-type and sulfur-rich crystals are p-type. On
the basis of these simple considerations, it is expected that the
electrical properties of the surface may be closely related to any
prior reactions which may have occurred.

This is more clearly illustrated by considering the constant
current charging curve for the n-type electrode (Fig. 5) which was
taken simultaneously with the SPV curve of Fig. 3. Corresponding
segments of these are labeled with a common symbol to facilitate
comparison. The charging curve for the p-type sample was essentially the same as that for the n-type shown here, indicating that
the dominant reactions are the same on both electrodes.

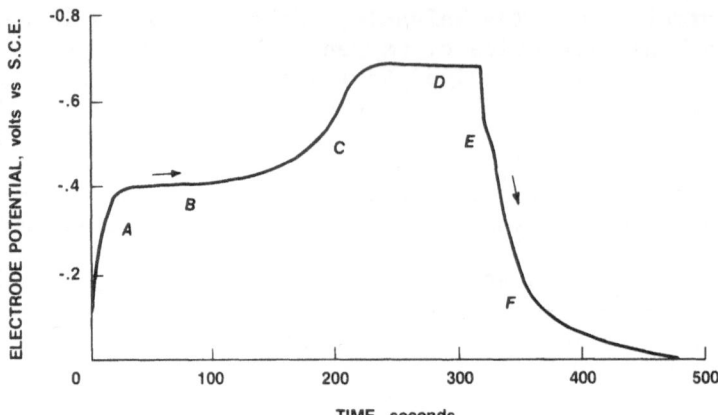

FIG. 5. – Constant current polarization curve for n-type electrode. I = 2.6 $\mu A/cm^2$, quiescent solution.

The possible reactions determining the potential-time dependences shown in this figure have been discussed in detail elsewhere[12]. The processes characterizing segments B and D are believed to result from, respectively, the reduction of elemental sulfur to hydrogen sulfide and the reduction of lead sulfide to elemental lead and hydrogen sulfide. Elemental sulfur is believed to be a product of prior anodic polarization and also of the open-circuit oxidation reaction which is clearly evident in the cleavage experiment. Segment C represents a transition between B and D and along this segment it is believed that the electrode consists predominately of a single phase, that is, lead sulfide, with varying amounts of dissolved lead or sulfur. The activity of sulfur would be expected to be unity along B and to decrease to zero along C, perhaps near the inflection, with the activity of lead then increasing from zero to unity at D.

The above interpretation of segment C is consistent with the initial potential observed shortly after cleaving a sample. Assuming the oxidation process and associated changes in the stoichiometry are slow, the potential 10 msec after cleavage is probably of lead sulfide with no elemental phases of lead or sulfur present. On this basis, one would expect the electrode potential at cleavage to have a value somewhere in the potential range of segment C. This is in agreement with the experimentally observed potentials.

A comparison of Figs. 3 and 5 shows that the SPV changes most rapidly with polarization over segment C of the reduction cycle, that is, where the stoichiometry is probably changing. In contrast, the SPV along A, B, and E is nearly insensitive to large changes in electrode potential. Over these segments, the concentration of sulfur (A, B) and lead (E) are believed to exceed their solubility limits in lead sulfide, with the result that the stoichiometry is nearly constant.

ELECTRONIC STRUCTURE OF THE INTERFACE

From the standpoint of the electronic structure of the interface, as reflected in the SPV measurements, the simplest way to treat the changing stoichiometry is to associate acceptor-like surface states with excess sulfur and donor-like states with excess lead, within their solubility limits in lead sulfide. This is consistent with the known[19] experimental behavior of these excess constituents in bulk lead sulfide. With normal diffusion rates in solids, the variations in the chemical composition of the electrodes as a result of electrochemical reactions are expected to be localized over a few atomic layers at the surface, that is, over distances much less than the width of the space-charge layer.

The association of surface states with changes in stoichiometry is amenable to analysis using the theory of the small signal photovoltage in the presence of surface states which has been developed by Frankl and Ulmer[13]. In our experiments, the density of surface levels can be considered to be variable and related to past electrochemical history. To carry out an accurate calculation would require a knowledge of the density of surface states along a charging curve, and a knowledge of their rate constants for charge exchange with the bulk levels. These properties are not known. However, it is instructive to calculate the theoretical photoresponse using reasonable estimates for the surface state parameters and to compare the calculated response with experiment. Following closely the treatment and notation of these authors, the results of these calculations are shown in Fig. 6 for the n- and p-type electrodes, where we have used the values of the bulk and surface state parameters listed in Table 1. We have assumed a shallow acceptor-like surface state characterized by four parameters: the density, N_t; the energy, E_t, and the rate constants r_c and r_v for transitions to and from the conduction and valence bands respectively. The trapping rate constants are those expected for a center having a capture cross-section of atomic dimensions.

Consider the calculated curves for the p-type sample. For $N_T = 0$, the SPV approaches zero at the flat-band potential. At the extremes of large positive and negative surface potentials, the SPV saturates. The ratio of the saturation values is given by[18] $p_b/n_b \simeq 8.6 \times 10^3$. Hence, on a p-type electrode, the photovoltage under strong cathodic bias is expected to be much larger than under

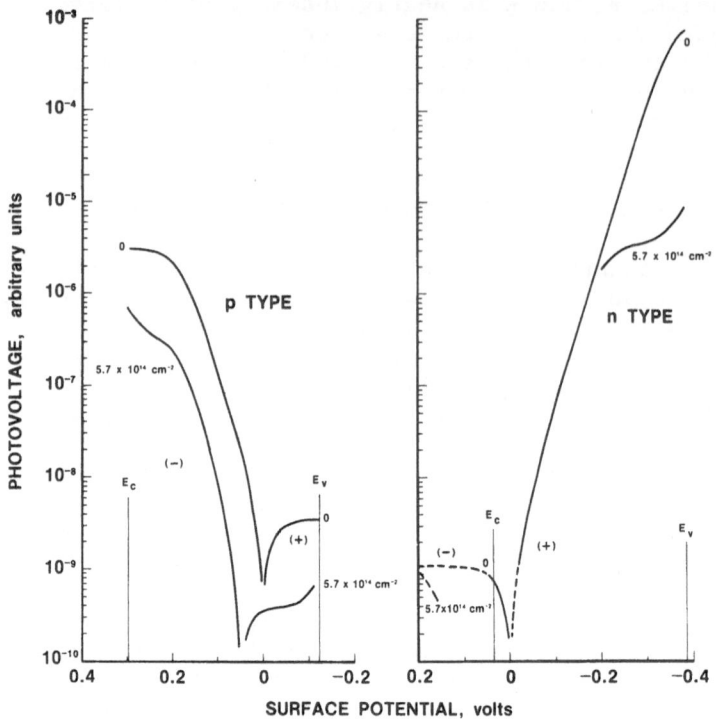

FIG. 6. – Theoretical SPV for the p- and n-type electrodes. The
surface-state densities are indicated on the curves. The
lines labeled E$_c$ and E$_v$ show when the Fermi level at the
surface enters the conduction and valence bands, respectively. The effects of degeneracy have not been taken
into account.

strong anodic bias. In the presence of surface states, the zero
SPV is no longer at the flat-band potential[13]. In the case of an
acceptor-like state, the zero photovoltage is shifted towards a positive surface potential. Finally, shoulders appear on the curves
at both sides of the zero in photovoltage when surface states are
present. Similar considerations apply to the n-type sample with
the exception that the negative branch of the SPV is extremely small
and may not be observable experimentally.

The general features of the calculated response are observable
in our experimental curves. This is more easily seen by plotting
the experimental photovoltage on the traditional logarithmic scale
as shown in Fig. 7. Note that the abscissa on the experimental
curves is electrode potential. This scale is only compatible (with-

Table 1. Bulk and Surface Parameters

Parameters	Value	Source
E_g--width of forbidden gap	0.42 eV	(21)
τ_b--bulk lifetime	15 μsec	(22)
m*--effective mass	0.17	(23)
K--relative dielectric constant	160	(24)
α--optical absorption coefficient	2×10^4 cm^{-1}	(19)
n_i--intrinsic carrier concentration	2.1×10^{15} cm^{-3}	calculated
n_b--electron concentration, n-type	2.0×10^{18} cm^{-3}	measured
p_b--hole concentration, p-type	6.0×10^{16} cm^{-3}	measured
μ_n--electron mobility	590 cm^2/volt·sec	measured
μ_p--hole mobility	325 cm^2/volt·sec	measured
D--ambipolar diffusion constant	11 cm^2/sec	calculated
N_t--surface state density	see graphs	assumed
E_t--surface state energy	-0.2 eV	assumed
r_v--rate constant	1×10^{-9} cm^2/sec	assumed
r_c--rate constant	1×10^{-9} cm^2/sec	assumed

in a constant) with the surface potential scale when an increment in electrode potential produces an equal increment in surface potential. For the p-type sample, the behavior near the zero of photovoltage is qualitatively in agreement with the theoretical response for both directions of polarization and indicates that near the zero's a significant fraction of the changes in electrode potential occur across the sample. Theoretically, the ratio of SPV's at the limits should be 8×10^3 in the absence of surface states. This can be compared with an experimental ratio of about 10 for the shoulders during the reduction cycle. This suggests that the magnitude of the SPV along these shoulders is determined by recombination at the surface or in the space-charge layer.

For the n-type electrode the apparent flat-band potential is not as readily determined as it is for the p-type electrode because we do not observe the zero crossing. We have made measurements with a sensitivity of $\simeq$ 0.005 mV and observed a small ($\simeq$ + 0.01 mV) constant SPV which is independent of electrode potentials below -0.6 V. This is believed to be a Dember potential[25], arising from the different diffusion rates of electrons and holes. In Fig. 7, the dashed line shows the effect of subtracting this small contribution. It is obvious that a small uncertainty in the Dember potential can have a large effect on the apparent FBP.

In calculating the theoretical SPV, we have assumed a discrete acceptor-like state with a constant density. A more general calculation would include both donor and acceptor states with their

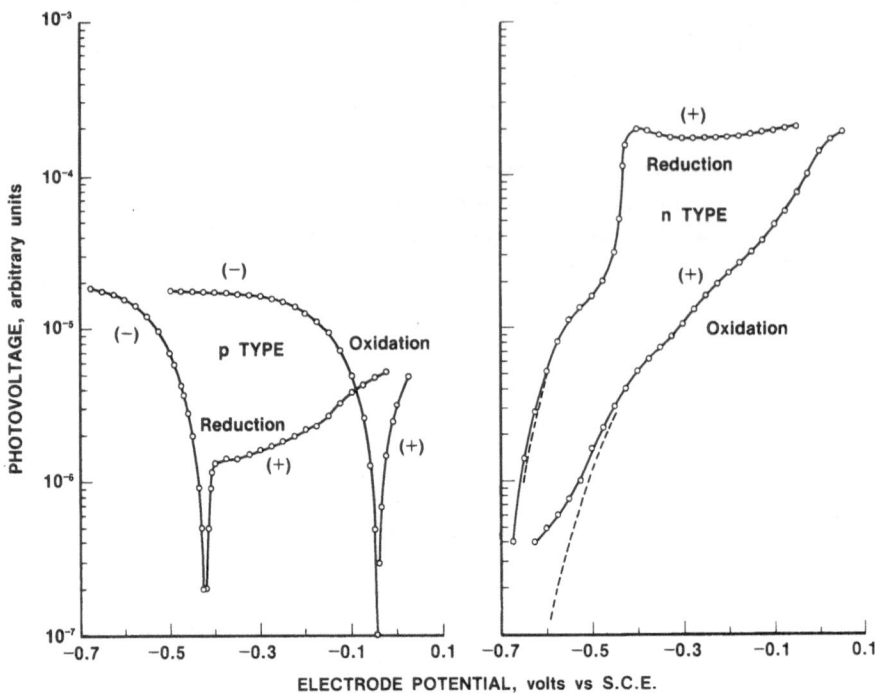

FIG. 7. - SPV versus electrode potential for p- and n-type samples.

densities varying with surface potential. The two sets of surface-
state parameters could then be used to adjust, independently, the
shoulders on either side of the FBP to fit the experiment results.
Additional structure could also be obtained in the SPV curves depending on the exact relationship between the surface-state density,
surface potential and electrode potential.

SUMMARY

1. A negative surface potential exists on both n- and p-type
galena electrodes at their rest potential. This potential barrier
is closely associated with prior oxidation reactions at the surface.
2. The apparent flat-band potentials have been determined for
the electrodes under dynamic polarization conditions and have been
found to depend on the bulk carrier concentrations and the past history of the electrodes.
3. Virgin surfaces produced by cleavage in deoxygenated borate
undergo spontaneous oxidation. The balancing cathodic process is
probably the cathodic reduction of oxygen. Immediately after cleav-

age, the surface potential is zero and subsequently becomes negative with spontaneous oxidation.

4. The results have been discussed in terms of the theoretical treatment of Frankl and Ulmer[13] using a model in which the surface potential and the SPV are determined primarily by surface states associated with the stoichiometry of the electrodes.

5. From the standpoint of flotation, it has been shown that the surface of galena is electron deficient after short times in alkaline solutions containing low concentrations of oxygen. This result is in agreement with the Plaksin school[1-5]. However, the proposed mechanism for the action of oxygen differs from earlier chemisorption models[2].

REFERENCES

1. I. N. Plaksin, and R. Sh. Shafeev, Acad. of Sci. of the USSR Proc., Phys. Chem. Sec., 132, No. 2 (1960) 421-423.

2. I. N. Plaksin, and R. Sh. Shafeev, Trans. Inst. Min. Met., 72 (1963) 715-722.

3. I. N. Plaksin, Trans. AIME, 214 (1959) 319-324.

4. I. N. Plaksin, VIII Int. Miner. Processing Congr., Leningrad, USSR (1968) paper S-3, 1-8.

5. I. N. Plaksin, Proc. Int. Miner. Processing Congr., Group III, No. 13 (1960) 253-392.

6. P. Eadington and A. P. Prosser, Trans. Instn. Min. Metall., 78 (1969) C74.

7. E. E. Maust, Jr., P. E. Richardson, and G. R. Hyde, "A Conceptual Model for the Role of Oxygen in Xanthate Adsorption on Galena," Bureau of Mines RI to be published.

8. A. Granville, N. P. Finkelstein, and S. A. Allison, Trans. Inst. Min. Met., 81 (1972) C1-C30.

9. R. Woods, J. Phys. Chem., 75 No. 3 (1971) 354-362.

10. H. T. Minden, J. Chem. Phys., 25 (1956) 241-248.

11. G. W. Poling, and J. Leja, J. Phys. Chem., 67 (1963) 2121-2126.

12. P. E. Richardson, and E. E. Maust, Jr., accepted for publication, Proc. A. M. Gaudin Memorial Symp., 105th AIME Meeting, Las Vegas, Nevada, February 1976.

13. D. R. Frankl, and E. A. Ulmer, Surface Sci., 6 (1966) 115-123.

14. S. Schuldiner, T. B. Warner, and B. J. Piersma, J. Electrochem.
     Soc., 114, No. 4 (1967) 343-349.

15. P. J. Boddy, and W. H. Brattain, Annals New York Acad. of Sci.
     101 (1963) 683.

16. J. F. Dewald in H. C. Gatos (Ed.), The Surface Chemistry of
     Metals and Semiconductors, John Wiley and Sons, Inc., New
     York, 1960, p. 205.

17. H. Gerischer, J. Electrochem. Soc., 113 (1966) 1174-1182.

18. C. G. B. Garrett, and W. H. Brattain, Phys. Rev., 99, No. 2
     (1955) 376-387.

19. W. W. Scanlon in F. Seitz and D. Turnbull (Eds.), Solid State
     Phys., Academic Press, New York, 9 (1959) 83-137.

20. J. B. Brodie, "The Electrochemical Dissolution of Galena,"
     Ph.D. Dissertation, The University of British Columbia,
     March 1969.

21. R. B. Schoolar, and J. R. Dixon, Phys. Rev., 137, 2A (1965)
     A667-A670.

22. M. Moldovanova, and S. Dimitrova, Phys. Status Solidi, 8, 1
     (1965) 173-176.

23. A. K. Walton, T. S. Moss, and B. Ellis, Proc. Phys. Soc., 79
     (1962) 1065-1068.

24. M. M. Sokoloski, and P. H. Fang, "The Dielectric Constant of
     Lead Salts," NASA N66-22249, April 1965, 4.

25. A. Many, Y. Goldstein, and N. B. Grover, Semiconductor Sur-
     faces, North-Holland Publishing Co., Amsterdam (1965) 276.

CORROSION - AN ANTICORROSION VIEWPOINT

Edmund C. Potter

CSIRO, Division of Process Technology

Delhi Road, North Ryde, Australia, 2113

INTRODUCTION

The Opening Address to this Conference has provided a theme, Electrochemistry for a Future Society, that cannot fail to appeal to savants and men of affairs alike. A panoramic view of electrochemistry in action to enrich our lives and mould our destiny ought perhaps to be supplemented by glimpses of detail from different parts of the field. This paper aims to illuminate in context one such area, namely, the science of metallic corrosion.

There can be very few people anywhere who are unaware of the phenomenon of metallic corrosion. Our languages abound with words to describe the familiar manifestations of corrosion (tarnishing, scaling, rusting, wastage, pitting, and so on), and the nuisance caused by loss and deterioration of metals in everyday use is so commonplace that most sufferers do not realise that corrosion is worth fighting. Paradoxically almost, specialists in metallic corrosion frequently encounter the enquirer who expects from them miracles of metallic rejuvenation long after a vital loss of metal has taken place from some everyday item such as a pipe, a tank, or an engine. These unsatisfactory liaisons between the corrosion specialist and the public no doubt owe much to the easy conviction that something so strong and refractory as a structural metal would surely not suffer rapid perforation and wear on contacting substances as smooth and yielding as water and air. Sadly, the epithet "stainless" has been both a hindrance and a help, since it is untrue on an embarrassing number of occasions but justified more often than not.

JUSTIFYING CORROSION SCIENCE STUDY

Metallic corrosion spoils the quality of life by making metals shabby and useless. It mystifies some people how anyone would want to study the gradual demise of a corroding metal without also diligently looking about for ways to halt or remedy such an unwanted process. Corrosion science study, of course, cannot proceed without financial support, but it stands apart from most other study topics in that finance originates from bodies whose members will certainly have been personally inconvenienced by metallic corrosion and will possibly be suppressing a sense of grievance. Unconsciously or otherwise, the study projects that find favour are those showing promise of overcoming a corrosion problem, especially a costly one.

It is understandable that a phenomenon as widespread as metallic corrosion should excite curiosity as to its cost to the community, for the chances are that some large and newsworthy sums will be involved. From the point of view of the corrosion researcher seeking financial support, the larger the community cost appears to be the better, since the opportunities for saving money then seem greater. Two principal methods of estimating corrosion costs have thereby emerged, one giving a considerably higher cost than the other.

The method that yields the higher cost is the more prevalent (1,2), dating from the first time corrosion costs were made about 1920. The procedure consists of estimating the direct cost of metallic corrosion losses (often metal by metal or by industry or application) and adding to these the costs of protective measures. Sometimes, more indirect costs are added on, such as lost production, lost orders, and even lost goodwill. The resulting figure is the TOTAL COSTS OF CORROSION and for industrialized countries it runs to about $3\frac{1}{2}\%$ of the gross national product according to the best information (3). Updating an earlier estimate (4), the total costs to Australia now exceed one thousand million dollars per annum, a magnificent sum on any scale of thinking.

Let us be quite clear what these total costs represent. They are the money that would not have to be spent if the phenomenon of corrosion did not exist at all. If you like, it is the money saved if all our common metals happened to be as noble as platinum. Well, of course, all our common metals are thermodynamically unstable in natural or man-made terrestrial conditions, and kinetically they usually lose little time reverting to their ores.

In my experience, whenever these total corrosion costs have been seriously discussed, it has always emerged that it was too easy to mistake such costs for money that would be saved but for some investment in metals research. All would agree at these discussions how misleading and unrealistic such a situation was, but it was not

corrected until the second principal method of corrosion costing (3)
appeared in the Report of the Hoar Committee on Corrosion and
Protection published in Britain in 1971. The Committee agreed that
only a proportion of the total annual costs could be saved, and
estimated, averaging all classes of metals used, that this proportion was 23%, say just under a quarter. For Britain in 1971, the
estimated savings were £310M per annum, which is probably around
£500M per annum today - five years later (5). The figure represents
THE AVOIDABLE COSTS OF CORROSION. In Australia these avoidable
costs are not less than $250M per annum, so that this is the annual
wastage that will be reduced by the successful anticorrosion worker.
It is this sum that can be used to persuade those able to finance
scientific and technological endeavour to make grants.

The main line of argument so far, may be summarized by the
following sequence of five statements:

1.   Electrochemistry embraces the science of corrosion

2.   Corrosion wastes metals, a non-renewable resource.

3.   Metals, especially the common non-ferrous metals, are
      being corroded and dissipated over the globe and are
      needed by a future society.

4.   A future society will thank us for an anticorrosion view-
      point of corrosion science.

5.   Electrochemistry for a future society.

The reader might think that these words would be most appropriate to end with, but this is, I submit, not so; because one
should consider which particular anticorrosion studies merit prime
attention. This is a difficult area since researchers naturally
wish to tackle that which arouses their curiosity and which is a
challenge, whether it has any broader relevance or not. Furthermore, one has to recognize that specific problems are always
arising in corrosion science, problems that must be tackled as a
matter of industrial urgency or because the professor or board of
directors requires such action. What, therefore, I am seeking is a
view, personal it must be, concerning the areas that should be
receiving more attention because they appear to offer opportunities
for anticorrosion progress.

In seeking such a view I have considered a number of current
research topics, For example, the mechanism of stress corrosion,
the detailed steps in anodic dissolution of clean metal surfaces,
instantaneous and on-line corrosion rate measurement, high-temperature
scaling, pitting phenomena, thermodynamic portrayal of metals
behaviour, and others. However, all of these topics, important
though they are, seem less significant to me than two others which
I feel should receive priority and additional attention.

ALLOYING FOR STAINLESSNESS

The first priority topic is an understanding of the factors controlling the corrosion behaviour of alloys, particularly the circumstances leading to useful passivity or "stainlessness". Knowing what we do of the corrosion behaviour of iron, chromium, and nickel, I do not think one would be able to deduce that alloying iron with over about 10% chromium makes an enormous difference to the atmospheric stability of iron and that the addition of a comparable additional amount of nickel confers a further remarkable extension of stability to many common aqueous conditions. Yet 18/8 chromium nickel steel does fail in a considerable number of ionic solutions, and over the years modifications of the basic alloy have appeared, many with some improved corrosion resistance. Thus, alloys with relatively minor amounts of titanium, molybdenum, vanadium, and various other elements have been invented with, it seems to me, not much to guide the inventor towards superior corrosion resistance, the goal he often seeks. The tedium of testing innumerable variants of alloys with perhaps six or more compositional variables can easily be imagined. What we should be seeking, I submit, is a rationale for stainlessness that will cut short the tedious quest for corrosion resistance based upon empirical preparation of new alloys. A few years ago Uhlig, a metallurgical electrochemist, made some praiseworthy approaches to such a rationale (6), but as far as I know, no one has taken the matter much further. Tomashov (7) states: "At present we are not able to calculate theoretically the optimum composition of a corrosion-resistant alloy on the basis of the physical and chemical characteristics of its individual components". It may be, of course, that such a task is too difficult for us, and so I pass to the second area where I believe more effort should be made. In some ways this area is related to the first I have just mentioned.

PROTECTIVENESS OF SURFACE FILMS

It has been the custom in corrosion science for several decades to explain and interpret the corrosion resistance or otherwise of metals in terms of the formation of surface films of corrosion product, very often oxide films. This is a very sound scheme, but much more could be made of it I believe.

We have known for a long time, for example, that the good atmospheric corrosion resistance of the very unstable metal aluminium is attributable to its carrying a self-healing oxide film. I do not think one could have anticipated such a favourable event, and, moreover, our present day knowledge of the columnar morphology of the oxide film (8) might lead one to doubt its ability to protect aluminium as well as it does. Iron also forms a thin self-healing oxide film on its surface in air at ordinary temperature, but the

protective properties of this film are relatively poor, as the
ubiquitous atmospheric rusting of iron demonstrates. I submit that
we do not have sufficient rationale for the protectiveness of surface
films on metals. Most of our common constructional and industrial
metals and alloys rely upon surface films of some sort for much of
any corrosion resistance they possess. Why is rust on ordinary
steel not a good protector? Put a small amount of copper in the
steel and the rust then generated is changed in composition and
exhibits an obviously superlative protective quality in the atmosphere. In spite of much valuable research on this very matter, I
submit that we do not understand this most relevant effect to that
extent where the effects of other alloying additions to the same or
to other metals can be anticipated in terms of the protectiveness
of the oxide films produced. Who could have anticipated that copper
would grow a most useful and pleasing platina on its surface in a
slightly-polluted atmosphere, and who could deduce beforehand whether
aluminium or one of its alloys might withstand sea-water – a most
relevant matter if you sail the seas? One could not foresee that
iron grows a splendidly protective film of magnetite in hot air-free
water, and in many aqueous solutions, yet it would be arrogant to
assert that we know precisely why that protectiveness sometimes gives
way or that we can suggest what slight variations might be made to
the composition of the metal so that even better protection is
achieved. At present, engineers designing large, expensive, and
important structures or plants often have no suspicion that they may
be asking the available metals and alloys to perform beyond their
capabilities under the conditions that will prevail when construction
is complete. I have a fair example of this.

The original work on the formation of the thin protective oxide
film on iron was done in air or oxygen at atmospheric pressure over
a wide temperature range (9-12). Doubtless this work generated a
great proportion of the subsequent theoretical interest in oxide
film growth, and this is most meritorious. To my surprise, however,
there seems to be no comparable published work on the growth of the
oxide film on iron in atmospheres containing less oxygen than in air,
and moreover there is no detailed work on the combined filming
effects of oxygen, carbon dioxide, and water vapour, and no clear
knowledge of the balance of antagonism between these film-formers
and potential film-destroyers such as hydrogen chloride, chlorine,
sulphur oxides, and nitrogen oxides. Yet gaseous environments
containing trace amounts of such film destroyers are common in
industry and even in urban atmospheres, with the result that
unnecessary losses of metal and equipment are taking place for want
of essential knowledge.

The position is especially grievous with steel plant that
encounters the mixture of gases arising from the incineration of
urban refuse. Here all the aforementioned gases are present at once
and in variable concentrations depending on the operational regime

of the plant.   Some idea of the complexity of the film forming
processes on mild steel at a steady 300°C is presented in Figure 1,
which illustrates (13) that the oxide film withstands the introduction of hydrogen chloride (at relevant levels) provided sufficient
film-former is present (oxygen and/or water vapour).   Traces of
hydrogen chloride themselves form a protective non-oxide film, but
the development of this film is thwarted at a level of introduced
oxygen around 1% and accelerated attack is suffered until a mixed
oxide/non-oxide film accumulates to a protective thickness.   Thus,
the balance between film protectiveness and otherwise is rather
delicate, so that proper procedures for start-up, operation, and
close-down of incinerator plant will have to be worked out and
observed if steel is to perform adequately in this application.

CONCLUSION

    Metal-care begins at the design stage of structures, plant,
and equipment, and continues throughout the subsequent phases of
construction, operation, and maintenance.   The electrochemist
applying his talents to progress in corrosion science should heed
the costly consequences of metal failure and direct his studies
from an anticorrosion viewpoint.

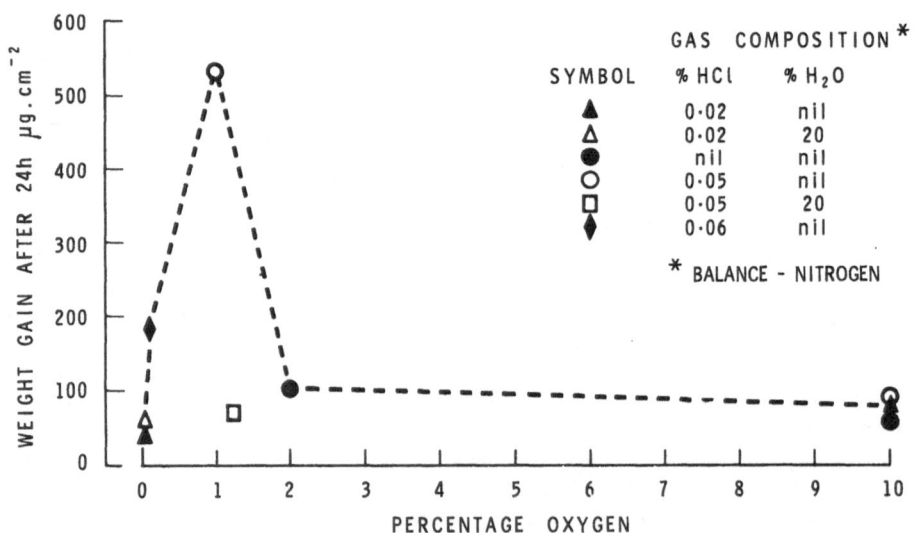

EFFECT OF OXYGEN CONCENTRATION ON IRON OXIDATION AT 300°C

Fig. 1   Effect of Oxygen Concentration on Iron Oxidation at 300°C
         (Nancarrow (13)).

REFERENCES

1. H.H. Uhlig, Corrosion, 6 (1950) 29.
2. W.H.J. Vernon, in *The Conservation of Natural Resources*, Inst. Civil Engrs (London), 1957.
3. Report of the Committee on Corrosion and Protection, London HMSO, 1971.
4. E.C. Potter, Australasian Corrosion Engng. 16 (1972) 3.
5. T.P. Hoar, Proce. R. Soc. Lond. A, 348 (1976) 1.
6. H.H. Uhlig, *Proceedings of the Second International Congress on Metallic Corrosion*, New York 1963, published by Nat. Ass. Corrosion Engrs, Houston, 1966, pp 1-8.
7. N.D. Tomashov, *Proceedings of the Third International Congress on Metallic Corrosion*, Moscow 1966, published by Mir Publishing House, Moscow, 1969, Vol I, p.37.
8. G.C. Wood in J.W. Diggle (Ed.), *Oxides and Oxide Films, Volume II*, Dekker, New York, 1973, p. 167.
9. H.A. Miley, Carnegie Schol. Mem., 25 (1936) 197.
10. W.H.J. Vernon, Trans. Faraday Soc., 31 (1935) 1670.
11. W.H.J. Vernon, E.A. Calnan, C.J.B. Clews, and T.J. Nurse, Proc. R. Soc. Lond. A, 216 (1953) 375.
12. D.E. Davies, U.R. Evans, and J.N. Agar, Proc. R. Soc. Lond. A, 225 (1954) 443.
13. P.C. Nancarrow, private communication.

THE EFFECT OF OXIDE FILMS ON CEMENTATION REACTIONS

G.A. Hope, I.M. Ritchie and J.E. Wajon

Department of Physical and Inorganic Chemistry
University of Western Australia
Nedlands, W.A.

ABSTRACT

Metal displacement reactions take place at the surface of a
metal in a solution or melt. This paper describes the effect of
oxidizing the metal surface on the rate of a displacement reaction.
In the first part of the paper, the deposition of copper from
copper(II) solutions on a wide variety of metals coated with an air-
formed film of oxide was investigated. It was found that most of
the metals displaced copper, the exceptions being aluminium, nickel,
and the titanium, vanadium and chromium sub-groups. The relationship
between these results and the mechanisms of metal oxidation reactions
is considered. The second part examines the effect on the copper(II)/
zinc reaction of oxides which have been grown both thermally and
anodically. The reaction rate was measured at a rotating disc. It
was found that the rate was controlled by the diffusion of copper(II)
ions to the metal surface largely irrespective of whether there was
a surface oxide present or not. However, the oxide did have some
second order effects on the hydrodynamics of the system and hence
indirectly on the reaction rate.

INTRODUCTION

When a metal is placed in a molten salt or solution containing
the cations of a second less electropositive metal, a metal
displacement or cementation reaction can take place. The first metal
is oxidized, its ions going into solution, while the cations of the
second metal are reduced, depositing it on the surface of the first.
Reactions of this type are used widely in industry: in the winning

341

of metals, in the purification of electrolyte streams, and in
several other special applications such as plating[1]. This paper
describes an investigation of the effect of oxide films on the
kinetics of a number of metal displacement reactions, in particular
the reaction between copper(II) ions and zinc metal

$$Cu^{2+} + Zn \rightarrow Cu + Zn^{2+}$$

which is of importance in the manufacture of zinc by the electrolytic
method.

Nearly all metals are thermodynamically unstable in oxygen, and
react rapidly in air to form an oxide film on the metal surface. In
most cases, the air-formed oxide grows to a thickness of about 3 nm,
upon attaining which, the reaction virtually ceases. Thicker oxides
can be grown by heating the metal in oxygen or anodizing it. Now the
inevitable presence of an oxide film on the precipitant metal can
influence the rate of a displacement reaction in two ways.

The first way is by altering the chemical steps at the metal/
solution interface. In particular, the oxide can act as a barrier
separating the reactants. Unless the oxide film is ruptured or
dissolved in the reactant solution, the cementation process can only
be sustained provided ions and electrons pass freely through the
oxide. The ions moving through the oxide go into solution; the
electrons are required for the reduction of the less electropositive
species. The effectiveness of the oxide film formed at room temperature in air as a barrier to the metal displacement reaction on a
wide variety of elements is investigated qualitatively.

The second way in which an oxide film can take part in a metal
displacement reaction may be termed hydrodynamic. In a reaction
such as that between copper(II) ions and zinc metal,where the slow
step has been reported to be the diffusion of the reactant cations to
the metal surface[2], the reaction rate depends on the hydrodynamic
conditions there. For this reason, the kinetic investigations
described here on the copper(II)/oxidized zinc system have been
undertaken using a rotating zinc disc. However, in order to be able
to calculate the flux of cations to the disc, it is necessary to have
laminar flow across the surface. It is known that the reaction rate
at a rough disc can exceed that calculated for a smooth one[2], and so
oxidation, which can directly alter the roughness of the surface,
might influence the reaction rate. In addition, by partially blocking
the surface, oxidation can affect the disposition, and hence roughness, of the cementation deposit. Since it has been found[3] that some
insight into the mechanism of cementation reactions can be obtained
by considering the appropriate Evans diagram, i.e. a diagram
constructed from the polarization curves of the cathodic and anodic
processes, we have also examined the effect of oxide films on the
copper(II) ions/zinc metal system from this point of view.

So far as we are aware, this is the first study of its type. That oxide films do play a role in cementation processes has been noted by other authors. For example, Tsvetkov and Zarechnyuk[4] found that chloride ions helped break down the oxide film on aluminium during the deposition of copper onto that metal from ethanolic solutions.

EXPERIMENTAL

Qualitative Experiments: To test the effectiveness of the air-formed oxide on various metals as a barrier to the displacement of copper from solution, metal samples (0.1 - 1.0g) were placed in 25 cm^3 of a 0.1 mol dm^{-3} copper(II) sulphate solution which had been made up from the A.R. grade solid and distilled water. Where necessary, the metal samples were degreased, acid etched, washed thoroughly in distilled water, dried and allowed to oxidize for 24 hours in a desiccator. The metals used were as follows, their form and purity where known being specified: aluminium (wire, 5N); antimony (shot, 6N); bismuth (shot, 6N); cadmium (shot, 5N8); calcium (granules); chromium (pellets, 5N); cobalt (wire, 4N5); copper (sheet, 5N); gadolinium (powder, 3N); indium (wire); iron (wire, 4N8); lead (shot, 6N); magnesium (ribbon); manganese (flake, 4N); nickel (wire, 5N); scandium (ingot, 3N); sodium (lumps, 2N8); tantalum (sheet); tellurium (lumps, 5N); thallium (rod, 5N); tin (granules, 2N); titanium (crystals, 3N); tungsten (wire); vanadium (lumps, 3N); yttrium (lumps); zinc (bar, 4N); zirconium (bar, 3N).

Quantitative Experiments: Two types of experiment were carried out on the copper(II)/zinc system. These were kinetic and electro-chemical.

In the kinetic experiments, the reactant solution was a 1.57 x 10^{-4} mol dm^{-3} (10 p.p.m.) solution of copper sulphate made by dissolving the A.R. grade solid in distilled water. The zinc samples were machined from ingots having a purity of 99.99% in the form of discs, 3 cm dia., which could be press-fitted into a Delrin holder of the type described previously[5]. Usually, the surface of the zinc was prepared by abrasion with 400 grade carborundum paper followed by 600 grade paper. It was then washed with water, acetone and dried. In some cases, the zinc surface was electropolished in a solution composed of equal volumes of 85% orthophosphoric acid and ethanol. During the polishing process, the disc was mounted about 1 cm below a sheet nickel cathode in the electropolishing solution and a potential difference of about 2 V applied between the electrodes. After passing a current for about 30 minutes, a bright shiny surface formed.

Kinetic experiments were carried out with freshly cleaned discs, with some discs which had been thermally oxidized, and others which had been anodized. The thermal oxidation was carried out by placing the disc in an open-air tube furnace whose temperature was held constant at the desired value to within ±3°C. After an appropriate period of oxidation, usually 14-16 hours, the disc was cooled, either rapidly by removing it from the furnace and placing in a desiccator, or slowly by leaving it in the furnace which was turned off. Various conditions were used for anodizing the zinc, the guides being the papers by Huber[6] and Fry and Whitaker[7]. For the production of a black anodic oxide, an anodizing solution of 1 mol dm^{-3} sodium hydroxide was used. If the anodizing solution was 0.15 mol dm^{-3} in sodium hydroxide and 0.05 mol dm^{-3} in sodium carbonate, the anodic oxide layer was white rather than black. In all cases, the cathode was a nickel sheet positioned 1 cm above the zinc surface.

The kinetic runs were carried out in a 2 dm^3 reaction vessel which was sufficiently large to avoid interfering with the hydro-dynamic flow near the rotating disc[8]. Before introduction of the rotating disc, the copper sulphate solution, 1 dm^3 in volume, was purged for 3 hours using oxygen-free nitrogen which reduced the oxygen content of the solution to less than 1 p.p.m. The solution was subsequently maintained under a nitrogen atmosphere. After allowing the solution to come to thermal equilibrium in a water bath kept at 25±0.5°C, the disc was immersed and rotated at 750 r.p.m. using a constant speed motor. The angular velocity was checked using a tachometer which had been calibrated stroboscopi-cally. The reaction was followed by taking 5 cm^3 samples at regular time intervals and analysing the samples for copper content by atomic absorption spectrophotometry using the line at 327.4 nm.

Reacted discs were examined under a microscope which had a 35 mm camera attachment.

In addition to the kinetic studies, some electrochemical measurements were undertaken. Anodic polarization curves for zinc discs with different surface preparations were determined ampero-statically in a deaerated solution containing 0.1 mol dm^3 zinc sulphate. A platinum sheet was made the cathode. The polarization current was measured using a calibrated ammeter. The potential of the zinc electrode was determined relative to a saturated calomel electrode via a Luggin capillary located axially and about 1 mm below the disc surface.

RESULTS AND DISCUSSION

Qualitative Experiments: The results of the qualitative
experiments are presented in the modified periodic table of Fig. 1,
together with the most likely composition of the air-formed oxide,
where this is known. It can be seen that the majority of the metals
investigated, all of which are thermodynamically unstable in
0.1 mol dm^{-3} copper sulphate, displaced copper from solution. The
exceptions were the elements of the titanium, vanadium and chromium
subgroups, nickel and aluminium. We suggest that in these cases,
the oxide on the metal is acting as a barrier against the reaction.
Passive iron is also known to inhibit the cementation of copper[9].
That it is the oxide which is preventing the reaction from taking
place can be demonstrated for the metals nickel, aluminium and
passivated iron. When the oxide skins on these metals are broken
by scratching, the deposition of copper takes place at a measurable
rate. However, this is not true of the titanium, vanadium and
chromium subgroup metals, perhaps because their oxides or hydroxides
reform more rapidly than the copper can be deposited. Not only is
this group of metals unstable in water[10], but for many of them,
their cations are highly charged, easily hydrolysed and polymerize[11]
ultimately to form the hydrated oxide.

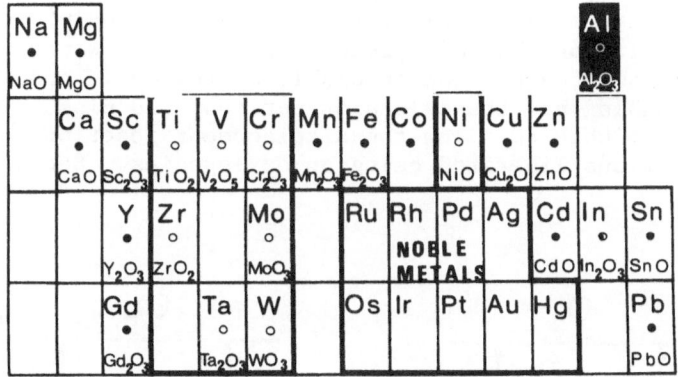

Fig. 1 Modified periodic table showing which elements displace
copper from solution. Key: ●, reactive elements; O, unreactive
elements; ◑, slowly reacting elements. Copper was judged to be
reactive since it displaced silver from silver nitrate solution.
Also listed are the oxides most likely to be present on the metal
surface.

The growth to a limiting thickness of aluminium oxide and chromium oxide on the two parent metals when these are heated in oxygen at temperatures above room temperature (aluminium >50°C; chromium >200°C) has been investigated[12,13]. In both cases, it was concluded that a Mott-Cabrera type of mechanism was operative, i.e. the slow step in the growth process was the field assisted movement of cations across the oxide film, the assistance of the field decaying to a negligible quantity when the limiting thickness was reached. If such a mechanism governs the growth in air of the oxide film formed at room temperature, it is not unreasonable to suppose that cation transport across these same films during the displacement reaction with copper(II) ions will also be small. The similarity between the oxidation and cementation reactions is shown in Fig. 2. It seems possible that the oxides of the titanium, vanadium and chromium subgroup metals, with the exception of Cr_2O_3, also impede the transport of cations to the oxide/solution interface during the displacement reaction. It is known that the oxides of these metals are all n-type oxygen-deficient semiconductors in which oxygen transport is likely to dominate[14]

If the oxide-covered metal allows the deposition of copper to take place on its surface, as is the case for the majority of the metals investigated here, then the oxide does not constitute an effective barrier to the reaction. Three reasons may be advanced for this. Firstly, the oxide may dissolve in water, the metals sodium, magnesium, calcium and thallium falling into this category. Deprived of their oxide coating, these metals not only precipitate copper from solution, but also react directly with water to liberate hydrogen. In some cases, the insoluble hydroxide is precipitated. It seems probable that the oxides of antimony and bismuth are also sufficiently soluble to allow the displacement reaction to take place, the cations liberated being an oxy-species. Secondly, the

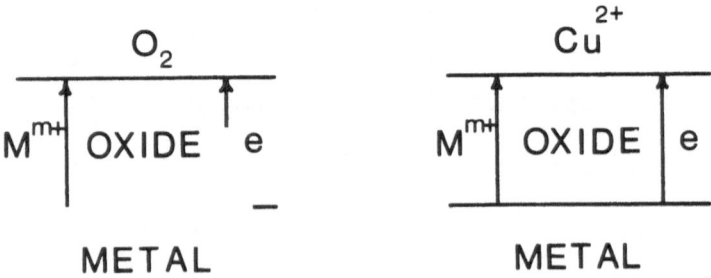

Fig. 2   Diagram showing the common electronic and ionic transport processes in the oxidation of a metal, and a metal displacement reaction across an oxide film.

oxide may rupture, perhaps as the result of internal stresses
generated by the conversion of the oxide to an insoluble hydroxide.
Stresses of this type have been shown to occur when the surface of
a metal is converted to a substance of different volume ratio, this
being the volume of the substance divided by the volume of metal
from which it was formed[14]. Under these circumstances, metal
deposition would be observed at the points where the oxide had
broken down mechanically. However, conversion to the hydroxide does
not seem very probable for most systems. The change in free energy
for the conversion of the oxide to the hydroxide is generally quite
small. For example, ΔG_{298}° values for the hydration of Sc_2O_3, CdO,
MnO and NiO are 64, -9, -10, 0 kJ mol^{-1} respectively[15].

Thirdly, transport of cations and electrons may proceed freely
across the oxide layer. This is apparently the case for the metals
scandium, yttrium, gadolinium, manganese, iron, cobalt, copper,
zinc, cadmium, tin and lead. The question then arises as to why
these metals show the phenomenon of oxide growth to a limiting
thickness in air. By reference to Fig. 2, it can be seen that this
phenomenon must be related to some slow step at the oxygen/oxide
interface, since the diffusion step is common to both the oxidation
and displacement reactions. Consistent with this, it has been shown
that the growth on zinc in oxygen of a zinc oxide film to a limiting
thickness between 300 and 375°C is controlled by a surface reaction[16].

Two apparent anomalies may be noted. It is surprising that
nickel, alone of the series manganese to zinc, has an oxide which
behaves protectively towards the deposition of copper. This may be
due to the high resistivity of the nickel oxide formed, nickel oxide
being a "hopping" type semiconductor[17]. The other unexpected
result is that scandium cements copper, whereas the chemically
similar element aluminium does not, and indium reacts only slowly.

Quantitative Experiments: The rate of displacement of copper(II)
ions from solution is equal to the net flux (J) of ions to the zinc
surface:

$$-VdC/dt = J \qquad (1)$$

where V is the volume of reactant solution whose concentration is
C at time t. For a diffusion-controlled reaction, such as the
copper(II)/zinc reaction[2], the flux to a smooth rotating disc of
area A under conditions of laminar flow is given by the Levich
equation[18]:

$$J = 0.62 \; D^{2/3} \; \nu^{1/6} \; \omega^{1/2} \; CA \qquad (2)$$

where D is the diffusion coefficient, ν is the kinematic viscosity
and ω is the angular velocity. Combining Eqs. (1) and (2), and

integrating with the assumptions that V, A and D are constant we get

$$\ln(C_o/C) \quad = \quad kAt/V \tag{3}$$

where C_o is the initial concentration of copper ions. In fact, V is not constant during the course of the reaction, samples of volume v being removed for analysis at regular time intervals Δt. The volume correction factor is small and may be accommodated[19] by using a modified time scale t*:

$$t^* \quad = \quad \Delta t \sum_{1}^{n} 1/\{1-(v/V_o)(n-1)\} \tag{4}$$

where n is the number of samples taken from an initial volume V_o.

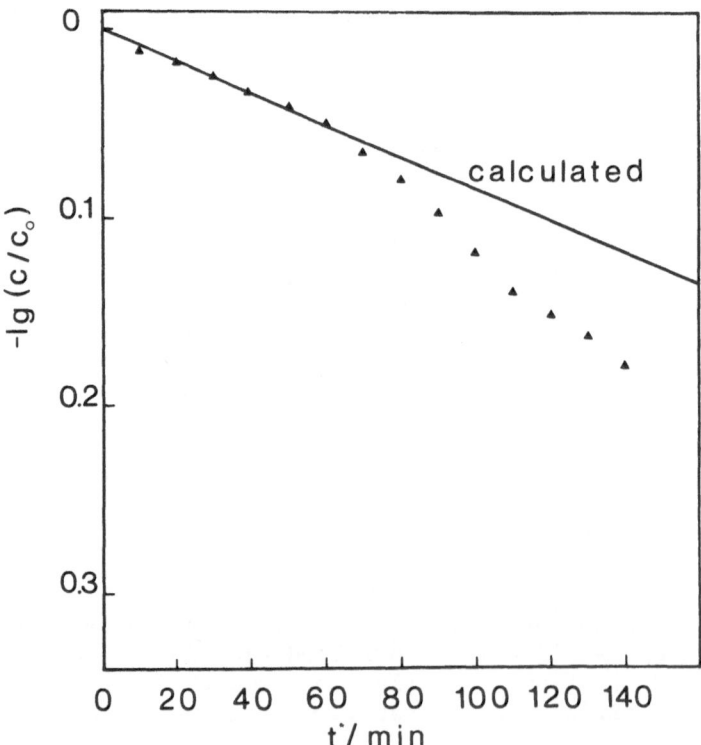

Fig. 3. The kinetics of a copper(II)/zinc reaction at 25°C and 750 r.p.m. plotted according to Eq.(3) with t replaced by t*.

Thus, the reaction rate should conform to Eq.(3) with t replaced by t*, the rate constant being given by

$$k = 0.62D^{2/3} \nu^{-1/6} \omega^{1/2} \tag{5}$$

Taking D to be equal to 7.3×10^{-10} m^{2}s^{-1} [2] and ν to be equal to 8.9×10^{-7} m^{2}s^{-1} [15], we obtain a value for k of 4.47×10^{-5} m s^{-1} at 25°C and 750 r.p.m., which were the conditions of our experiments.

Fig. 3 shows the results of a typical experiment plotted according to Eq.(3) for an unoxidized zinc disc. It can be seen that the reaction initially conforms to the expected behaviour, but then becomes faster. Similar curves have been obtained by Strickland and Lawson[2]. In this work, we have neglected the enhanced second stage of the reaction, which has been ascribed to deposit roughening[2], and calculated rate constants for the first stage using a least squares method and assuming Eq.(3) to be valid. Individual and average ($\bar{k}$) rate constants for unoxidized zinc discs are presented in Table I. The errors quoted are two times the standard deviation.

These results are, within the limits of experimental error, all equal to, or greater than the calculated rate constant, which is what one would expect for a reaction which is diffusion controlled, but which might show some increase in reaction rate due to deposit roughening.

In all cases, the copper deposit was black, sometimes showing the spiral pattern characteristic of reactions at rotating discs.

TABLE 1

Rate Constants for the Copper(II)/Unoxidized Zinc System at 25°C and 750 r.p.m.

Surface Preparation	$10^5 k$/m s^{-1}	$10^5 \bar{k}$/m s^{-1}
Mechanically polished[20] (10 results)		6.0 ± 1.7
Mechanically polished	4.5 ± 0.7 4.2 ± 0.7 5.8 ± 0.8 4.8 ± 0.7	4.8 ± 1.4
Electropolished	4.7 ± 0.3 4.5 ± 0.3	4.6 ± 0.5
Calculated		4.5

TABLE II

Rate Constants for the Copper(II)/Oxidized Zinc System at 25°C
and 750 r.p.m.

Surface Preparation	Oxide Thickness/nm	Cooling	$10^5 k/m\ s^{-1}$
Mechanically polished	10	fast	4.8 ± 0.7
	10		5.2 ± 0.5
	10		4.0 ± 0.7
	140		0.8 ± 0.3
	140		3.8 ± 0.7
	140	slow	8.7 ± 0.7
	140		5.7 ± 0.7
Electropolished	140	fast	3.5 ± 0.7
	140	slow	3.3 ± 0.7
Calculated			4.5

In general, the appearance of the mechanically polished discs
after thermal oxidation was a dull grey. The effect of this oxide
on the kinetics of the copper displacement reaction for a selection
of oxidation conditions is shown in Table II. The approximate
oxide thicknesses listed in this Table were estimated from the data
of Vernon *et al.*[21] Similar results were obtained at other rotation
speeds. From an inspection of Table II, it can be seen that the
majority of the rate constants are close to the calculated value,
i.e. the reaction may still be considered to be under chemical
control. Only in one run was a rate reduction observed which could
clearly be ascribed to the oxide acting as a barrier to the reaction.
The cause of this isolated result is not known. However, it is
possible for parts of the zinc surface to be covered with a highly
resistive oxide on which copper will not deposit without affecting
the reaction rate greatly. Power[22] has considered some simple
geometries for the reduction in active area of a rotating disc,
and found that the percentage reduction in rate can be significantly
less than the reduction in area.

Macroscopically, the copper deposit appeared black as before.
Under the microscope it was sometimes found that the copper had
been laid down preferentially on certain grains or along grain
boundaries. Preferential deposition at a grain boundary is shown
in Fig. 4. Clearly, the disc cannot be regarded as equiaccessible[18]
under these circumstances, as the surface of an ideal rotating disc
should be. Since the reaction rate can be expected to depend on the
size of the copper deposits, and their distribution over the metal
surface, some scatter in the results is to be anticipated.

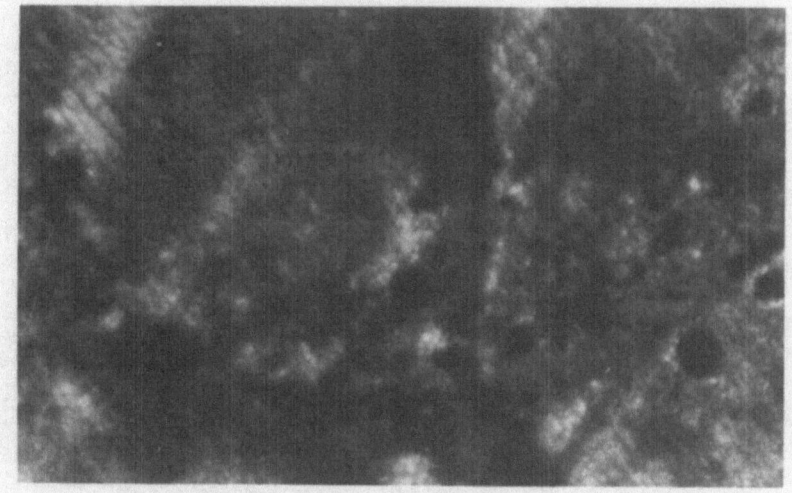

Fig. 4. Micrograph showing the preferential deposition of copper along the grain boundaries of a zinc sample oxidized in air at 400°C for 14 hours.

Consistent with this, the rate constants on oxidized surfaces were found to be much more variable than those for unoxidized surfaces, $\bar{k}$ being equal to $4.9 \pm 3.5 \times 10^{-5}$ m s^{-1}.

The effect of several different anodic oxide films on the rate of the copper displacement reaction is shown in Table III. The oxide thicknesses have been estimated from the magnitude and duration of the anodizing current. These must be regarded as upper estimates

TABLE III

Rate Constants for the Copper(II)/Anodized Zinc System at 25°C and 750 r.p.m.

Type of Film	Oxide Thickness/μm	$10^5 k$/m s^{-1}
White	1	5.2 ± 0.7
	4	3.0 ± 0.7
	4	5.2 ± 0.8
	4	5.2 ± 0.8
Black	1	$\cdot 4.5 \pm 0.7$
	3	7.5 ± 0.8
Calculated		4.5

since gas was evolved during anodization, and the current efficiency
cannot have been 100%. As with the thermally oxidized discs, the
oxides did not appear to constitute an effective barrier to the
reaction, $\bar{k}$ being 5.1 ± 2.8 m s^{-1}.

Before cementation, the black films appeared smooth and reflecting, while the white appeared uniformly dull. Neither type of film
appeared to be porous. In general, the copper deposits on each were
black and uniform, unless flaking of the anodic film occurred.

Anodic polarization curves were measured for polished, thermally oxidized and anodized zinc discs. For polished and thermally
oxidized samples, the potential-current characteristics were very
similar showing rotation speed independence, an equilibrium potential of about -0.8 V with respect to a standard hydrogen electrode,
and a Tafel slope of around 30 mV per decade of current density.
In contrast, the equilibrium potential of a white anodized disc was
about -0.53 V and the Tafel slope was about 170 mV per decade.
This latter figure is high, but this is not uncommon for electrode
reactions on oxide surfaces[23]. It was difficult to obtain measurements for the black anodic films which reacted during the passage
of the measuring current. Possibly the zinc particles which are
supposed to impart the blackness[6] to the films were dissolving.

Because the presence of an oxide coating does not raise the
potential of the zinc electrode very much, the cathodic curve for
the reduction of copper ions to copper metal intersects the anodic
curves of the oxidized zinc samples in the region in which the
cathodic current is diffusion limited. As a result, we would
expect[3] the rate of the metal displacement reaction between
copper(II) ions and zinc metal to be controlled by rate of diffusion
of copper ions to the zinc surface, irrespective of whether the zinc
is covered by an oxide or not. This finding is consistent with
the kinetic results.

REFERENCES

1.   G.P. Power and I.M. Ritchie in B.E. Conway and J.O'M. Bockris
      (Eds.), *Modern Aspects of Electrochemistry, Volume 11*, in press.

2.   P.H. Strickland and F. Lawson, *Proc. Aust. Inst. Min. Met.*,
      236 25 (1970).

3.   G.P. Power and I.M. Ritchie, submitted to *Aust. J. Chem.*

4.   N.S. Tsvetkov and O.S. Zarechnyuk, *Zhurn. Prikl. Khim.*, 33
      636 (1960).

5.   J.T.T. Pang, I.M. Ritchie and D.E. Giles, *Electrochim. Acta*,

in press.

6.   K. Huber, *J. Electrochem. Soc.*, 100   376 (1953).

7.   H. Fry and M. Whitaker, *J. Electrochem. Soc.*, 106   606 (1959).

8.   D.P. Gregory and A.C. Riddiford, *J. Chem. Soc.*, 3756 (1956).

9.   U.R. Evans, *J. Chem. Soc.*, 130   1020 (1927).

10. M. Pourbaix, *Atlas of Electrochemical Equilibria in Aqueous Solutions*, Pergamon, Oxford, 1966.

11. D.L. Kepert, *The Early Transition Metals* Academic Press, London, 1972, Chap. 1.

12. G.L. Hunt and I.M. Ritchie, *J. Chem. Soc., Farad. Trans. I*, 68   1413 (1972).

13. G.A. Hope and I.M. Ritchie, *Thin Solid Films*, in press.

14. O. Kubaschewski and B.E. Hopkins, *Oxidation of Metals and Alloys* Butterworths, London, 1962, p.24.

15. R.C. Weast, (Ed.) *Handbook of Chemistry and Physics*, 52nd. Edn., The Chemical Rubber Company, Cleveland, 1971.

16. I.M. Ritchie and R.K. Tandon, *Surface Sci.*, 22   199 (1970).

17. A.T. Howe and P.J. Fensham, *Quart. Rev.*, 21   507 (1967).

18. V.G. Levich, *Physicochemical Hydrodynamics*, Prentice-Hall, Englewood Cliffs, N.J., 1962, Chap.2.

19. G.P. Power and I.M. Ritchie, unpublished results.

20. P.H. Strickland, *Ph.D. Thesis*, Monash University, Victoria, Australia, (1972).

21. W.H.J. Vernon, E.I. Akeroyd and E.G. Stroud, *J. Inst. Metals*, 65 301 (1939).

22. G.P. Power, *Ph.D. Thesis*, University of Western Australia, Australia, (1975).

23. A.K. Vijh, *Electrochemistry of Metals and Semiconductors*, Marcel Dekker Inc., New York, 1973, Chap. 6.

ELECTROWINNING COPPER FROM CHLORIDE SOLUTIONS

K. J. Cathro

CSIRO Division of Mineral Chemistry

P.O. Box 124, Port Melbourne, Vic. 3207, Australia

I. INTRODUCTION

There are several references to electrowinning copper from chloride solutions (e.g. 1-4), the copper being recovered in powder form. Ashcroft (5) and Hazen (6) have mentioned the possibility of obtaining firm coherent deposits, while the conditions necessary for obtaining such deposits were studied by Mitter et al. (7) and Gokhale (8). Cuprous chloride solutions are used in order to obtain acceptable current efficiency, and it is necessary to maintain a large excess of chloride ion to retain the cuprous salt in solution as a chloro-complex. Gokhale (9) has given data on the solubility of cuprous chloride in sodium chloride and sodium chloride/hydrochloric acid solutions.

This study forms part of a larger project on the recovery of copper from chalcopyrite concentrates by sulphur activation, cupric chloride leaching, and electrowinning of copper (10, 11). The work on electrowinning has been carried out to confirm and extend the data of previous investigators, some attention being paid to the co-deposition of impurities, and the quality of the cathode copper obtained.

II. EXPERIMENTAL

Preliminary experiments were made using a 250 ml pot cell employing a 10 cm^2 electrode of copper for the cathode, and National Carbon type CS graphite for the anode. A description of this cell has been published previously (10). Briefly, the cell consisted of a tall form beaker, the anode compartment consisting of a glass tube closed at the immersed end with a fine porosity glass sinter, and provision being made for a Luggin capillary to the cathode, thermometer, and agitation by a magnetically driven stirring bar.

A short series of tests was made using a two-compartment diaphragm cell, also described elsewhere (10). Although some useful data were obtained using this cell, it proved difficult to operate over extended time periods, so a three-compartment diaphragm cell was designed. This was made sufficiently large that a few hundred grams of cathode copper could be obtained during a run lasting about 100 h, thus permitting sufficient metal to be obtained to allow evaluation of the product.

The general arrangement of this cell is shown in Fig. 1. It was covered with a lid which carried the electrodes, inlet and outlet tubes for electrolyte circulation, thermometer pocket, and provision for adding fresh feed to both cathode and anode compartments. The cathode was usually of 0.1 mm copper foil, the anodes of type CS graphite $100 \times 80 \times 19$ mm. The initial cathode active area was 158 cm^2, both sides, and the interelectrode distance 45 mm. The diaphragm material was of polypropylene felt (American Felt Co. "Feutron", usually of grade PO-8114). Electrolyte temperature was not controlled, but averaged 35°C.

In operation the fresh feed, and any gelatin solution required, was fed to the central cathode compartment at a rate sufficient to maintain a copper concentration of 25-30 g/l in the catholyte. There was no separate overflow from the cathode compartment, all the solution passing through the diaphragms to the anode compartments. Fresh feed was added also directly to the anode compartments, the amount being regulated to a value which ensured that very nearly all the copper was reoxidized to the cupric state. Since the cell voltage increases sharply as the cuprous ion concentration is reduced, it was possible to devise a control system, responding to the value of the cell voltage, which controlled the rate of addition of fresh feed to the anode compartments. The electrolyte was circulated by use of Gormann-Rupp oscillating pumps, separate pumps being used for anolyte and catholyte, the circulation rate being chosen to give a 1 min turnover time.

There were several changes in the details of the electrical system during the course of the tests, the final form employing a dual constant-current supply, the two negative outputs being connected to the common cathode, while the positive outputs were connected, via a current measuring resistor, each to an anode. This arrangement was adopted to enforce equal current sharing between anodes. The potential drops across the current measuring resistors were added in an analog summing circuit, and the total charge passed calculated by integrating the output of this device. Cell voltage,

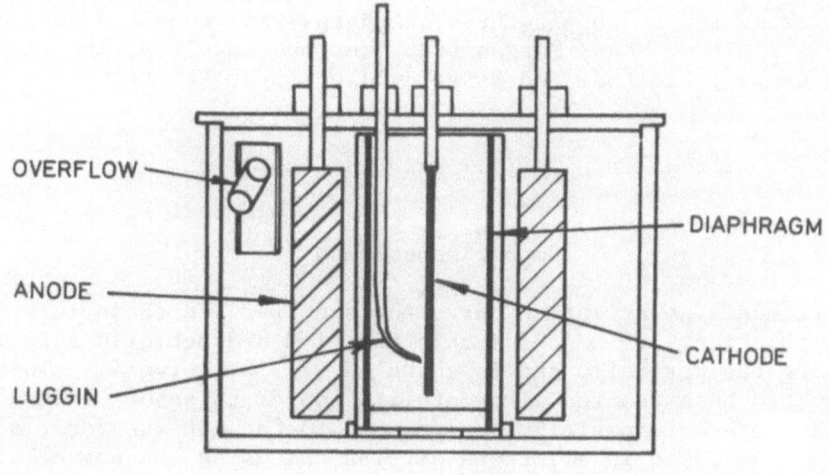

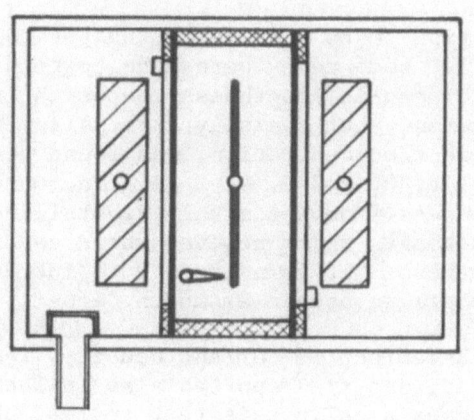

FIG. 1. Diaphragm cell

cathode potential, and electrolyte temperature were measured and recorded on a multi-point recorder.

Most of the work was done using solutions made up from laboratory reagent grade chemicals. However, four electrolyses were made using leach solutions obtained by the sulphur activation cupric chloride leaching process described earlier (11). The chalcopyrite

concentrates used to obtain these leachates were supplied by Mount
Isa Mines Limited, Mount Morgan Ltd, The Mount Lyell Mining and
Railway Company Limited, and Bougainville Copper Pty Ltd.

III.   RESULTS

Copper Deposition

A cathodic polarization curve for a copper electrode in 0.5 M
cuprous chloride, 3.5 M sodium chloride, 1 M hydrochloric acid was
made using the pot cell, and is shown in Fig. 2, curve A.   This
solution had been reacted with metallic copper to ensure that any
cupric ion present initially was reduced to the cuprous form, a
precaution followed in all tests carried out using the pot cell.
The curve shows a Tafel region up to a current density of 500 A/m^2,
the Tafel slope being 0.033 V per decade.   Above 500 A/m^2 some
concentration polarization is evident.

Copper deposition tests made using the pot cell are reported in
Table 1.   Initially, some tests were made to find an addition agent
which would improve deposit smoothness, deposits from the unmodified
solution being coherent, but coarsely crystalline and uneven (10).
Gelatin, in sufficient concentration, was found to be suitable.
As shown by runs U, V, W, and J, at a gelatin concentration of 0.5
g/l the deposit changes from a coarsely crystalline uneven texture
to a finer, even deposit, which however shows severe nodule growth
at the electrode edges.   It is necessary to increase the gelatin
concentration to 2 g/l to obtain a smooth deposit, free of nodules.

In addition the effect of current density, temperature and
acidity were studied, and are reported also in Table 1.   For the
short deposition times used in this test series, about 1 h, good
deposits were obtained with a current density as great as 350 A/m^2.
Increase of temperature (tests P and R) and increase of acidity
(tests S and T) favoured smoother deposits.   The only solution
impurity studied was ferrous iron, present in run X at 30 g/l.
This addition had no affect on deposit quality.   Current efficiency
averaged 93% over the whole series, without showing any noticeable
trend with change in test parameters.   The fall in efficiency below
100% is probably due to small, variable, amounts of cupric ion diffusing into the catholyte from the anode compartment.

The results of tests made using the two-compartment diaphragm
cell are shown in Table 2.   In all cases current density was 150
A/m^2, and the electrolyte contained 3.5 M sodium chloride, 1 M
hydrochloric acid, and 2 g/l gelatin.   These results show a much
wider range in current efficiency than for the pot cell runs, and it

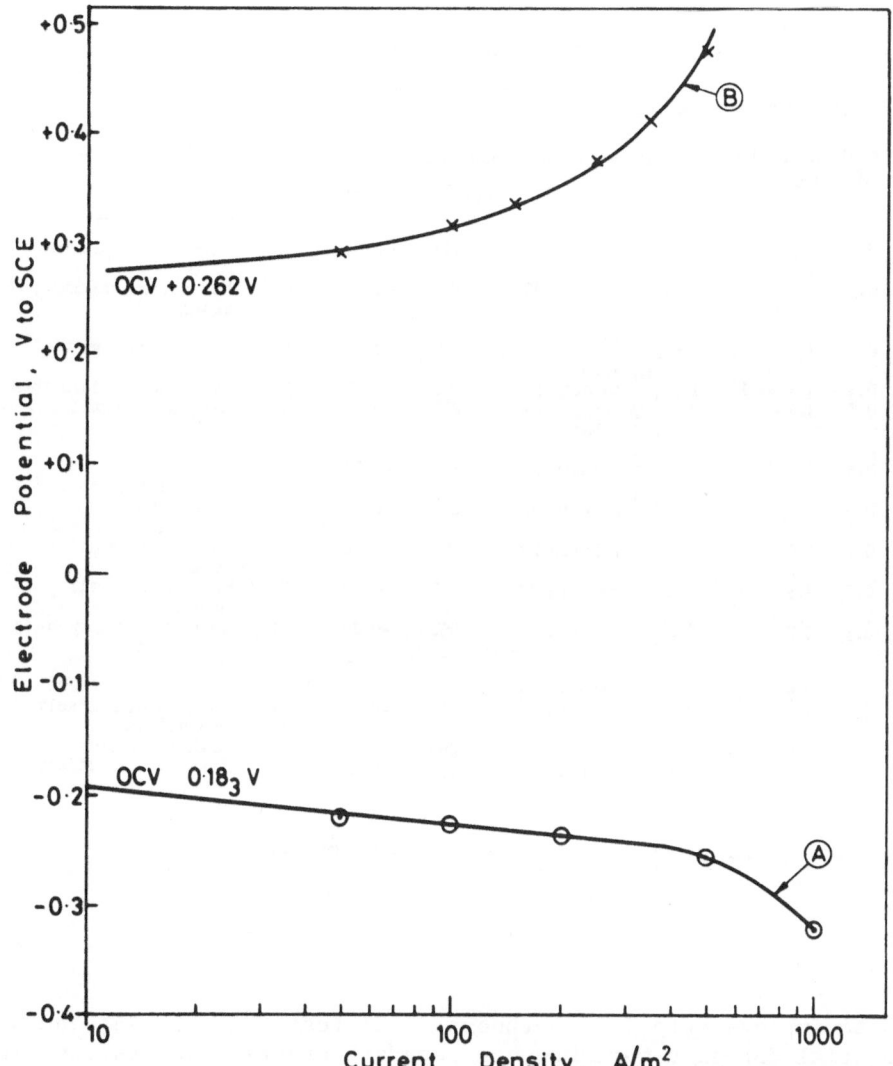

FIG. 2. Polarization curves.   Curve A: cathodic polarization of
10 cm² copper foil electrode in 0.5 M CuCl.   Curve B: anodic
polarization of 10 cm² type CS graphite electrode in 0.5 M CuCl.
The solution contained 3.5 M NaCl and 1 M HCl in both cases.

is possible to relate current efficiency with cupric ion concentration, as shown in Fig. 3, current efficiency decreasing with increase
in cupric concentration.   Increasing cupric concentration seemed

TABLE 1

RESULTS OF ELECTROLYSES USING POT CELL

Test No.	Electrolyte composition					Cathode performance			
	CuCl (M)	NaCl (M)	HCl (M)	Additives	Temp (°C)	Current[+] density (A/m^2)	Pot'l v S.C.E. (V)	Curr. eff.(%)	Quality of deposit
B	0.5	3.5	1	none	25	250	-0.21	94	Coarsely crystalline, uneven
U	0.5	3.5	1	none	22	150	-0.21	95	Coarsely crystalline, uneven
V	0.5	3.5	1	0.5 g l^{-1} gelatin	26	150	-0.25	95	Severe nodule growth at edges
W	0.5	3.5	1	1 g l^{-1} gelatin	26	150	-0.25	93	Some nodules, less than V
X	0.5	3.5	1	2 l^{-1} gelatin, 30 g l^{-1} Fe^{++}	21	250	-0.29	96	Even, a few small nodules
J	0.5	3.5	1	2 g l^{-1} gelatin	25	150	-0.27	90	Even, finely crystalline
K	0.5	3.5	1	2 g l^{-1} gelatin	28	100	-0.25	90	Even, finely crystalline
L	0.5	3.5	1	2 g l^{-1} gelatin	23	200	-0.28	88	Even, a few small nodules
M	0.5	3.5	1	2 g l^{-1} gelatin	28	250	-0.29	97	Even, a few small nodules
N	0.5	3.5	1	2 g l^{-1} gelatin	28	350	-0.29	90	Even, nodule growth bottom edge
P	1.0	3.5	1	2 g l^{-1} gelatin	23	250	-0.25	95	Even, a few nodules
R	0.5	3.5	1	2 g l^{-1} gelatin	70	250	-0.25	93	Even, smooth, finely crystalline
S	0.5	4.5	0.1	2 g l^{-1} gelatin	28	250	-0.25	92	Dendrites at edges
T	0.5	none	4.5	2 g l^{-1} gelatin	28	250	-0.27	94	Even, smooth, finely crystalline

Note: [+] calculated for a one-electron deposition reaction

also to improve deposit smoothness.   In tests 3A, 3B, 4A, and 4B sequential deposits were made on separate starting sheets, and in each case the second deposit (corresponding to a higher catholyte cupric ion concentration) was free of nodules.   The results of tests 5 and 6 confirm this, the deposit made in test 6, where the cupric ion concentration was three times that of test 5, being noticeably the better.   Although not conclusive, these results suggest that the presence of a moderate cupric ion concentration (of the order of 0.1 M in a total copper concentration of 0.65 M) minimizes nodule growth.

Energy consumption averaged 0.45 kWh/kg of copper, the variations observed being due mainly to variations in current efficiency.   There is also a secondary effect due to variations in the cell voltage, which are in turn a function of the cupric/cuprous ratio of the anolyte.

TABLE 2

RESULTS OF ELECTROLYSES USING
TWO-COMPARTMENT DIAPHRAGM CELL

Test No.	Average copper concentration				Cathode Pot'l versus S.C.E. (V)	Curr. eff.+ (%)	Cell		Quality of deposit
	Catholyte		Anolyte				EMF (V)	Energy (kWh/kg)	
	Cu^+ (M)	Cu^{++} (M)	Cu^+ (M)	Cu^{++} (M)					
1	0.61	0.03	0.64	0.075	-0.25	95.0	0.88	0.39	Good deposit, few small nodules on edges
2	0.60	0.050	0.65	0.091	-0.23	90.4	0.87	0.41	Good deposit, slight buildup on edges
3A	0.59	0.051	0.33	0.30	-0.23	83.2	0.91	0.46	Rough & some nodules on side facing anode
3B	0.52.	0.086	0.28	0.36	-0.23	76.3	0.93	0.51	Very smooth deposit
4A	no assays, 40 g 1^{-1} Fe^{++}				-0.24	89.7	0.96	0.45	Some nodules on side facing anode
4B	present as $FeCl_2$				-0.24	84.2	0.97	0.49	Very smooth deposit
5	0.66	0.029	0.50	0.11	-0.24	94.7	0.88	0.39	Good deposit, a few small nodules on edge
6	0.51	0.092	0.44	0.16	-0.24	73.3	0.88	0.51	Very smooth deposit

Notes:

+ calculated for one-electron deposition reaction

All runs made at 150 A/m^2, 40°C

Again, the test made with an appreciable concentration of
ferrous iron in the electrolyte showed no significant change in
cathode quality as compared to the deposits obtained from iron-free
solutions.   Analysis of the acid-washed cathode material showed an
iron content of 6 ppm.

One major purpose of the tests made using the three-compartment
diaphragm cell was to produce sufficient copper to allow evaluation
of the cathode metal.   Therefore tests were run for from 10 to 100
h, giving cathodes from 67 to 378 g.   The heaviest of these deposits
corresponds to a cathode thickness of 5.6 mm, and the thicker
deposits were always quite brittle.   In two trials deposition was
made onto a titanium cathode, in order to examine the feasibility
of preparing starting sheets.   The deposits obtained were coherent,
and could be stripped easily from the titanium sheet, but were of
uneven thickness.   The composition of the feed solutions used is
given in Table 3, and the results of the electrowinning runs in
Table 4.

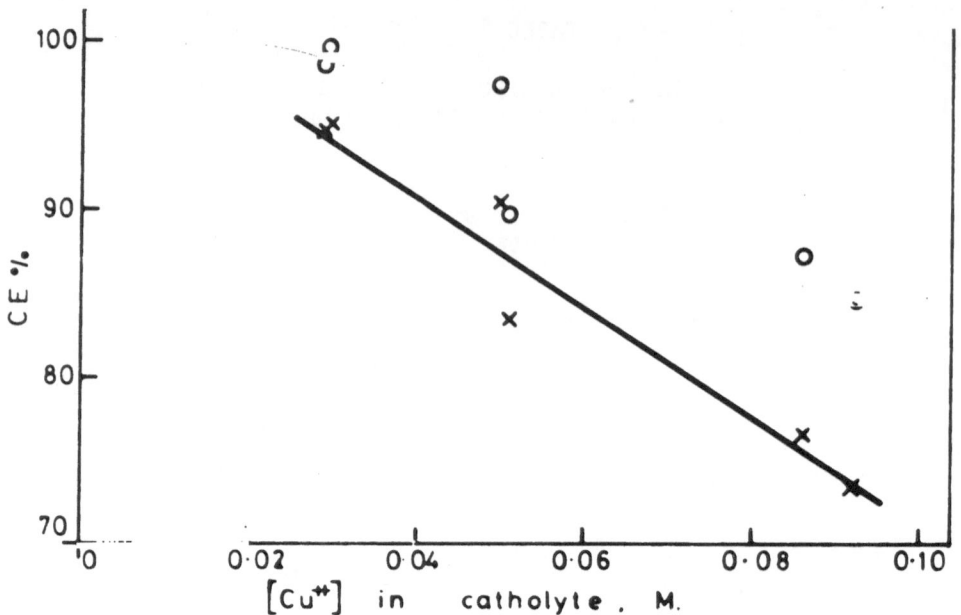

FIG. 3. Current efficiency vs. cupric concentration. ×,
calculated on one-electron basis; o, corrected for Cu⁺⁺ present.

Cathode quality was acceptable at current densities up to 235
A/m^2, except that slightly powdery deposits were obtained when the
catholyte copper concentration fell to a low level (runs 11 and 12),
or in the presence of arsenic and molybdenum together (run 9), this
latter combination causing a very poor deposit.   Current efficiencies were low, 60% to 70% on a one-electron basis;   reasons for
this will be discussed below.

Anodic Oxidation of Cuprous Ions

Fig. 2, curve B, shows an anodic polarization curve for the
oxidation of cuprous ions to cupric, on a type CS graphite electrode,
in a solution containing 0.5 M cuprous chloride, made using the pot
cell described above.   There is no clearly defined Tafel region,
the plot being curved above 50 A/m^2.   This is due in part to
concentration polarization, as the over-potential is decreased by
increasing agitation.   Fig. 4 shows the effect of cuprous ion

TABLE 3

FEED SOLUTIONS TO ELECTROWINNING

Three Compartment Cell

Run No.	Origin of Solution	Fe++ g/l	Pb mg/l	Bi mg/l	Element As mg/l	Mo mg/l	Te mg/l	Ag mg/l
1	Synthetic:- Lab Grade Reagent	Nil	Nil	Nil	Nil	Nil	Nil	Nil
2	"	"	"	"	"	"	"	"
3	"	28.4	"	"	"	"	"	"
4	"	22.3	2200	"	"	"	"	"
5	"	26.8	Nil	12	"	"	"	15
6	"	15.1	"	430	"	"	"	Nil
7	"	26.8	"	Nil	1400	"	"	"
8	"	27.4	"	"	1500	1300	"	"
9	"					ND		
10	Leachate:- Bougainville	33.6	32	2	27	ND	3	10
11	Leachate:- Mt Lyell	32.6	66	1	19	0.3	2	4
12	Leachate:- Mt Morgan	34.8	2	1	24	ND	3	2
13	Leachate:- Mt Isa	30.7	415	74	79	ND	ND	5

Notes:

Nil – not added to synthetic feed solutions

ND – not detectable by analytical method used

Other Elements – Total Cu 1.15 M, Cu+/total Cu = 0.74
H+ 0.6 M, Cl- 6.4 M (Average Figures)

TABLE 4

RESULTS OF ELECTROWINNING RUNS

Three Compartment Cell

Run No.	Catholyte Total Cu M	Ratio Cu+/total	Anolyte Total Cu M	Ratio Cu++/total	Current Density A/m²	Cell EMF volts	Energy Kwhr/kg	Current Efficiency Cathode %	Anode %	Quality of Deposit
1	0.66	0.80	0.93	0.72	144	1.10	0.75	62	72	Smooth massive deposit
2	0.62	0.79	0.96	0.67	235	1.26	0.81	66	75	Smooth, a few nodules at edges
3	0.59	0.80	0.94	0.65	285	1.43	0.87	69	72	Poor, rough with dendritic growths
4	0.59	0.84	0.87	0.87	144	1.17	0.71	69	76	Smooth massive deposit
5	0.69	0.86	0.87	0.87	220	1.40	0.91	65	71	Fair, some large nodules
6	0.60	0.83	0.84	0.88	151	1.1	0.70	67	75	Smooth massive deposit
7	0.61	0.83	0.85	0.86	148	1.10	0.75	62	72	Smooth massive deposit
8	0.84	0.74	0.90	0.72	152	1.0	0.92	44+	49+	Smooth massive deposit
9	0.58	0.84	0.82	0.79	153	1.2	0.8	65	72	Poor, rough & powdery
10	0.60	0.84	0.76	0.98	148	0.95	0.69	58	59	Good deposit, a few nodules
11	0.39	0.80	0.64	0.94	150	1.07	0.71	64	69	Fair, slightly powdery
12	0.40	0.74	0.67	0.95	149	0.94	0.65	61	65	Fair, slightly powdery
13	0.45	0.84	0.76	0.82	151	1.04	0.60	73	82	Good deposit

Notes:

(i) + A more open type of diaphragm was used this run

(ii) Current efficiencies calculated on a one electron basis

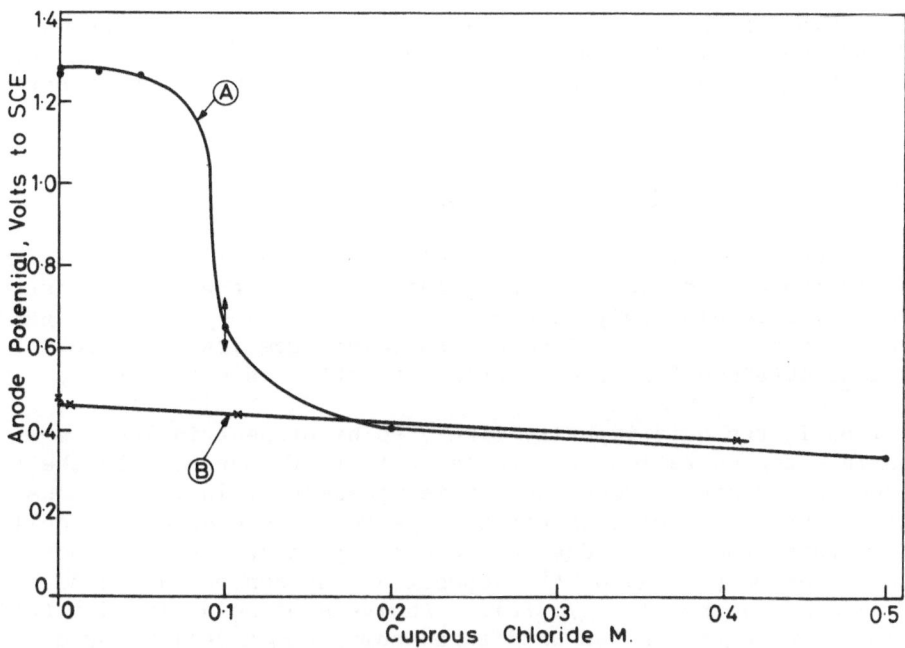

FIG. 4.   Anode potential vs. cuprous concentration.   Curve A:
total copper 0.5 M, cuprous copper concentration as shown, no Fe[++]
present.   Curve B: total copper 0.5 M, cuprous copper concentration as shown, Fe[++] 0.5 M.   In both cases the current density was
150 A/m², temperature approximately 25°C, and the solution contained 3.5 M NaCl and 1 M HCl.

concentration on anode potential, at a constant current density of
150 A/m², for a constant total copper concentration of 0.5 M.   The
effect of adding 0.5 M ferrous ion is shown also (curve B).

     The effect of appreciable ferrous iron in the anolyte was
observed also during operation of the three-compartment cell.
When using electrolytes free of iron (e.g. runs 1-3 of Table 3), if
the feed rate was adjusted such that anolyte cuprous concentration
fell below about 0.1 M - the precise value being agitation dependent
- the cell voltage would rise from approximately 1 V to nearly 2 V,
and the odour of chlorine was noticed.   In the presence of ferrous
iron (0.3-0.6 M, runs 4-13) this did not occur, the cell voltage
remaining steady at about 1.1 V.   If the feed rate is chosen
correctly, nearly complete reoxidation of cuprous copper to cupric

is possible, as shown by runs 10-12.   The one set of anodes was
used throughout all work with the three-compartment cell, a total
of 1122 h operation, and showed no evidence of any attack or deterioration.

Co-deposition of Impurities

The co-deposition of arsenic, bismuth, iron, lead, and silver
were studied by addition of appropriate salts of these elements to
copper chloride electrolytes made up from laboratory grade reagents.
The co-deposition of tellurium was followed from its presence in
solutions obtained from the leaching of activated chalcopyrite.

Iron is the only impurity likely to be present in leach solutions in a concentration comparable to that of copper.   In the
presence of excess cuprous ion, it is necessarily in the ferrous
state.   The effect of iron has been tested with each of the cells
used in this study.   It does not cause any change in the deposition
potential or morphology of the deposit at any concentration up to
the maximum examined (40 g/l Fe).   Analysis of acid rinsed cathode
obtained from both the two- and three-compartment cells showed iron
figures of between 5 and 15 ppm, using a precipitation-polarographic
method.   The figures of 7-67 ppm shown in Table 5 were obtained by
mass spectrometry on melted and swaged material.   The high and
somewhat erratic values so obtained may reflect iron pickup during
handling.   Some confirmation of this is given by the relatively
high iron analyses of samples 1 and 3, which were plated from an
iron-free electrolyte, whereas the lower values were shown by samples
8 and 10 where the electrolyte contained approximately 30 g/l of
iron.

Of the other elements added only silver co-deposited to any
significant extent, except that, when molybdate was added to a
solution containing arsenic, the cathode deposit is heavily contaminated with both (Table 5).   One further element, tellurium, was
found in the cathodes prepared from leach solutions.   It had not
been realized that this would present any problems, but it is
extracted by a cupric chloride leach and co-deposits readily with
the copper.

Samples of cathode material were melted into small ingots under
a borax-fluospar flux, swaged into 4.6 mm rods, and drawn into 2 mm
wire for conductivity tests.   The swaging and wiredrawing did not
present any difficulty, except where some flux inclusions required
some samples to be remelted.   The conductivities obtained were in
the range 97-102% of I.A.C.S., except for the material deposited
from the arsenic-molybdenum containing solution (run 9) and the
sample from run 10, which was accidentally contaminated with nickel
during the melting.   The detailed results are shown in Table 5.

TABLE 5

COMPOSITION OF CATHODE COPPER

| Element | Present in Run No, ppm. | | | | | | | | | |
	1	3	6	7	8	9	10	11	12	13
Bi	<0.1	<0.1	0.1	1.5	<0.1	1.0	<0.1	1.0	0.1	0.5
Pb	<0.1	0.6	2.7	5.0	1.3	1.0	1.6	2.3	1.7	2.0
Te	ND	ND	ND	ND	ND	ND	41	27	36	ND
Sb	0.2	<0.1	0.1	0.1	0.1	10	5.0	5.4	0.4	0.9
Sn	<0.1	<0.1	0.3	3.2	0.7	<0.1	<0.1	<0.1	<0.1	<0.1
Ag	5.5	196	400	21	21	200	170	59	24	52
Se	0.4	0.6	<0.1	0.3	<0.1	0.7	<0.1	<0.1	<0.1	<0.1
As	<0.1	<0.1	<0.1	0.2	9.1	8300	1.8	8.0	2.6	1.4
Zr	<2	<2	2	4	2	<2	<2	2	4	5
Ni	0.8	0.7	0.3	35	0.1	0.1	+	0.8	1.0	0.7
Co	ND	ND	ND	ND	ND	ND	+	ND	ND	ND
Fe	17	21	42	38	66	12	7	24	30	67
Mn	ND	ND	ND	ND	ND	ND	ND	ND	ND	ND
Cr	ND	ND	ND	ND	ND	ND	ND	ND	ND	ND
P	<1	<1	<1	<1	<1	<1	<1	<1	<1	<1
Na	<1	<1	<1	<1	<1	<1	<1	<1	<1	<1
O	320	220	130	50	260	400	670	390	330	1350
S	3	37	75	57	97	3	26	25	72	25
Cl (1)	–	–	145	54	117	1300	57	160	295	164
Cl (2)	5.6	4.5	0.6	1.1	0.9	2.4	3.6	3.6	1.4	2.1

Conductivity*

	1	3	6	7	8	9	10	11	12	13
	101.8	101.8	99.3	100.9	98.6	30.3	+	100.1	97.0	98.2

Origin	Synthetic Solutions prepared from lab. reagent chemicals	Boug.	Mt Lyell	Mt Morgan	Mt Isa

Notes:

*   % I.C.A.S.

+   Contaminated with nickel during melting, so these values omitted

(1)   Before melting

(2)   After melting. Runs 6, 9, 11, 12, 13 melted twice

ND   Not detectable by Mass Spectrometry

IV. DISCUSSION

The standard potential for the Cu^+/Cu equilibrium is -0.11 V to the saturated calomel electrode (SCE), while that of the Cu^{2+}/Cu^+ equilibrium is +0.29 V vs. SCE (12), giving an E.M.F. of 0.40 V for the reaction

$$2CuCl \rightarrow CuCl_2 + Cu \hspace{4cm} (1)$$

where reactants and products are in their standard state. An open circuit potential of -0.18 V vs. SCE has been measured for the Cu^+/Cu system in a solution containing 0.5 M cuprous chloride, 3.5 M sodium chloride, and 1 M hydrochloric acid, this potential moving in the positive direction with increase of cupric ion concentration, and is -0.15 V vs. SCE for a cupric/total copper ratio of 0.1. The open circuit cell voltage with this ratio is approximately 0.46 V; with an anolyte having a cupric/total copper ratio of 0.9-0.98 the cell open circuit voltage is 0.56 V.

At a current density of 150 A/m^2, with a copper concentration of 0.6 M in each compartment, cupric/total copper being 0.1 in the catholyte and > 0.9 in the anolyte, the cathode potential is between -0.21 V vs. SCE (gelatin absent) and -0.26 V (gelatin 2 g/1). The cell voltage may vary however from 1.1 to approximately 2 V, due to variation in anode potential. Curve A of Fig. 4 shows that at low cuprous concentrations the anode potential can increase to 1.28 V vs. SCE. This sharp jump in potential is highly dependent upon agitation, indicating that the rise is due to concentration polarization of the anode with respect to cuprous oxidation, which forces the anode potential to that of chlorine evolution (E_o = +1.11 V vs. SCE). The presence of ferrous ions was stated above as eliminating this potential jump (see curve B, Fig. 4), and this is explicable if the anode potential rises, when concentration polarization towards cuprous oxidation occurs, to the ferrous/ferric oxidation potential rather than to chlorine evolution. The ferric ions so produced will react rapidly with cuprous ions in the bulk of the solution, regenerating the ferrous, that is, the ferrous salt is acting as a redox buffer. When working with the three-compartment cell at a current density of 150 A/m^2 the cell voltage was typically 1.05 V with the cupric/total copper ratios stated above, and 0.3-0.6 M ferrous iron. With a cathode potential of -0.26 V and a measured ohmic drop of 0.34 V, the anode potential must be +0.45 V vs. SCE. This is quite comparable to the standard potential for Fe^{+++}/Fe^{++} (0.50 V vs. SCE), or the value measured at 150 A/m^2 during pot cell tests, using a 0.4 M ferrous 0.1 M ferric solution, namely +0.49 V vs. SCE.

The results for cathode current efficiency obtained using the two-compartment diaphragm cell showed that the most significant factor reducing current efficiency (one-electron basis) was the presence of cupric ion. Part of the deficit is due to calculation of the efficiency on a one-electron reduction basis, whereas the reduction of cupric ion, whether direct or indirect (13), requires two electrons. Assuming that the fraction of the copper deposited from the cupric state is equal to the cupric/total copper ratio of the catholyte, a second current efficiency value can be calculated which should be more representative of the actual distribution of charge. The values so calculated are shown as the upper line of Fig. 3, and it is clear that even with this correction there is still a decrease in current efficiency with cupric concentration. One possible explanation of this is that the rate of dissolution of copper in cupric chloride is much more rapid than its redeposition from the cuprous solution so formed, but there are not sufficient data available in the present series to test this suggestion.

When using the three-compartment cell, the average current efficiency (one-electron basis) was 65%, while the average anode efficiency, based on a Cu^+/Cu^{++} conversion, was 72%. This excludes the data of run 8, which was made with a very porous diaphragm. These low values are caused by the method of cell operation, the fresh feed being added to the cathode compartment containing 20 g/l of cupric ion, whereas the average catholyte composition is 6.5 g/l of cupric ion. Thus, an appreciable amount of charge is required to reduce cupric copper to the cuprous state. Unlike the two-compartment cell where there were separate anolyte and catholyte streams, there was no separate catholyte overflow in these runs, a volume of catholyte equal to the volume of feed solution to the cathode compartment being displaced to the anode compartment, where it is reoxidized to the cupric state. The charge involved in reducing 20 g/l cupric to 6.5 g/l in a volume of solution equal to the feed, and its subsequent reoxidation, appears as a loss in faradaic efficiency. If this quantity of charge is calculated and used to correct the apparent current efficiency figures quoted, 90% of the charge to the cathode can be accounted for, on average, as can 96% of that to the anode. If a further correction is made to the cathode current efficiency for the effect of the two-electron reduction of cupric ions, as was done above, 100% ± 7% of the charge to the cathode can be accounted for. This is somewhat higher than was obtained with the two-compartment cell, where at a similar cupric concentration only about 90% of the charge was accounted. The difference may be more apparent than real in view of the poor precision of the results for the three-compartment cell. However, the agitation level was higher in the two-compartment cell, which might enhance the redissolution of the cathode copper by cupric chloride.

The average power consumption at a current density of 150 A/m^2 was 0.72 kWh/kg, while at 230 A/m^2 it was 0.86 kWh/kg, when using

the three-compartment cell.   These values are probably representative of what could be obtained in practice, and are about one-third
of the consumption achieved in electrowinning copper from a sulphate
electrolyte.

The most significant impurities co-deposited with copper are
tellurium and silver.   If powder copper is available, as for
example if it were produced in the process of stripping copper from
the catholyte bleed (11), it could be used to cement silver from
the electrolyte prior to electrolysis (6, 14).   Removal of tellurium, if it were present in excessive quantity, would probably
require a separate fire refining stage, when the cathode material
was melted prior to casting (15).   Chlorine, present at appreciable
levels in the cathode material, is volatilized during melting.
Finally, it has been observed that the presence of molybdenum in
solution brings about the co-deposition of arsenic probably via the
formation of an arsenomolybdate complex.   Fortunately, molybdenum
is not extracted from chalcopyrite concentrates by a cupric chloride
leach (11), and therefore causes no difficulty in practice.

ACKNOWLEDGEMENTS

The assistance of Mr. R. Mitchell of Copper Refineries Pty Ltd
in the analysis of copper samples, and Mr. A. Westrope of BHP
Research Laboratories in the preparation of these samples is
acknowledged, as is the supply of chalcopyrite concentrate samples
by The Mount Lyell Mining and Railway Company Limited, Mount Morgan
Ltd, Bougainville Copper Pty Ltd, and Mount Isa Mines Limited.

REFERENCES

1   P.R. Kruesi, E.S. Allen and J.L. Lake, CIM Bull., 66 (1973) 81.
2   R.S. Olsen, D.H. Yee, G.L. Hundley, R.E. Mussler and F.E. Block,
     Trans. AIME, 254 (1973) 301.
3   M.M. Wong, F.P. Haver and R.D. Baker, 104th AIME Meeting, Feb.
     1975, New York.
4   W.L. Chambers and R.W. Chambers, US Patent 3,692,647 (1972).
5   E.A. Ashcroft, Trans. Electrochem. Soc., 18 (1933) 23.
6   W.C. Hazen, US Patent 3,767,543 (1973).
7   G.C. Mitter, B.K. Bose, S.G. Dighe, Y.W. Gokhale and B.P.
     Choudhury, J. Sci. Ind. Res. (India), D20 (1961) 114.
8   S.D. Gokhale, J. Sci. Ind. Res. (India), B10 (1951) 316.
9   S.D. Gokhale, J. Univ. Bombay, Sect. A, 20(3) (1951) 53.
10   K.J. Cathro, Australas. Inst. Mining Met. Proc., No. 252 (1974) 1.
11   K.J. Cathro, International Symposium on Copper Extraction and
     Refining, Las Vegas, Feb. 1976.
12   W.M. Latimer, Oxidation Potentials, 2nd edn, Prentice Hall Inc.,
     New Jersey, 1952, p. 186.

13   R. Astakhova and B. Krasikov, Russ. J. Appl. Chem., 44 (1971)
     356.
14   E.M. Atadan, US Patent 3,776,826 (1973).
15   H.J. Miller, in A. Butts (Ed.), Copper, the Science and
     Technology of the Metal, its Alloys and Compounds (ACS Mono-
     graph), Reinhold, New York, 1954, p. 297.

IMPROVEMENTS TO THE HALL-HEROULT PROCESS FOR ALUMINIUM ELECTROWINNING

Kai Grjotheim, Conrad Krohn and Harald Øye

Institute of Chemistry, University of Oslo, The Foundation
of Scientific and Industrial Research at the University
of Trondheim and Laboratory of Inorganic Chemistry, The
Norwegian Institute of Technology, University of Trondheim

The Hall-Heroult process has several disadvantages, the main
ones listed in order of importance are:

(i)   the energy efficiency is relatively low ($\approx$40%);

(ii)   pure and expensive raw materials have to be used,
        especially the aluminium oxide and anode carbon;

(iii)   the cost of an alumina reduction plant is high, due to
        the large number of relatively small production units;

(iv)   the frequent discontinuities in the cell operation that
        occur cause current losses as well as being costly
        because of the corrective labour and lost time involved;

(v)   the electrolytic cell life is extremely sensitive to
        operational faults;

(vi)   cell emissions give rise to environmental pollution
        problems.

One of the limiting factors in the development of new industry today
is the availability of suitable energy. The present Hall-Heroult
process is extremely demanding on electrical energy resources and it
is therefore appropriate to discuss the energy efficiency of the
process as a basis for considerations of future technological
developments.

ENERGY REQUIREMENTS

Alumina can be extracted from bauxite by several methods, the most common of which is the Bayer process.

The Gibbs free energy, ΔG^o, for the decomposition of one mole of alumina to aluminium and oxygen amounts to 1280 kJ per mole of alumina at 1000^oC (1). However, the total energy requirement is the enthalpy, ΔH, and as $\Delta S^o > 0$, then $\Delta H^o > \Delta G^o$. The value of ΔH^o is 1690 kJ per mole of alumina at the same temperature. This corresponds to a theoretical energy requirement of 8.7 kWh per kg Al produced from alumina at 1000^oC. The actual energy needed to produce 1 kg of aluminium from alumina at room temperature will include the energy to heat the alumina. Extracting the alumina from a bauxite or a clay demands even more energy, as the oxide is chemically bound to other elements in the ore.

In principle, reduction may be achieved either by the use of a chemical reductant, or by electrolysis. The Hall-Heroult method combines these two principles, as the carbon anode acts as a co-reductant during electrolysis.

The dominant cell reaction of the Hall-Heroult process is (2):

$$Al_2O_3 \text{ (dissolved)} + \frac{3}{2} C(s) = 2Al(1) + \frac{3}{2} CO_2(g) \qquad (1)$$

Assuming the electrolyte to be saturated with oxide, and the pressure of carbon dioxide is 1 atm, the enthalpy of this reaction is around 1100 kJ per g atom Al at 1000^oC. Thus the theoretical energy requirement to produce 1 kg of Al at this temperature is 5.6 kWh per kg of Al, as compared with 8.7 kWh per kg of Al for decomposition of the oxide without carbon as a reducing agent. Electrical energy is saved at the expense of carbon.

The current efficiency of the process usually ranges from 85 to 92%. Most of the loss in current efficiency (CE) is due to the reaction:

$$2 \ Al(1) + 3 \ CO_2(g) = Al_2O_3 \text{ (dissolved)} + 3 \ CO(g) \qquad (2)$$

whereby a fraction (1-x) of the metal formed is re-oxidized. To express the energy requirement for the real process, taking the reduced current efficiency into account, it is convenient to consider the thermodynamic cycle (3) given in Figure 1. ΔH^o (tot) represents the enthalpy change for the transformation of the reactants at 298K to the reaction products at T, K. ΔH^o (reac,T) denotes the enthalpy change for the reduction at the specified reaction temperature T. H^o is the enthalpy of the various components at the temperature stated. Thus, it is seen that the enthalpy requirement per g atom

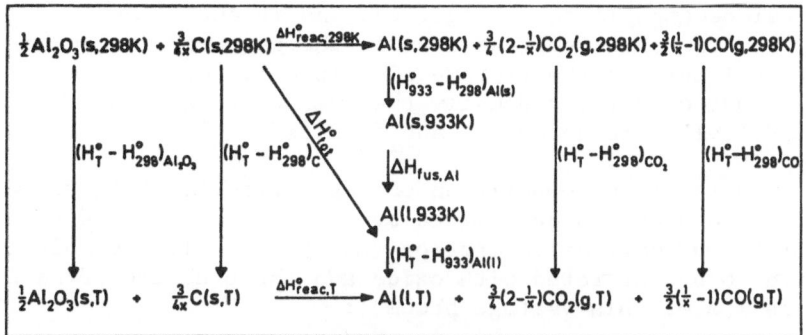

Fig. 1   Cycle for the Hall-Heroult process.

of aluminum is:

$$\Delta H^{o}_{tot} = \Delta H^{o}_{reac,T} + \frac{1}{2}(H^{o}_{T} - H^{o}_{298})_{Al_2O_3} + \frac{3}{4x}(H^{o}_{T} - H^{o}_{298})_{C}$$

$$= \Delta H^{o}_{reac,298} + (H^{o}_{933} - H^{o}_{298})_{Al(s)} + \Delta H_{fus,Al} +$$

$$(H^{o}_{T} - H^{o}_{933})_{Al(l)} + \frac{3}{4}(2 - \frac{1}{x})(H^{o}_{T} - H^{o}_{298})_{CO_2(g)} +$$

$$\frac{3}{2}(\frac{1}{x} - 1)(H^{o}_{T} - H^{o}_{298})_{CO(g)} \qquad (3)$$

Schmidt-Hatting (4) included a term containing the heat of solution of Al_2O_3 in molten cryolite. As seen from the cycle, however, this term does not contribute to the net energy requirement.

Inserting the various energy terms (3) we obtain the total minimum energy requirement as a function of the current efficiency:

$$\Delta H^{o}_{tot} = \frac{139.18}{x} + 477.24 \text{ kJ g atom Al}^{-1}$$

$$= \frac{1.43}{x} + 4.91 \text{ kWh kg Al}^{-1} \qquad (4)$$

In practice, the energy demand is substantially higher, usually between 12 - 18 kWh per kg of Al. Extra energy must be supplied to maintain the high temperature of the cell, heat the effluent cell gases, and provide energy for the other side reactions and the anode effect. The real cell voltage, correspondingly, is a sum of the

theoretical decomposition voltage, the anodic and cathodic overvoltages and the ohmic voltage drop in the bath and the busbars.
An additional term is the average rise in cell voltage due to the
periodic anode effects. Usually the average total operational cell
voltage, V (cell), is between 4 and 5 volts.

The reversible decomposition voltage, V(rev), of Al_2O_3 in the
Hall-Heroult process is determined by the change in Gibbs free
energy of the primary cell reaction (Eq. [1]). If the molten bath
is assumed to be saturated with oxide and the pressure of carbon
dioxide is 1 atm, this voltage becomes:

$$V_{rev} = 1.307 - \frac{0.125}{x} \text{ volts} \qquad (5)$$

at $977^{\circ}C$ (1250 K) (3). Actual measurements of this voltage in
operating cells, however, have given higher values (about 1.7 V) due
to an anodic overpotential of about 0.5 to 0.6 Volts (2).

The *electrical* energy required for the electrolysis is:

$$\Delta G = n \, F \, V_{rev} \qquad (6)$$

where ΔG is the change in Gibbs free energy of the reduction process.
The *total* energy requirement, however, as previously mentioned, is
ΔH, which is larger than ΔG. Hence, the cell would be cooled if
heat were not supplied externally. When the reaction is to take
place *isothermally*, the electrical energy will be ΔH, and not ΔG.

The *total energy input* per mass unit of aluminium produced is:

$$E_{input} = \frac{1}{x} n \, F \, V_{cell} \qquad (7)$$

Only about one quarter of this is expended as electrical work ΔG.
The larger part of it results in ohmic heat due to the electrical
resistance of the cell.

The energy balance is then:

$$E_{input} = \Delta G + E_{ohmic} = \Delta H_{tot} - T \, \Delta S + E_{ohmic} \qquad (8)$$

where E(ohmic) is the overpotential and the ohmic energy developed.
The energy difference (Eq. [9]) is dissipated as heat to the
surroundings in order to keep the cell temperature constant.

$$E_{input} - \Delta H_{tot} = E_{ohmic} - T \, \Delta S = Q \qquad (9)$$

We now define the energy efficiency (EE) as the sum of the
enthalpy of reaction, $\Delta H(reac,T)$, at the temperature T, and the
energy needed to heat the raw materials, $\Sigma_i (H^{o}_T - H^{o}_{298})_i$, divided
by the energy input to the cell:

$$EE = \frac{\Delta H^o_{tot}}{E_{input}} = \frac{\Delta H_{reac,T} + \Sigma_i (H^o_T - H^o_{298})_i}{\frac{1}{x} \, n \, F \, V_{cell}} \tag{10}$$

$$EE = x \left(\frac{H^o_{react,T} + \frac{1}{2} (H^o_T - H^o_{298})_{Al_2O_3} + \frac{3}{4x} (H^o_T - H^o_{298})_C}{n \, F \, V_{cell}} \right) \tag{11}$$

Inserting the previously calculated sum in the numerator and the numerical values for n and F, we have the following expressions (3) for the energy efficiency as a function of the current efficiency and the cell voltage:

$$EE = \frac{0.481 + 1.65 \, x}{V_{cell}} \tag{12}$$

Average values are 90% (x = 0.9) for the current efficiency and 4.5 V for the cell voltage. In that case the energy efficiency is around 44%.

FACTORS GOVERNING THE ENERGY EFFICIENCY

The rather low value of the EE can be improved by either increasing x (the CE) or decreasing the cell voltage. The CE in industrial cells is, in principle, governed by the following factors:

(i)   metal re-oxidation by the anode gas;
(ii)   metal re-oxidation by air;
(iii)   anodic oxidation of dissolved metal;
(iv)   evaporation of metal-containing species;
(v)   electronic conduction losses through the electrolyte, by short-circuits between the anode and cathode, and through the parallel current path of the top crust;
(vi)   metal losses into the cell lining, carbide formation, etc.

The cell voltage is determined by the decomposition potential, the ohmic voltage drops and the over-voltages, the latter two being dependent on current density.

Some of the factors that influence the EE are determined by cell construction, while others may be varied during the operation of the cell.

The following parameters are primarily determined by the cell design and construction:

(i)   the resistance in the external conductors and in the
         busbar system;
(ii)   the resistance in the carbon anodes, which varies with
         the size and with the nature of the connection to the
         holders; the resistance in the prebaked anode decreases
         as the anode is consumed during operation;
(iii)   the resistance in the carbon cathode; this resistance
         changes with the age of the cell, but not in a predictable
         manner;
(iv)   the arrangement of the cell superstructure such as the
         position of the current leads, the thermal insulation,
         the covering of the cell, and the equipment for collect-
         ing the anode gases;

Other parameters that vary during the cell operation include:

(i)   inter-electrode distance;
(ii)   composition of the bath;
(iii)   amount and distribution of aluminium oxide on the crust;
(iv)   crust breaking cycle;
(v)   metal tapping cycle;
(vi)   depth of immersion of the anode in the bath;
(vii)   thickness of the metal layer;
(viii)   current density and total current;
(ix)   magnetic field from the current leads.

The energy balance shows that more than 50% of the energy is
dissipated to the surroundings. Part of this is removed as heat
in the tapped metal and the effluent gases, but a major part is
radiated from the cell through the crust and cell casing.

It is very essential to have a cell in thermal balance in order
to obtain smooth operation and a long life time. The temperature
influences several parameters, e.g. the rate of the back reactions
and the electrolyte resistance. One attempts to keep the temperature in the cell at $970 - 975°C$, which is $10 - 15°C$ above the melting point of the electrolyte.

Temperature fluctuations will occur during cell operation, with
more important disturbances occurring during feeding of aluminium
oxide, tapping of the metal, and anode adjustments. Changing of
anodes, addition of salts other than aluminium oxide, (e.g. sodium,
aluminium, lithium or calcium fluoride) and the occurrence of the
anode effects also disturb cell temperature.

Even with proper cell operation these discontinuities may cause
fairly large temperature fluctuations. During the anode effect
temperatures higher than $1100°C$ have been recorded. Feeding of
the aluminium oxide has caused temperature drops larger than $30°C$,
thus demonstrating the importance of continuously feeding the cells.

POSSIBILITIES FOR IMPROVEMENTS

Large improvements have been made during the 90 years that the Hall-Heroult process has been operated, but there is room for further improvement.   Although the process may appear to be simple theoretically, it is really extremely complex and the basic requisite for its improvement still relies on a fundamental knowledge of the theoretical background.

Areas of possible improvements are the following:

(i)   improved physical properties of the electrolyte;
(ii)   overvoltage reduction by choice of anode material;
(iii)   factors prolonging cell life;
(iv)   cell construction;
(v)   cell operation.

Some possible developments in these fields will be discussed in the following sections.

Physical Properties of the Electrolyte

The properties to be considered are: electrical conductivity, wettability, surface and interfacial tensions, density, solidus temperature, and aluminium solubility in the electrolyte.

A metal salt would be useful as an additive if its price were moderate, and the cation less noble than aluminium in the fluoride-oxide environment, provided it increased the conductivity and lowered the density of the salt melt.   The surface properties should be altered to destabilize small aluminium droplets in the melt and then enhance the electrolyte wetting of the anode.   It should further not give rise to operating difficulties, or produce unwanted by-products during electrolysis.

In a previous paper (5) the requirements for an "ideal salt additive" together with a discussion of a number of possibilities were presented.   Only the results of experiments on industrial cells will be mentioned here.

Recently (6) the effect of AlF_3 and CaF_2 additions on the current efficiency have been studied experimentally in commercial cells at Ardal and Sunndal Verk.   Figure 2 gives statistical results for AlF_3 addition.   The results confirm the beneficial effect of this addition.   An increase from 5 to 7 wt % AlF_3 was found to give an increase in the current efficiency from 88 to 90% in the particular cells studied.   A comparison with literature data confirms that this observed effect of AlF_3 addition is representative for large cells.

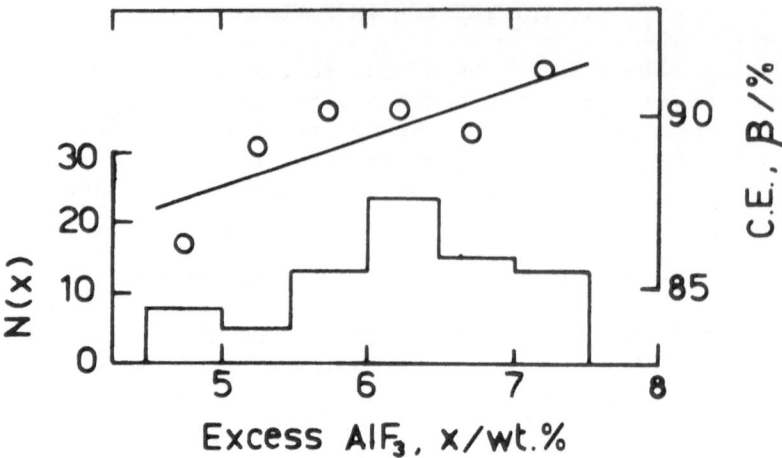

Fig. 2   The Effect of AlF$_3$ content on the CE of 150kA pre-baked
          anode cells.

Anode Overvoltage Reduction

        The overvoltage implies a primary formation of carbon dioxide
at the anode, even though carbon monoxide should be formed at the
temperature of the electrolysis from thermodynamic data.   It has
been shown that the overvoltage increases with increasing graphitiza-
tion of the anode.   Theoretically, it should be possible to lower
the overvoltage by changes in the anode material (catalysts, inhib-
itors, structural treatment, etc.), but hitherto no dramatic
improvements have been revealed.   The same also applies to progress
on the permanent anode concept.   Although the latter would increase
the amount of electrical energy needed, a successful development
would result in simplifications in construction and operation of
the cells.   A lot of work on this has been done and is still going
on at various research laboratories.   Possibilities of lowering
the anodic overvoltage are also under consideration in these efforts.

Factors Prolonging Cell Life

One of the limiting factors governing the cell life has been the chemical reactions which occur in cracks and crevices in the carbon lining. One such reaction that shortens cell life is the formation of Al_4C_3 in the cells. Hard carbide particles in the liquid metal also present a severe problem in some applications of the metal, such as rolling thin sheet for printing purposes. Carbide formation mainly occurs at the bottom of the cells, as the side walls are usually covered with a ledge of solidified electrolyte.

From a thermodynamic point of view, Al_4C_3 should readily be formed from the elements at $1000^\circ C$ by the reaction:

$$4 \text{ Al}(1) + 3 \text{ C}(s) = 4 \text{ Al}_4\text{C}_3(s) \tag{13}$$

since the standard Gibbs free energy change is -170 kJ. However, in spite of this it is known that liquid aluminium and graphite can remain in contact without any reaction occurring. This has been attributed to the formation of an oxide layer on the aluminium surface, making the reaction rate dependent on diffusion through the aluminium oxide layer (7). When this oxide layer dissolves in molten cryolite, aluminium and carbon react more readily (8).

Recently, it has been found (9) that the carbide formation in alumina - cryolite melts at low alumina concentration is strongly dependent on the alumina content. The results are shown in Figure 3. The important feature of the results is that in pure cryolite about 3 times as much Al_4C_3 has been formed compared to alumina - cryolite melts containing more than 4 wt % Al_2O_3. In the concentration range 0-4 wt % Al_2O_3 the carbide formation decreases with the alumina content. These experiments give an additional reason for preventing the cells running too low in alumina content. There also are indications that minor amounts of some compounds strongly influence the carbide formation.

With a suitable treatment of the cathode carbon some improvement should be possible by minimizing side reactions that limit cell life. Also the use of new and better cathode materials may be of importance in the future.

Cell Design

As most development towards optimum cell design has to be done on the industrial scale, only a few comments will be presented here. One of the more surprising features is the range of variations in cell construction in the industry. With our present knowledge of how the strong magnetic fields influence the convection rate and

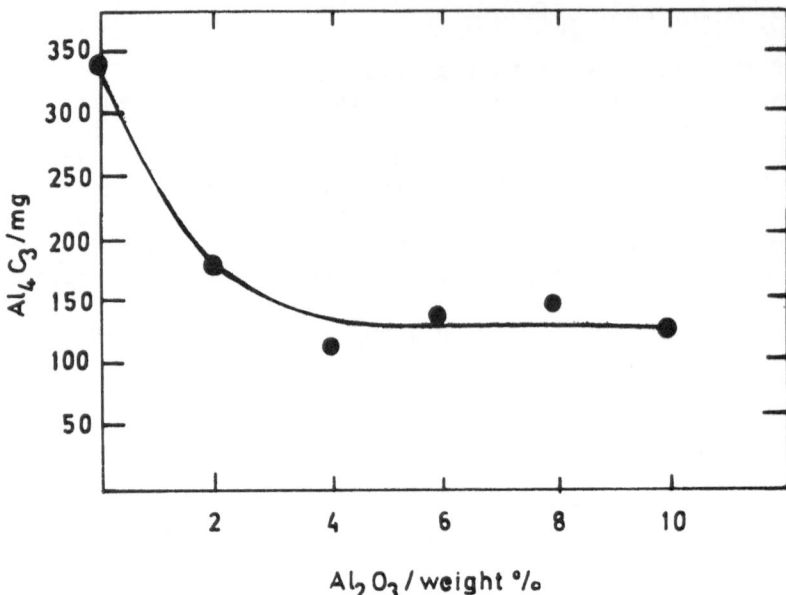

Fig. 3   The effect of the concentration of alumina in cryolite on
         the reaction between liquid aluminium and graphite (AGR –
         Union Carbide) at 1050°C.   Reaction time 4 h.

pattern in cells, causing tilting of the metal surface, it is to be
expected that this factor would be better accomodated in cell design.

As mentioned, not even carbon is entirely corrosion resistant
during electrolysis.   Therefore, the inside of the carbon lining
of the cell should be covered with a ledge of frozen electrolyte
during normal operations.   With the large temperature fluctuations
in the cell, the configuration of this ledge will vary during
operation.   The geometrical form and size of this ledge heavily
influences the convection pattern, and thereby the CE of the cell.

The convection pattern also influences the dissolution and
transport of alumina in the cell.   Usually the convection rate is
lowest in the centre of the cell.   This should be taken into
account in deciding the period of the cycle and the positioning of
the automatic oxide feeding devices.

The frozen electrolyte ledge is also effective in directing
the electrical current through the bottom of the cell, and not
through the walls.   If the ledge, because of the conditions of the
thermal insulation, should creep under the metal and partly over the

carbon cathode, this will change the current distribution pattern and its magnetic field.   This in turn will influence the operation of the cell in a problematicable way.

In our opinion, innovations are still needed for more positive improvements in the design of the cell.

Cell Operation

It is important in cell operation to know the optimum combination of cell parameters for a specific type of cell.   An economic optimization of the process would also require a knowledge of manual labour requirements as a function of cell size, investment costs, etc.

The CE of a large number of 150 kA pre-baked anode cells has been studied experimentally.   By carrying out multiregression analysis, the contribution of some of the operating parameters have been evaluated.

The CE can be described by the following equation (10):

$$CE(\%) = -0.1388 \, t + 0.59 \, X_{AlF_3} + 58.9 \sin(3h) - 0.132 \, A + 163.7 \tag{14}$$

where t is the bath temperature (^{o}C), X_{AlF_3} excess AlF_3 (wt %), h height of metal pad (cm), and A cell age (months). (Multiple correlation coefficient = 0.70).

In more modern plants some of the cell operations such as feeding of aluminium oxide, anode adjustments (voltage control) and termination of the anode effect have been automated.   However, in spite of the increased automation, operator attention is still required for anode changing and adjustment, metal tapping, bath additions and extinguishing anode effects when not automatically terminated.

Today, operators have to supervise more cells than in the past making the newer cells more sensitive to operational faults. Therefore, aluminium electrolysis is still recognized as one of the most operator-dependent metallurgical processes, requiring both high skill and constant attention by conscientious operators.

THE FUTURE OF THE HALL-HEROULT PROCESS

The establishment of an alumina reduction plant on a new site represents a far-ranging and long-term commitment of a country's

resources.   Before making such a decision, it is therefore appropriate to consider the possibility that future technological development may reduce the expected future benefits from this venture.

Over the past three decades, electric power requirements for
smelting aluminium have been cut by nearly half.   Prior to World
War II, 22 kWh were needed to produce one kg of aluminium.   Today,
the average smelters use 16 kWh and the most efficient ones are
down to 13-14 kWh.   At the same time the average life time of
cells has also increased from about 2-3 years to more than double
that figure.   It seems reasonable to assume that research could
further improve both the cell life and the energy efficiency.

The search for new, less energy consuming processes for primary
aluminium production has in recent years been intensified (11).
The savings indicated by possible new processes would have to be
sufficient to compensate for the cost of the new plants as well as
for the loss of capital due to the replacement of old plants.
Therefore, even if new process development accelerates, one may
draw some careful conclusions from the fact that the Hall-Heroult
process presently dominates.   Some of the new processes might
attain industrial importance in the future, but it is unlikely that
any will be able to compete with the Hall-Heroult process in the
coming 15-20 years.

ACKNOWLEDGEMENTS

Extensive financial support from the Royal Norwegian Research
Council for Scientific and Industrial Research (Grant B 581) is
gratefully acknowledged.   Thanks are further extended to A/S Ardal
and Sunndal Verk for financial support and permission to do research
on commercial cells in their plants.

REFERENCES

(1) JANAF Thermochemical Data, compiled and calculated by the Dow
     Chemical Company, Thermal Laboratory, Midland, Michigan, 1973.
(2) K. Grjotheim, C. Krohn, M. Malinovsky, K. Matiasovsky and
     J. Thonstad: "Aluminium Electrolysis", Al-Verlag, Düsseldorf
     1976, at press.
(3) D. Bratland, K. Grjotheim and C. Krohn: in Light Metals 1976,
     AIME, New York, 1976, at press.
(4) W. Schmidt-Hatting: Erzmetall. 21 (1968) 317.
(5) K. Grjotheim, J.L. Holm, C. Krohn and K. Matiasovsky: Svensk
     Kemisk Tidskr. 78 (1966) 10.
(6) B. Berge, K. Grjotheim, C. Krohn, R. Naeumann and K. Tørklep:
     in Light Metals 1975, AIME, New York 1975.

(7) R.C. Dorward: Met. Trans. 4 (1973) 386.
(8) R.C. Dorward: Aluminium 49 (1973) 686.
(9) K. Grjotheim, S. Jørgensen, R. Nikolic and H.A. Øye: Metall
     30 (1976), at press.
(10) B. Berge: Thesis, Institute of Inorganic Chemistry, The
     Technical University of Norway, Trondheim 1975.
(11) K. Grjotheim, C. Krohn and H. Øye: Aluminium 51 (1975) 697.

AUTHOR INDEX

Numbers that are underlined show the page on which the complete reference is listed. Contributing authors and page numbers pertaining to this volume are presented in script.